Applied Classical Electrodynamics

Applied Classical Electrodynamics

Volume I: Linear Optics

F.A. HOPF

G.I. STEGEMAN

Optical Sciences Center
University of Arizona, Tucson

A Wiley-Interscience Publication

JOHN WILEY & SONS, New York • Chichester • Brisbane • Toronto • Singapore

Library of Congress Cataloging in Publication Data:

Hopf, F. A.
Applied classical electrodynamics.

"A Wiley-Interscience publication."
Includes index.
Contents: v. 1. Linear optics.
1. Electrodynamics. 2. Optics.
I. Stegeman, G. I. II. Title.

QC631.H73 1985 537.6 85-12043
ISBN 0-471-82788-6 (v. 1)

Printed in the United States of America

10 9 8 7 6 5 4 3 2 1

Preface

This is the first book of a two volume work on the classical physics of optical interactions. The two volumes deal with linear and nonlinear optics, respectively. These volumes are teaching texts developed for the academic program at the Optical Sciences Center of the University of Arizona. They are designed to give a background on issues in optical physics that relate to material science and laser applications. The emphasis on classical physics reflects the fact that most practical applications involve the classical limit. Matter obeys quantum mechanics and the interaction of radiation with matter cannot be developed with complete consistency from a model of electrons on springs. Nonetheless, once the conceptual difficulties resolved by quantum mechanics are dealt with, nearly all remaining cases in applied optics can be modeled classically. Classical mechanics leads to a useful phenomenology of considerable breadth of application.

When first written, Volume I consisted of thirteen chapters. In the present version, the chapter on multipole fields is Appendix B. For teaching or reading purposes, it can be inserted anywhere. We have found that many students prefer to get comfortable with the dipole interactions before dealing with higher order terms. The design of the volumes is modular, and there is considerable flexibility in the order in which the text can be read or taught. In the Table of Contents, the chapters, and in some cases the sections, contain, in parentheses, a reference back to the basic material needed for the chapter. Most basic material is in the first three chapters. Some supplemental techniques are discussed in Chapter 10. Otherwise one can read the book in almost arbitrary order. In addition, each topic is developed starting from a fairly elementary level. One need not master all of the background material if all one is interested in is an overview of the ideas. The problem sets are designed to tie the book together into a coherent whole, and deal with conceptual issues that might otherwise become diversionary. Those who wish to do the problems should be aware that the degree of difficulty varies substantially. Each problem has been tried at least once by one of the authors (the first author has answered all problems to within factors of ε_0).

The book is designed to allow a reader with a background in classical mechanics and electromagnetism to learn enough to continue on with texts that specialize in the specific topics. Thus each topic is not covered in great detail, and we have tried to judge which areas are adequately covered in existing texts to avoid duplication. In particular, the material in Chapter 4 is dealt with in a number of texts, so we have covered the subject in much less detail than it deserves. Nearly every topical area in optics has its own conventional notation. We have attempted throughout to use notation that is standard for the topic so that one can move directly to specialized texts. Some symbols are used in many different ways in optics while others are not used at all, so we have occasionally compromised. Appendix A discusses some of the more difficult notational problems.

This book is printed in camera ready form from a personal computer. The software package we use is still under development by M. Sargent III and colleagues. The experimental software imposes some defects and constraints.

We wish to acknowledge the help of many students who proofread this material and helped with the figures. We gratefully acknowlege the useful suggestions of G. Salamo, E. van Stryland and A. Smirl who have been using preliminary versions of these volumes in their courses.

Tucson Arizona
March 18, 1985

Frederic A. Hopf
George I.A. Stegeman

Contents

[1] Indicates chapter needed as background.

1
Review of Electromagnetic Theory

1.1 MAXWELL'S EQUATIONS

In this section we review the key elements of electromagnetic theory that are needed for this book. We cover the subject briefly, since the material is treated in detail in existing texts. We formulate the theory in terms of macroscopic field variables. The electric and magnetic forces are defined in terms of the electric field **E** and magnetic field **B**. In addition, there are displacement fields **D** and **H** which arise from real charges ρ and currents **J**. Maxwell's equations read

$$\nabla \cdot \mathbf{D} = \rho \tag{1.1}$$

$$\nabla \times \mathbf{H} = \mathbf{J} + \frac{\partial \mathbf{D}}{\partial t} \tag{1.2}$$

$$\nabla \cdot \mathbf{B} = 0 \tag{1.3}$$

$$\nabla \times \mathbf{E} = -\frac{\partial \mathbf{B}}{\partial t}. \tag{1.4}$$

We are almost exclusively concerned with cases of electromagnetic fields that arise in charge- and current-free space, i.e., in which $\rho = 0$ and $\mathbf{J} = 0$. In that case Eqs. (1.1) and (1.2) read

$$\nabla \cdot \mathbf{D} = 0 \tag{1.5}$$

$$\nabla \times \mathbf{H} = \frac{\partial \mathbf{D}}{\partial t}. \tag{1.6}$$

The displacement currents and fields are related through polarizations **P** and magnetizations **M**, whose calculation is one of the main functions of this book. These equations read

$$\mathbf{B} = \mu_0(\mathbf{H} + \mathbf{M}) \tag{1.7}$$

and

$$\mathbf{D} = \varepsilon_0\mathbf{E} + \mathbf{P}. \tag{1.8}$$

The forces acting on the microscopic charges that give rise to **P** and **M** are described by the Lorenz force

$$\mathbf{F} = e\mathbf{E} + e\mathbf{v} \times \mathbf{B}, \tag{1.9}$$

where e is the charge and **v** is the velocity in the charge.

The Poynting vector **S** measures the flux of a radiation field, and is given by

$$\mathbf{S} = \mathbf{E} \times \mathbf{H}. \tag{1.10}$$

One must always be careful in using Eq. (1.10) at a single point in space. The flux can vary rapidly as a function of position, and one should always integrate over a small surface area about the point of interest, to minimize the chance of artifacts.

1.2 WAVE EQUATIONS

Let us now consider the various forms of wave equations that are useful in this book. We begin by letting the current **J** be nonzero for reasons described below. First take the curl of Eq. (1.4), substitute Eq. (1.7) to eliminate **B**, and then use Eq. (1.2). This gives

$$\nabla \times (\nabla \times \mathbf{E}) + \mu_0 \frac{\partial^2\mathbf{D}}{\partial t^2} = -\mu_0 \frac{\partial \mathbf{J}}{\partial t} - \mu_0 \nabla \times \frac{\partial \mathbf{M}}{\partial t}. \tag{1.11}$$

First let us eliminate **D** from this equation using Eq. (1.8). We also use the general vector relation written in terms of an arbitrary vector field **V** as

$$\nabla \times (\nabla \times \mathbf{V}) = -\nabla^2\mathbf{V} + \nabla(\nabla \cdot \mathbf{V}) \tag{1.12}$$

so that

$$-\nabla^2\mathbf{E} + \nabla(\nabla \cdot \mathbf{E}) + \frac{1}{c^2}\frac{\partial^2\mathbf{E}}{\partial t^2} = -\mu_0\frac{\partial \mathbf{J}}{\partial t} - \mu_0\frac{\partial^2\mathbf{P}}{\partial t^2} - \mu_0 \nabla \times \frac{\partial \mathbf{M}}{\partial t}, \tag{1.13}$$

where $c = (\mu_0\varepsilon_0)^{-1/2}$ is the velocity of light in vacuum.

Most derivations in standard electromagnetic texts are given in terms of real currents **J**. However, most light-matter interactions are expressed in terms of polarization **P** and magnetization **M**. We note that the terms $-\mu_0\ \partial\mathbf{J}/\partial t$, $-\mu_0\ \partial^2\mathbf{P}/\partial t^2$ and $-\mu_0\ \nabla \times \partial\mathbf{M}/\partial t$ all appear as source or driving terms on the right-hand side of Eq. (1.13). Therefore we can write

$$\mathbf{J}_{\text{eff}} \equiv \frac{\partial \mathbf{P}}{\partial t} + \nabla \times \mathbf{M} \tag{1.14}$$

as an effective current source. By this device, we define a fictitious current in terms of polarization and magnetization source terms that can be used in place of real currents. As a consequence, we can make use of mathematical formulae developed in electromagnetic texts. It is a simple exercise to define an effective charge density and to show that these relations are consistent with Maxwell's equations and charge continuity.

For most cases of practical interest, Eq. (1.13) is used with $\mathbf{M} = \mathbf{J} = 0$. We need this equation often, so for later reference we write Eq. (1.13) as

$$-\nabla^2\mathbf{E} + \nabla(\nabla \cdot \mathbf{E}) + \frac{1}{c^2}\frac{\partial^2 \mathbf{E}}{\partial t^2} = -\mu_0 \frac{\partial^2 \mathbf{P}}{\partial t^2}. \tag{1.15}$$

For a weak optical field in an optically isotropic media, $\mathbf{P} = \varepsilon_0 \chi \mathbf{E}$, where χ is a scalar susceptibility. In this case

$$\frac{1}{c^2}\frac{\partial^2 \mathbf{E}}{\partial t^2} + \mu_0 \frac{\partial^2 \mathbf{P}}{\partial t^2} = \frac{n^2}{c^2}\frac{\partial^2 \mathbf{E}}{\partial t^2}, \tag{1.16}$$

where n is the refractive index given by $n^2 = 1 + \chi$.

It is occasionally of value to write the wave equation in terms of **D**. One obtains this by substituting Eq. (1.8) into Eq. (1.13) to eliminate **E**. Then (with $\mathbf{J} = 0$)

$$-\nabla^2\mathbf{D} + \frac{1}{c^2}\frac{\partial^2 \mathbf{D}}{\partial t^2} = \nabla \times (\nabla \times \mathbf{P}) - \frac{1}{c^2}\nabla \times \frac{\partial \mathbf{M}}{\partial t}. \tag{1.17}$$

The advantage of using Eq. (1.17) is that the vector relationships on both sides are explicitly orthogonal to the propagation wave vector of an electromagnetic plane wave.

In scattering problems it is useful to express the answers in terms of the vector potential **A**. The magnetic field is given by

$$\mathbf{B} = \nabla \times \mathbf{A}$$

and the electric field by

$$\mathbf{E} = -\nabla\phi - \frac{\partial \mathbf{A}}{\partial t},$$

where ϕ is the scalar potential. The scalar and vector potentials are related by

$$\nabla \cdot \mathbf{A} + \frac{1}{c^2}\frac{\partial \phi}{\partial t} = 0,$$

in which case it is left as an exercise to show the conventional result (use $\mathbf{P} = \mathbf{M} = 0$) that

$$- \nabla^2 \mathbf{A} + \frac{1}{c^2} \frac{\partial^2 \mathbf{A}}{\partial t^2} = \mu_0 \mathbf{J}. \tag{1.18}$$

If one wishes to use Eq. (1.18) in general, it is necessary to learn about gauges which are discussed in standard texts. We always use Eq. (1.15) to compute the radiation pattern far from the source $\mathbf{J}_{eff}$ (which is, in practice, usually an induced polarization in Eq. (1.15)). It is then not necessary to worry about the gauge, since we compute results in a free space region where $\mathbf{B} = \mu_0 \mathbf{H}$ and $\mathbf{D} = \varepsilon_0 \mathbf{E}$, thus obtaining a gauge-independent answer.

1.3 PLANE WAVES

In free space, any arbitrary field can be written as a superposition of plane waves

$$\mathrm{E}(\mathbf{r},t) = \int \mathbf{dk} \int d\omega \, \mathrm{E}(\mathbf{k},\omega) \, e^{i(\mathbf{k}\cdot\mathbf{r} - \omega t)}. \tag{1.19}$$

For a single monochromatic plane wave

$$\mathrm{E}(\mathbf{r},t) = \frac{1}{2} \boldsymbol{E}(\mathbf{k},\omega) \, e^{i(\mathbf{k}\cdot\mathbf{r} - \omega t)} + \mathrm{cc}, \tag{1.20}$$

where $\mathbf{k}$ is the wave normal, ω is the angular frequency, $\mathrm{E}(\mathbf{k},\omega)$, $\boldsymbol{E}(\mathbf{k},\omega)$ are vector amplitudes, and $i = \sqrt{-1}$. Throughout this book, the convention for amplitudes is the following: field amplitudes are written as italics; fields and Fourier amplitudes are written in ordinary type. In each case, the independent variables indicate which amplitude is meant. Field amplitudes are related to the Fourier amplitudes through Dirac delta functions. These relations are explored in detail in Chapter 12. The symbol cc stands for complex conjugate. With this convention, total fields are real and amplitudes are complex. These conventions apply to all fields, matter or electromagnetic, except when the use of greek symbols for the field calls for some flexibility. The conventions are restricted to fields and do not apply to properties of individual atoms and molecules. The position coordinate $\mathbf{r}$ denotes the vector (x,y,z) which is used throughout this book as the coordinate system in which electromagnetic propagation problems are solved. When it is necessary to distinguish the (x,y,z) axes from others (e.g., crystal axes), they are referred to as the propagation axes.

In free space, $\nabla \cdot \mathrm{E} = 0$ and

$$\nabla^2 \boldsymbol{E}(\mathbf{k},\omega) \, e^{i(\mathbf{k}\cdot\mathbf{r} - \omega t)} = - k^2 \, \boldsymbol{E}(\mathbf{k},\omega) \, e^{i(\mathbf{k}\cdot\mathbf{r} - \omega t)}. \tag{1.21}$$

So from Eq. (1.15) with $\mathrm{P} = 0$ (i.e., in vacuum)

$$\frac{\omega}{k} = c. \tag{1.22}$$

In Eqs. (1.21) and (1.22) the symbol k stands for the magnitude of the vector **k** and is defined by $\mathbf{k} = k\hat{\mathbf{k}}$. The caret symbol over a vector denotes a unit vector, i.e., $|\hat{\mathbf{k}}| = 1$. When the same letter denotes a vector (the letter may be upper or lower case) and a unit vector (usually the letter is lower case) the unit vector is understood to be parallel to the vector. In this book, an absolute sign around a vector (which is often used elsewhere to denote vector amplitudes) means the following; for an arbitrary vector **V**

$$|\mathbf{V}|^2 = V_x V_x^* + V_y V_y^* + V_z V_z^*, \tag{1.23}$$

where for real a and b, $(a+ib)^* = (a-ib)$. Hence the absolute sign signifies simultaneously an absolute value and a vector amplitude. In free space, Eq. (1.5) with $\mathbf{D} = \varepsilon_0 \mathbf{E}$ implies

$$\mathbf{E} \cdot \mathbf{k} = 0, \tag{1.24}$$

and one has in general that

$$\mathbf{D} \cdot \mathbf{k} = 0. \tag{1.25}$$

Each electric field is associated with a magnetic field, written as

$$\mathbf{H}(\mathbf{r},t) = \frac{1}{2}\, \boldsymbol{H}(\mathbf{k},\omega)\, e^{i(\mathbf{k}\cdot\mathbf{r}-\omega t)} + cc, \tag{1.26}$$

in which case Eq. (1.4) with $\mathbf{B} = \mu_0 \mathbf{H}$, and the identity

$$\nabla \times \left(\boldsymbol{E}(\mathbf{k},\omega)\, e^{i(\mathbf{k}\cdot\mathbf{r}-\omega t)} \right) = i\, \mathbf{k} \times \boldsymbol{E}(\mathbf{k},\omega)\, e^{i(\mathbf{k}\cdot\mathbf{r}-\omega t)} \tag{1.27}$$

gives

$$i\, \mathbf{k} \times \boldsymbol{E}(\mathbf{k},\omega) = i\omega\mu_0\, \boldsymbol{H}(\mathbf{k},\omega). \tag{1.28}$$

The **H** vector is thus perpendicular to **k** and **E**. Using Eq. (1.28), we can obtain a more useful form for the Poynting vector **S**. In so doing, we are not interested in the short-time oscillations of **S**, but rather its average in time over an interval of one optical cycle, i.e.,

$$\overline{\mathbf{S}} = \frac{\omega}{2\pi} \int_0^{2\pi/\omega} \mathbf{S}\, dt. \tag{1.29}$$

Any quantity like $\exp(-in\omega t)$, where n is a nonzero integer, gives zero when averaged in this fashion. Using Eqs. (1.10) and (1.28) gives

$$\overline{\mathbf{S}} = \frac{1}{2}\left(\frac{\varepsilon_0}{\mu_0}\right)^{1/2} |\boldsymbol{E}(\mathbf{k},\omega)|^2\, \hat{\mathbf{k}}. \tag{1.30}$$

It is usually the case that we work with scalar rather than vector amplitudes. In that case, we define two orthogonal unit vectors $\hat{\mathbf{e}}_a$ and $\hat{\mathbf{e}}_b$ such that

$$\hat{\mathbf{e}}_a \cdot \hat{\mathbf{e}}_b = 0, \quad \hat{\mathbf{e}}_a \cdot \hat{\mathbf{k}} = 0, \quad \hat{\mathbf{e}}_b \cdot \hat{\mathbf{k}} = 0. \tag{1.31}$$

One can then express an arbitrary polarization of the optical field as linear combinations of these vectors $a\hat{\mathbf{e}}_a + b\hat{\mathbf{e}}_b$, where

$$|a|^2 + |b|^2 = 1. \tag{1.32}$$

Thus we write the general complex vector amplitude $\boldsymbol{E}(\mathbf{k},\omega)$ in terms of a scalar amplitude such that

$$\mathbf{E} = \frac{1}{2}\,(a\,\hat{\mathbf{e}}_a + b\,\hat{\mathbf{e}}_b)\, E(\mathbf{k},\omega)\, e^{i(\mathbf{k}\cdot\mathbf{r}-\omega t)} + \mathrm{cc}, \tag{1.33}$$

where we henceforth suppress the dependence on k and ω for notational convenience, i.e., $E \equiv E(\mathbf{k},\omega)$. It is occasionally of value to write the complex scalar amplitude in terms of the conventional amplitude of the wave, which is denoted as $|E|$ and the phase ϕ, i.e.,

$$E = |E|\, e^{i\phi}. \tag{1.34}$$

The various senses of polarizations of the light come from different choices of a and b, for example:

Linear polarization, a and b are real;
Circular polarization, $a = 1/\sqrt{2}$, $b = \pm\, i/\sqrt{2}$.

In optical media, arbitrary waves can be written as the superposition of orthogonally polarized plane waves, but the particular plane waves used in this superposition are not arbitrary. In many cases of interest in this book, only special choices of directions of polarizations and propagation can be written in the form of Eq. (1.33). These special choices are given the name "proper vectors" or eigenvectors, which signifies that they are valid solutions of the wave equation. An eigenvector has the property that it has a unique velocity v, such that

$$v = \frac{\omega}{k}. \tag{1.35}$$

We can always associate an index of refraction with each eigenvector such that

$$v = \frac{c}{n}, \tag{1.36}$$

i.e.,

$$\frac{\omega}{k} = \frac{c}{n}. \tag{1.37}$$

In general, n is a function of both ω and k, and in nonlinear optics it can also be a function of $|\boldsymbol{E}|$.

1.4 REFLECTION AND TRANSMISSION AT BOUNDARIES

Whenever an optical field encounters a discontinuity in the optical properties of matter, it is partially reflected and partially transmitted. For a plane surface, the frequency ω and the component of the wave vector parallel to the surface are conserved across the interface. The wave vector component normal to the boundary is not conserved. These conservation conditions are a consequence of translational symmetry parallel to the surface, but not normal to the surface.

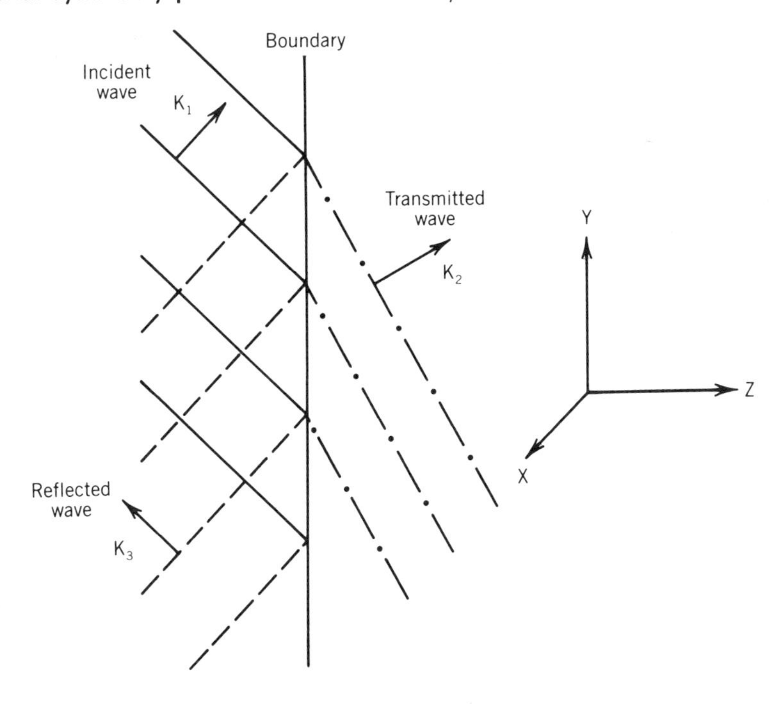

Figure 1.1. Reflection and transmission of a wave incident on a boundary.

For the boundary problem sketched in Figure 1.1, one defines the angles of incidence and refraction with respect to the wave normals **k** (note that **S** is not necessarily parallel to **k**). All three waves are assumed to be eigenvectors. The appropriate boundary conditions are that the transverse components (i.e., E_x, E_y) of E are continuous at the boundary, as well as the normal component of **D** (i.e.,

D_z). These conditions require that

$$k_{x1} = k_{x2} = k_{x3} \tag{1.38a}$$
$$k_{y1} = k_{y2} = k_{y3} \tag{1.38b}$$
$$\omega_1 = \omega_2 = \omega_3. \tag{1.38c}$$

Equations (1.38a) and (1.38b) lead directly to the fact that the angle of incidence equals the angle of reflection (for optically isotropic media), and to Snell's Law

$$\frac{\sin\theta_i}{\sin\theta_t} = \frac{n_t}{n_i}, \tag{1.39}$$

where n_t and n_i are the refractive indices on the transmission and incidence side of the boundary respectively.

1.5 RADIATION FROM A DIPOLE SOURCE

The various vector quantities needed for this calculation are illustrated in Figure 1.2. Define **R** as a vector to any point in the source. Let **J**(**r**',t) be a current field within the source, where we can use Eq. (1.14) to replace **J** by **P** for

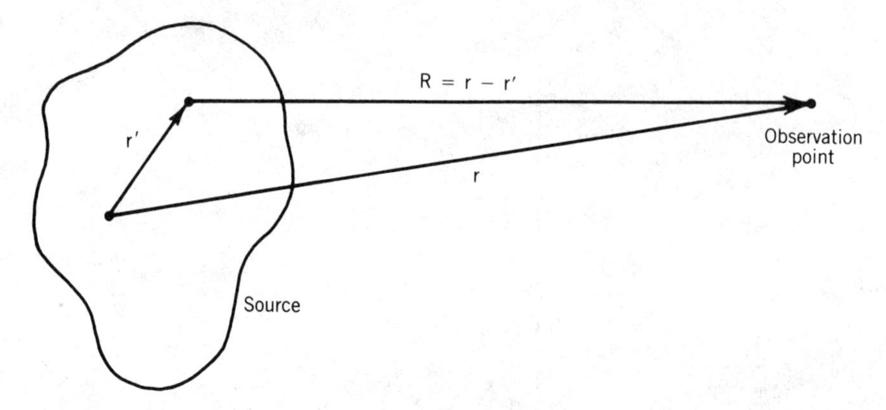

Figure 1.2. Illustration of vectors and geometry used in calculating the radiation from a source.

the case of induced dipoles. Then the standard solution of Eq. (1.18) for the vector potential is

$$\mathbf{A}(\mathbf{r},t) = \frac{\mu_0}{4\pi} \int \mathbf{dr'} \int dt' \, \frac{\mathbf{J}(\mathbf{r'},t')}{R} \, \delta(t' + R/c - t). \tag{1.40}$$

When the limits of integration are not explicitly denoted, one integrates over all space and time. Let us digress temporarily and write down some expressions that are useful for the future. First let us take the current to read

$$\mathbf{J}(\mathbf{r}',t') = \frac{1}{2}\,\mathcal{J}(\mathbf{r}')\,e^{-i\omega t'} + \text{cc}, \tag{1.41}$$

and write the vector potential as

$$\mathbf{A}(\mathbf{r},t) = \frac{1}{2}\,\mathcal{A}\,\frac{e^{i(kr-\omega t)}}{r} + \text{cc}, \tag{1.42}$$

which is valid only in the far field (i.e., r >> r'). Then

$$\mathcal{A}\,\frac{e^{ikr}}{r} = \frac{\mu_0}{4\pi}\int d\mathbf{r}'\,\frac{\mathcal{J}(\mathbf{r}')}{R}\,e^{ikR}.$$

This equation is not valid until we also make the far field approximation for the integrand. We approximate this expression by noting the following

$$R = |\mathbf{r} - \mathbf{r}'| = (r^2 + r'^2 - 2\mathbf{r}\cdot\mathbf{r}')^{1/2}$$

$$\simeq r - \frac{\mathbf{r}\cdot\mathbf{r}'}{r} = r - \hat{\mathbf{r}}\cdot\mathbf{r}',$$

Note that the use of the absolute symbol in these expressions is legitimate since all the vectors are real. The far field approximation is valid if $r'/r \ll 1$. One can further approximate $R \simeq r$ in the denominator. However, we are often interested in media whose dimensions are larger than a wavelength of light, in which case one cannot drop the term $\hat{\mathbf{r}}\cdot\mathbf{r}'$ inside the exponential. Hence this term is left in the integrand, and we obtain

$$\mathcal{A} = \frac{\mu_0}{4\pi}\int d\mathbf{r}'\,\mathcal{J}(\mathbf{r}')\,e^{-ik\hat{\mathbf{r}}\cdot\mathbf{r}'}. \tag{1.43}$$

Note that $\mathcal{A}$ is a function of $\hat{\mathbf{r}}$, which means that it is a function of angle, but not of distance to the source.

The problem we are interested in at present is the case

$$\mathcal{J}(\mathbf{r}') = -i\omega\,\mathbf{p}_0\,\delta(\mathbf{r}'),$$

where we have used Eq. (1.14) to associate the current with a point dipole of amplitude $\mathbf{p}_0$ written as

$$\mathbf{p} = \frac{1}{2}\,\mathbf{p}_0\,e^{i\omega t} + \text{cc}.$$

Note that the dipole amplitude is an atomic or molecular property, and so the amplitude convention discussed at the beginning of Section 1.3 does not apply here. We then have

$$\mathbf{A} = -\frac{i\omega\mu_0}{4\pi}\,\mathbf{p}_0 \tag{1.44}$$

or

$$\mathbf{A}(\mathbf{r},t) = -\frac{1}{2}\,\frac{i\omega\mu_0}{4\pi}\,\mathbf{p}_0\,\frac{e^{i(kr-\omega t)}}{r} + \text{cc.} \tag{1.45}$$

From this, the field can be computed using the gauge-invariant expression $\mathbf{B} = \nabla \times \mathbf{A}$, which gives

$$\mathbf{B}(\mathbf{r},t) = \frac{1}{2}\,\frac{k^2}{4\pi\,\varepsilon_0 c}\,(\hat{\mathbf{r}} \times \mathbf{p}_0)\,\frac{e^{i(kr-\omega t)}}{r} + \text{cc.} \tag{1.46}$$

We can then use Maxwell's equations in vacuum as a gauge-invariant procedure to compute E (see Problem 1.4). This gives

$$\mathbf{E}(\mathbf{r},t) = \frac{1}{2}\,\frac{k^2}{4\pi\,\varepsilon_0}\,(\hat{\mathbf{r}} \times \mathbf{p}_0) \times \hat{\mathbf{r}}\,\frac{e^{i(kr-\omega t)}}{r} + \text{cc.}$$

This expression can also be rewritten using standard vector identities into the more useful form

$$\mathbf{E}(\mathbf{r},t) = \frac{1}{2}\,\frac{k^2}{4\pi\,\varepsilon_0}\,[\mathbf{p}_0 - \hat{\mathbf{r}}(\hat{\mathbf{r}}\cdot\mathbf{p}_0)]\,\frac{e^{i(kr-\omega t)}}{r} + \text{cc.} \tag{1.47}$$

The vector direction of $\mathbf{E}$ is determined by dotting the dyad $\mathbf{I} - \hat{\mathbf{r}}\hat{\mathbf{r}}$, where $\mathbf{I}$ is the unit dyad (see Eq. (A.93)) into the dipole source. This dyad finds the projection of the dipole that is perpendicular to the line of sight $\hat{\mathbf{r}}$. This can be easily seen since $\hat{\mathbf{r}}\hat{\mathbf{r}}$ is the projection onto the line of sight, and the sum of the two projections must be the identity, clearly

$$(\mathbf{I} - \hat{\mathbf{r}}\hat{\mathbf{r}}) + \hat{\mathbf{r}}\hat{\mathbf{r}} = \mathbf{I}.$$

The projection operator is useful, and we denote it with a special symbol

$$\mathbf{O}(\hat{\mathbf{r}}) \equiv \mathbf{I} - \hat{\mathbf{r}}\hat{\mathbf{r}} \tag{1.48}$$

The vector relations involving the projection operator are shown in Figure 1.3. It is left as an exercise to show that the $\mathbf{B}$ field in Eq. (1.46) is orthogonal to both $\mathbf{E}$ and $\mathbf{r}$. The Poynting vector is

$$\overline{\mathbf{S}} = \frac{ck^4}{8\pi\,\varepsilon_0}\,\frac{|\hat{\mathbf{r}} \times \mathbf{p}_0|^2}{4\pi\,r^2}\,\hat{\mathbf{r}}. \tag{1.49}$$

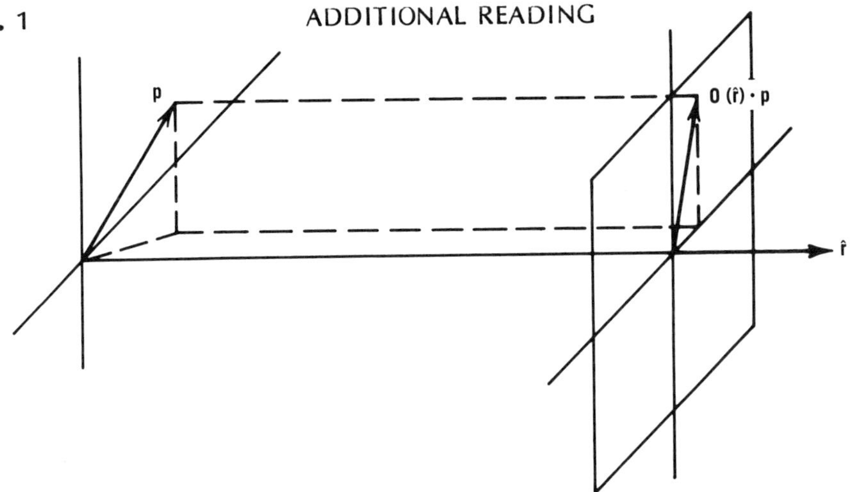

Figure 1.3. The relation between the dipole amplitude **p**, the line of sight $\hat{\mathbf{r}}$, and the projection $\mathbf{O}(\hat{\mathbf{r}})\cdot\mathbf{p}$.

Defining θ as the angle between $\hat{\mathbf{r}}$ and $\mathbf{p}_0$ (see Figure 1.3), we obtain

$$\overline{\mathbf{S}} = \frac{ck^4}{8\pi\varepsilon_0}\,\frac{\sin^2\theta\,|\mathbf{p}_0|^2}{4\pi r^2}\,\hat{\mathbf{r}}. \tag{1.50}$$

To find the total radiated power P_{rad} emitted by the dipole, construct a surface around the dipole and integrate the flux over the surface, so that

$$P_{rad} = \int \hat{\mathbf{r}}\cdot\overline{\mathbf{S}}\, d\sigma$$

or

$$P_{rad} = \int r^2\, d\phi \int \sin\theta d\theta\, \frac{ck^4}{8\pi\varepsilon_0}\,\frac{\sin^2\theta\,|\mathbf{p}_0|^2}{4\pi r^2},$$

which gives

$$P_{rad} = \frac{1}{4\pi\varepsilon_0}\,\frac{1}{3}\,ck^4|\mathbf{p}_0|^2. \tag{1.51}$$

When limits of integration are not denoted explicitly, an integral over a surface element $d\sigma$ is understood to run over the whole of the surface, integrals over ϕ run from 0 to 2π and integrals over θ run from 0 to π. Equation (1.50) is the basis for understanding the way dipoles scatter light. In addition, the dipoles both absorb and disperse light. These topics are taken up in Chapter 4, which gives a unified picture of dipole radiation.

ADDITIONAL READING

Electrodynamics

Jackson, J.D. *Classical Electrodynamics*, 2nd Edition, Wiley, New York, 1975. (or 1st Ed. 1962) Chapters 6 and 7.

Vector Identities

Portis, A.M. *Electromagnetic Fields: Sources and Media*, Wiley, New York, 1978. Appendices B and C.

PROBLEMS

1.1. Verify Eq. (1.12) and the following vector identities

$$\mathbf{A} \times (\mathbf{B} \times \mathbf{C}) = (\mathbf{A} \cdot \mathbf{C})\mathbf{B} - (\mathbf{A} \cdot \mathbf{B})\mathbf{C}, \tag{1.52}$$

$$\mathbf{A} \cdot (\mathbf{B} \times \mathbf{C}) = \mathbf{C} \cdot (\mathbf{A} \times \mathbf{B}). \tag{1.53}$$

1.2. Prove the following relationships for plane-wave fields of the form of Eq. (1.20)

$$\nabla \cdot \mathbf{E}(\mathbf{r},t) = \frac{1}{2}\, i\mathbf{k} \cdot \boldsymbol{E}(\mathbf{k},\omega)\, e^{i(\mathbf{k}\cdot\mathbf{r}-\omega t)} + \mathrm{cc}, \tag{1.54}$$

$$\nabla\, \mathbf{E}(\mathbf{r},t) = \frac{1}{2}\, i\mathbf{k}\, \boldsymbol{E}(\mathbf{k},\omega)\, e^{i(\mathbf{k}\cdot\mathbf{r}-\omega t)} + \mathrm{cc}, \tag{1.55}$$

$$\nabla^2\, \mathbf{E}(\mathbf{r},t) = -\frac{1}{2}\, k^2\, \boldsymbol{E}(\mathbf{k},\omega)\, e^{i(\mathbf{k}\cdot\mathbf{r}-\omega t)} + \mathrm{cc}, \tag{1.56}$$

$$\nabla(\nabla \cdot \mathbf{E}(\mathbf{r},t)) = -\frac{1}{2}\, \mathbf{k}\, (\mathbf{k} \cdot \boldsymbol{E}(\mathbf{k},\omega))\, e^{i(\mathbf{k}\cdot\mathbf{r}-\omega t)} + \mathrm{cc}. \tag{1.57}$$

1.3. Prove that for any scalar field ϕ or vector field **A**

$$\nabla \times (\nabla \phi) = 0, \tag{1.58}$$

$$\nabla \cdot (\nabla \times \mathbf{A}) = 0. \tag{1.59}$$

1.4. (a) Using Maxwell's equations in vacuum, verify that

$$\boldsymbol{E} = -\frac{c^2}{\omega}\, \mathbf{k} \times \boldsymbol{B}. \tag{1.60}$$

(b) Verify this expression from the vector amplitudes $\boldsymbol{E}(\mathbf{k},\omega)$, and $\boldsymbol{B}(\mathbf{k},\omega)$ obtained from Eqs. (1.46) and (1.47). Show explicitly that these amplitudes are orthogonal.

(c) Verify Eqs. (1.30) and (1.49).

1.5. Show that Eq. (1.40) is a solution of Eq. (1.18).

1.6. The Hertz vector $\mathbf{Z}$ can be used to define the vector and scalar potentials as

$$\mathbf{A} = \frac{1}{c^2}\frac{\partial \mathbf{Z}}{\partial t}, \ \phi = -\nabla \cdot \mathbf{Z}.$$

Show that these potentials obey

$$\nabla \cdot \mathbf{A} + \frac{1}{c^2}\frac{\partial \phi}{\partial t} = 0.$$

and derive the wave equation for $\mathbf{Z}$ in free space.

1.7. Show that the charge and current distributions

$$\rho = -e\,[\delta(x-\xi\cos\nu t - x_0)\,\delta(y-\eta\cos\nu t - y_0)\,\delta(z-\zeta\cos\nu t - z_0)$$
$$-\,\delta(x-x_0)\,\delta(y-y_0)\,\delta(z-z_0)], \tag{1.61}$$

$$\mathbf{J} = e(\xi\hat{\mathbf{x}}+\eta\hat{\mathbf{y}}+\zeta\hat{\mathbf{z}})\nu\,\sin\nu t\,\times$$
$$\delta(x-\xi\cos\nu t-x_0)\delta(y-\eta\cos\nu t-y_0)\delta(z-\zeta\cos\nu t-z_0), \tag{1.62}$$

obey the continuity equation for charge and current

$$\nabla \cdot \mathbf{J} + \frac{\partial \rho}{\partial t} = 0. \tag{1.63}$$

1.8. Find an effective charge such that it plus the effective current in Eq. (1.14) obey the continuity equation for current and charge.

1.9. A dipole moment is defined by the formula

$$\mathbf{p}(t) = \int d\mathbf{r}\ \mathbf{r}\ \rho(\mathbf{r},t), \tag{1.64}$$

where the integral runs over the volume containing the charge. Use this formula to determine the dipole moment of the charge distribution in Eq. (1.61).

1.10. (a) Show that if Eq. (1.16) is valid then $E \cdot \mathbf{k} = 0$ (i.e., $\nabla \cdot E = 0$) and that Eq. (1.37) is valid.

(b) Show that if $\boldsymbol{P}(\mathbf{k},\omega) = \varepsilon_0\chi(\omega)\ \boldsymbol{E}(\mathbf{k},\omega)$, then $E \cdot \mathbf{k} = 0$, $\omega/k = c/n(\omega)$, $n(\omega)^2 = 1+\chi(\omega)$.

2 Normal Modes of Matter

2.1 DEGREES OF FREEDOM

In this chapter we discuss the degrees of freedom and the associated normal modes that we use to describe the matter system. There are three position coordinates, i.e, degrees of freedom, for each particle in the system. In practice, few of these particles contribute significantly to the optical properties. Except for occasional applications in the ultraviolet, which we ignore, no more than one electron per atom, with charge $-e$, participates in optical interactions. The remaining particles of interest are the nuclei, a composite particle of charge $+e$ composed of the actual nucleus plus the remaining core electrons. For an N atom system, the 6N degrees of freedom (two particles per atom × three coordinates per particle) are denoted $\mathbf{r}_\alpha$, $\alpha = 1, 2, \cdots, 6N$. When the atoms are bound together into molecules, it is often sufficient to consider only one electron per molecule, in which case there are fewer than 6N degrees of freedom.

The particles can participate in three types of motion, translational, rotational, and vibrational. We are concerned with the motion of electrons which remain bound either to specific core atoms or to molecules. These motions cannot be described classically. However, if we restrict ourselves to cases where the electronic motion makes small departures from the quantum mechanical ground state, it turns out to be reasonable, if not always perfectly accurate, to describe the electronic motion as a vibration. We therefore treat all electrons as if they were point charges bound by springs to the atoms.

There remain two nuclear motions, nuclear vibration and rotation. Of these, rotations are somewhat special since they do not have restoring forces. In addition, the dependence of forces on angle is usually sinusoidal, rather than linear. Hence, while the angular motions can be treated by means analogous to those used in this chapter, the detailed algebra is different. The formulae developed in this chapter are restricted to vibrational cases.

Nuclear vibrations are motions of atoms relative to each other inside a molecule. This requires a minimum of two atoms. The atoms can be thought of as interconnected by springs, which act as interatomic forces. For small excursions from the equilibrium position, the motion is essentially that of a simple harmonic oscillator.

2.2 NORMAL COORDINATES

If one were to observe the motion of the atoms and electrons inside a molecule, a complex pattern would be found. A similar macroscopic case is a stick which is thrown so that it both rotates and translates. From standard texts on classical mechanics we know that the motion of the stick can be described as translation with components along three orthogonal axes, as well as rotation about the principal axes. The key point is that the motion can be resolved into six uncoupled motions. The equations describing these motions contain only one variable each (e.g., distance along the x axis, rotation angle about one of the principal axes). These decoupled motions are called normal modes and they are described mathematically by normal coordinates. There is one normal coordinate for each degree of freedom. In a molecule, the atoms move as if they were connected by springs, so that there are also vibrational normal coordinates. These are decoupled from each other and are decoupled from the translational and rotational normal coordinates.

Let us treat the translational and rotational normal coordinates first. Free translations present no difficulty since any rectilinear coordinate system decouples the motion. The rotations are similarly straightforward since the moment of inertia tensor can always be diagonalized to yield three principal axes (see Problem 2.1).

The bound structures, molecules, lattices, etc., require more careful attention. Let us assume that there is no net acceleration of the medium, and that the medium has a stable rest structure determining the position of the atoms and/or molecules. We define by r_α all the displacements that describe vibrations about the rest position. We can then write the potential energy as a Taylor's expansion about the rest positions, i.e.,

$$V = \sum_{\alpha,\beta} \zeta_{\alpha\beta}\, r_\alpha\, r_\beta + \sum_{\alpha,\beta,\gamma} \zeta_{\alpha\beta\gamma}\, r_\alpha\, r_\beta\, r_\gamma + \cdots, \tag{2.1}$$

where $\zeta_{\alpha\beta}$, $\zeta_{\alpha\beta\gamma}$, ... are constants. The terms in this expansion are all real. In addition each $\zeta_{\alpha\beta}$ is symmetrical under the interchange of each pair of indices. The higher indices have permutation symmetry, which is discussed later.

Let us now prove this symmetry for the quadratic (first) term. Suppose that $\zeta_{\alpha\beta}$ is initially chosen arbitrarily. The following symmetrization can always be made by an interchange of dummy variables:

$$V = \frac{1}{2} (\sum_{\alpha\beta} \zeta_{\alpha\beta} r_\alpha r_\beta + \sum_{\alpha\beta} \zeta_{\alpha\beta} r_\alpha r_\beta) = \frac{1}{2} (\sum_{\alpha\beta} \zeta_{\alpha\beta} r_\alpha r_\beta + \sum_{\alpha\beta} \zeta_{\beta\alpha} r_\beta r_\alpha)$$

$$= \frac{1}{2} (\sum_{\alpha\beta} \zeta_{\alpha\beta} r_\alpha r_\beta + \sum_{\alpha\beta} \zeta_{\beta\alpha} r_\alpha r_\beta) = \sum_{\alpha\beta} \frac{\zeta_{\alpha\beta} + \zeta_{\beta\alpha}}{2} r_\alpha r_\beta . \tag{2.2}$$

Note that it is always the case that

$$\zeta_{\alpha\beta} + \zeta_{\beta\alpha} = \zeta_{\beta\alpha} + \zeta_{\alpha\beta}.$$

So, whatever coefficients are used to start with, they can always be rearranged into a symmetrical form. Therefore $\zeta_{\alpha\beta}$ can, with complete generality, be chosen to be symmetrical in the first place, i.e.,

$$\zeta_{\alpha\beta} = \zeta_{\beta\alpha}. \tag{2.3}$$

If one considers the higher order terms in the potential, one sees that there are similar general considerations that lead to the following.

$$\zeta_{\alpha\beta\gamma} = \zeta_{\alpha\gamma\beta} \tag{2.4a}$$

$$\zeta_{\alpha\beta\gamma} = \zeta_{\gamma\alpha\beta} \tag{2.4b}$$

and

$$\zeta_{\alpha\beta\gamma\delta} = \zeta_{\alpha\beta\delta\gamma} \tag{2.5a}$$

$$\zeta_{\alpha\beta\gamma\delta} = \zeta_{\beta\alpha\delta\gamma} \tag{2.5b}$$

$$\zeta_{\alpha\beta\gamma\delta} = \zeta_{\delta\alpha\beta\gamma}. \tag{2.5c}$$

It is left as an exercise to show these symmetry properties, which underlie many basic optical and nonlinear optical properties of matter such as the symmetry of the dielectric tensor, the Manley-Rowe relations of nonlinear optics, and the existence of effective nonlinear coefficients.

The kinetic energy of the particles is written in a fashion similar to the potential energy, and reads

$$T = \sum_{\alpha\beta} t_{\alpha\beta} \dot{r}_\alpha \dot{r}_\beta, \tag{2.6}$$

where $t_{\alpha\beta}$ are constants. Following the same procedure that led to Eq. (2.3), one can show that $t_{\alpha\beta}$ can always be written in a symmetrical form, i.e.,

$$t_{\alpha\beta} = t_{\beta\alpha}. \tag{2.7}$$

One of the fundamental theorems of classical mechanics is that there exists a coordinate system that is a linear combination of the r_α's, i.e.,

$$q_\gamma = \sum_\alpha A_{\gamma\alpha}\, r_\alpha, \tag{2.8}$$

where $A_{\gamma\alpha}$ is a constant matrix defining a new coordinate system denoted q_γ such that $t_{\alpha\beta}$ and $\zeta_{\alpha\beta}$ are diagonal in this new coordinate system. The q_γ's are called the normal coordinates. When the motions are described by these coordinates, they are in the form of decoupled normal modes. It is in this set of coordinates that we work from now on. We rarely need to use the formal aspects of this development: it is usually easy to figure out what the normal coordinates are without formal development and to write down the appropriate equations by inspection.

We use a special notation for the tensors in this case. For example, for the kinetic energy

$$T = \frac{1}{2} \sum_\alpha M_{\alpha\alpha}\, \dot{q}_\alpha^2 \tag{2.9}$$

and for the potential energy

$$V = \frac{1}{2} \sum_\alpha k_{\alpha\alpha}\, q_\alpha^2 + \frac{1}{3} \sum_{\alpha\beta\gamma} k_{\alpha\beta\gamma}\, q_\alpha\, q_\beta\, q_\gamma + \frac{1}{4} \sum_{\alpha\beta\gamma\delta} k_{\alpha\beta\gamma\delta}\, q_\alpha\, q_\beta\, q_\gamma\, q_\delta + \cdots \tag{2.10}$$

where the $M_{\alpha\beta}$ and $k_{\alpha\beta}$ are diagonal. Therefore, in trying to find the normal coordinates, the goal is to express the first terms in the potential expansion in this form. Conversely, any complete expansion of the potential energy in this form, i.e., decoupled terms quadratic in some coordinates, implies that the coordinates are normal coordinates. We simplify the notation using

$$M_{\alpha\alpha} \equiv M_\alpha\,, \qquad k_{\alpha\alpha} \equiv k_\alpha, \tag{2.11}$$

where the M_α are the generalized masses and the k_α are the generalized restoring constants. From symmetry considerations

$$k_{\alpha\beta\gamma} = k_{\alpha\gamma\beta} \tag{2.12a}$$

$$k_{\alpha\beta\gamma} = k_{\gamma\alpha\beta} \tag{2.12b}$$

and

$$k_{\alpha\beta\gamma\delta} = k_{\alpha\beta\delta\gamma} \tag{2.13a}$$

$$k_{\alpha\beta\gamma\delta} = k_{\beta\alpha\gamma\delta} \tag{2.13b}$$

$$k_{\alpha\beta\gamma\delta} = k_{\delta\alpha\beta\gamma}. \tag{2.13c}$$

These relations, which are just special cases of Eqs. (2.4) and (2.5), are used in our discussion of electro-optics and nonlinear optics.

The equations of motion of the normal coordinates follow immediately from classical mechanics. The force on the α'th normal coordinate q_α is given by

$$F_\alpha = -\frac{\partial V}{\partial q_\alpha}. \tag{2.14}$$

The linear equations of motion of the normal coordinates are

$$M_\alpha \ddot{q}_\alpha = -k_\alpha q_\alpha, \tag{2.15}$$

i.e.,

$$\ddot{q}_\alpha + \omega_\alpha^2 q_\alpha = 0, \tag{2.16}$$

where

$$\omega_\alpha = \left(\frac{k_\alpha}{M_\alpha}\right)^{1/2} \tag{2.17}$$

is the resonant frequency of the oscillation. Equation (2.16) describes a classical simple harmonic oscillator with the solution

$$q_\alpha = A \sin\omega_\alpha t + B \cos\omega_\alpha t. \tag{2.18}$$

The constants A and B are determined by the initial position $q_\alpha(0)$ and velocity $\dot{q}_\alpha(0)$, i.e.,

$$q_\alpha = \frac{\dot{q}_\alpha(0)}{\omega_\alpha} \sin\omega_\alpha t + q_\alpha(0) \cos\omega_\alpha t. \tag{2.19}$$

Equation (2.19) describes the motion of a vibrational normal coordinate.

The translational and rotational normal coordinates can be recovered from Eq. (2.19) in the limit $\omega_\alpha \to 0$ (there are no restoring forces for translational or rotational motion). The solution is

$$q_\alpha = \dot{q}_\alpha(0)t + q_\alpha(0), \tag{2.20}$$

which corresponds to rotation if q_α is an angular coordinate, and to translation if q_α is a spatial coordinate.

2.3 NORMAL MODES

The rest of this chapter deals with the normal modes associated with each normal coordinate. The normal modes are the particle motions described by Eqs. (2.19) and (2.20). There are two limiting cases. In one, the relevant modes describe local motions. Each molecule constitutes a basic unit described by the q's, and the motions occur within the molecule. In contrast, sound waves involve forces acting between molecules, and the coordinates involve the entire medium. In this latter case the terminology collective motion or collective modes is used.

It is important to distinguish between these in optics since the motion associated with local modes occurs over distances small compared to a wavelength, whereas collective modes involve motions over distances large compared to a wavelength.

Since local modes require only intramolecular forces, they occur in all three phases of matter, gas, liquid, and solid. On the other hand, collective modes require intermolecular forces for motions to be coupled over long distances. Dilute gases in which intermolecular collisions occur with a frequency comparable to or less than the collective mode frequency cannot support collective modes. Dense gases, liquids, and solids all support collective modes.

2.3.1 Local Modes

We discuss local modes in terms of molecules of increasing complexity. A single atom has six relevant degrees of freedom, three for the atom-core electron combination and three for the optically active electron. There are three translational degrees of freedom as illustrated in Figure 2.1a. There are also three electronic degrees of freedom as shown in Figure 2.1b.

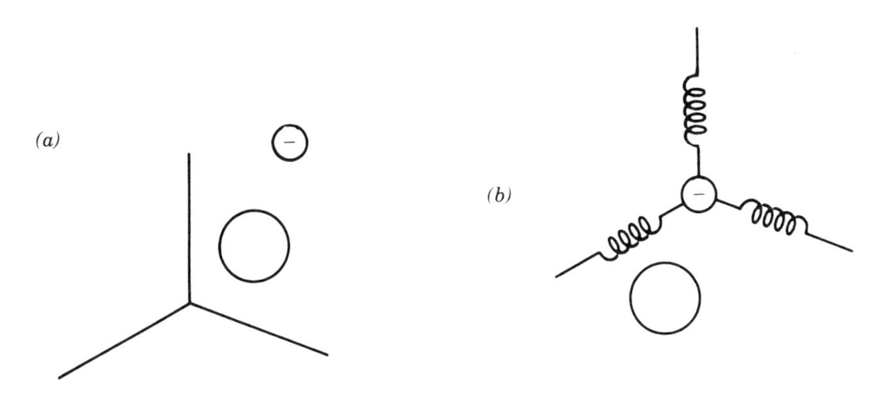

Figure 2.1. Three orthogonal degenerate degrees of freedom for a) translation and b) electronic vibration.

The electronic excitations are modeled as electrons on springs attached to the molecule. Since we have not yet included electric and magnetic fields, and since gravitational forces are too weak to play any role, the space in which the molecule sits has no preferred directions and is called isotropic. Any three orthogonal axes can be used as normal coordinates in which case the normal coordinates are said to be degenerate. Note that the spring system associated with electronic excitation also leads to degenerate normal coordinates because a single atom does not have a preferred orientation and hence all orientations of the springs are equivalent. This means that the springs necessarily have equal spring constants.

The next most complicated system is a diatomic molecule. In this case there are two atoms with their core electrons, and one optically active electron, i.e., nine degrees of freedom. These are sketched in Figure 2.2. Again there are three degenerate translational degrees of freedom and three electronic degrees of

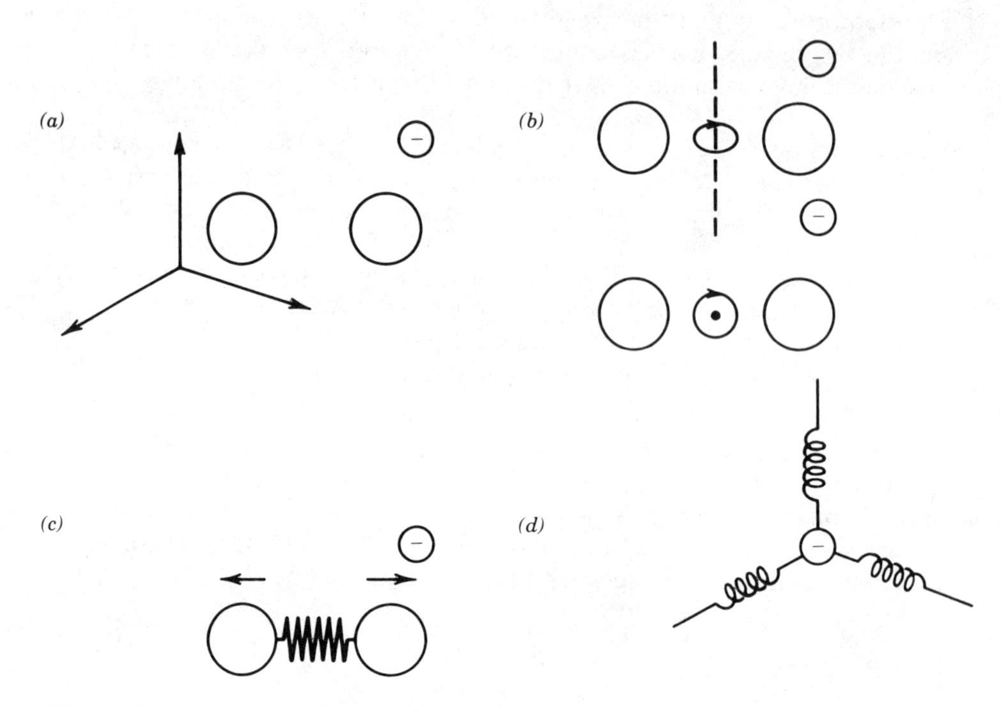

Figure 2.2. Degrees of freedom in diatomic molecule: a) three degenerate translations, b) two degenerate rotations, c) one vibration, and d) three electronic vibrations (two degenerate).

freedom. Note that all directions in a plane orthogonal to the molecular axis are equivalent and hence there is a twofold degeneracy in the electronic degrees of freedom; two spring constants for the electron are the same, and one (the one parallel to the molecular axis) is different. There are two rotational degrees of freedom (rotation about the molecular axis is part of the electronic motion) and one vibrational degree of freedom. Note that no coupling exists between the vibrational or rotational motion and the electronic coordinates. This is a consequence of the Born-Oppenheimer approximation, which asserts that the electrons follow adiabatically (i.e., essentially instantaneously) the motions of the nuclei.

There are two-cases within diatomics which need to be distinguished. One is for symmetrical molecules such as N_2, illustrated in Figure 2.3a, and the other is for asymmetric molecules such as HF, illustrated in Figure 2.3b. Asymmetric molecules can have permanent dipole moments. There can, however, be no permanent dipole in the rest structure of a symmetric diatomic. Imagine that one drew a dipole in the symmetric molecule in Figure 2.3a. By symmetry, one can interchange the nuclei without altering the picture. At this point, one can return the atoms to their original position by flipping both the atoms and the arrow (i.e., imagine looking at the picture in a mirror). This new picture must be as valid as the old, but the arrow has reversed direction. The only arrow that can point in more than one direction simultaneously is one of zero length. By similar arguments, one can show that in symmetrical molecules, all cubic terms in Eq. (2.10) are zero, while in the asymmetric molecule, cubic terms are allowed.

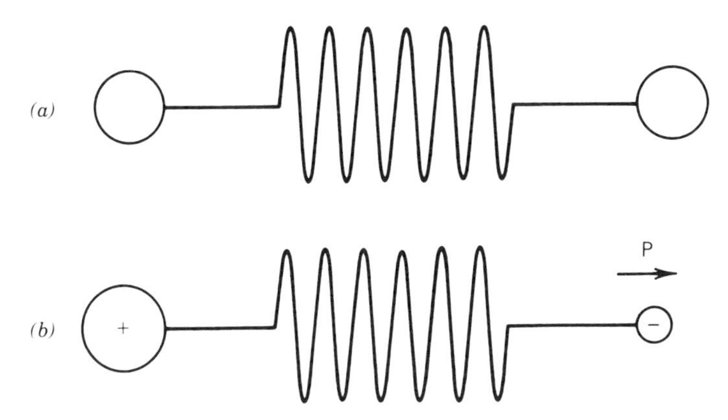

Figure 2.3. Illustration of a) symmetrical diatomic with no permanent dipole moment, and b) asymmetric diatomic molecule with a permanent dipole moment.

2.3.2. Collective Modes

Collective modes are a consequence of intermolecular forces in condensed matter. The simplest example is a crystalline solid which has well defined equilibrium atomic positions and restoring forces. The material is made up of unit cells. These cells correspond to the simplest configurations of atoms which repeat themselves along symmetry axes. Unit cells lose their translational and rotational degrees of freedom, which are replaced by collective excitations, i.e., sound waves (acoustic phonons). The simplest model which produces sound waves is a linear chain of atoms. Since it is a one-dimensional "crystal," it cannot reproduce all of the characteristics of sound waves in solids. Nevertheless, it does remarkably well.

Consider the one-dimensional chain of identical atoms shown at their equpilibrium positions in Figure 2.4. We denote the equilibrium position of the α'th atom in the chain by R_α and its instantaneous displacement from equilibrium by r_α. The atoms are separated by springs with force constant ζ, so the potential energy of the chain is given by

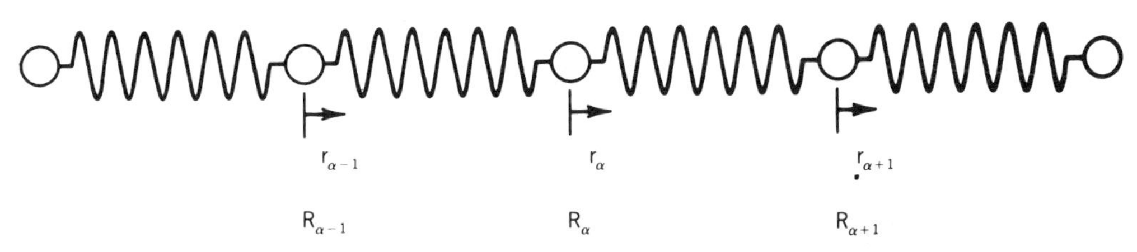

Figure 2.4. One-dimensional chain of identical atoms.

$$V = \sum_{\alpha} \zeta \, (r_{\alpha} - r_{\alpha+1})^2 \tag{2.21}$$

and the kinetic energy by

$$T = \frac{1}{2} m \sum_{\alpha} \dot{r}_{\alpha}^2. \tag{2.22}$$

From Eqs. (2.14) and (2.21) we obtain

$$m\ddot{r}_{\alpha} = - 2\zeta \, (2r_{\alpha} - r_{\alpha-1} - r_{\alpha+1}). \tag{2.23}$$

Equation (2.23) can be diagonalized by the transformation

$$r_{\alpha} = \frac{1}{2} \sum_{\beta} (d_{\beta} \, e^{i\,\alpha 2\pi\beta/N} + \mathrm{cc}), \tag{2.24}$$

which, when substituted into Eq. (2.23), gives

$$\ddot{d}_{\beta} = - \omega_{\beta}^2 d_{\beta} \, , \tag{2.25}$$

where

$$\omega_{\beta} = 2 \left(\frac{2\zeta}{m}\right)^{1/2} \sin(\beta\pi/N). \tag{2.26}$$

Since d_{β} is a complex number, it is not a normal coordinate; rather it is a linear combination of two normal coordinates. Our choice of the term $2\pi\beta/N$ in Eq. (2.24) is made obvious below.

The normal coordinates are found by solving Eq. (2.25) for pairs of sound waves traveling in opposite directions. Writing the solution as

$$d_{\beta} = A_{\beta} \, e^{i\omega_{\beta}t} + B_{\beta} \, e^{-i\omega_{\beta}t} \tag{2.27}$$

and substituting it into Eq. (2.24) gives

$$r_{\alpha} = \frac{1}{2} \sum_{\beta} \left(A_{\beta} \, e^{i(\alpha\beta \, 2\pi/N \, + \, \omega_{\beta}t)} + B_{\beta} \, e^{i(\alpha\beta \, 2\pi/N \, - \, \omega_{\beta}t)}\right) + \mathrm{cc}. \tag{2.28}$$

For N atoms separated by a distance a the position of the α'th atom is given by

$$R_{\alpha} = \alpha a, \tag{2.29}$$

and for a chain of length L

$$L = aN, \tag{2.30}$$

which gives

$$\alpha = \frac{R_\alpha N}{L}.$$

Therefore, defining the acoustic wave vector as

$$k_\beta = \frac{2\pi}{L}\,\beta \tag{2.31}$$

then

$$r_\alpha = \frac{1}{2}\sum_\beta \left(A_\beta\, e^{i(R_\alpha k_\beta - \omega_\beta t)} + B_\beta\, e^{i(R_\alpha k_\beta + \omega_\beta t)}\right) + \text{cc}. \tag{2.32}$$

The two terms correspond to oppositely traveling sound waves and the A_β and B_β are the normal coordinates, i.e., the normal mode amplitudes. The acoustic velocity is $v_\beta = \omega_\beta / k_\beta$.

The relationship between k_β and ω_β is called the dispersion relation. For this case

$$\omega_\beta = 2\left(\frac{2\zeta}{m}\right)^{1/2} \sin(k_\beta L/2N). \tag{2.33}$$

Unique frequencies are obtained for

$$0 < \frac{k_\beta L}{2N} < \frac{\pi}{2}. \tag{2.34}$$

[Negative frequencies are automatically included when we write our solutions in the form of Eq. (2.27).] The dispersion relation is sketched in Figure 2.5. For $k_\beta L/2N \ll 1$,

$$v_\beta = \frac{\omega_\beta}{k_\beta} = \left(\frac{2\zeta a^2}{m}\right)^{1/2}, \tag{2.35}$$

i.e., the acoustic velocity is a constant, independent of ω_β and k_β. This is called the hydrodynamic limit, and the acoustic modes with which light interacts belong to this regime. Deviations in linearity and the upper regions of the dispersion curve are studied, for example, by neutron scattering.

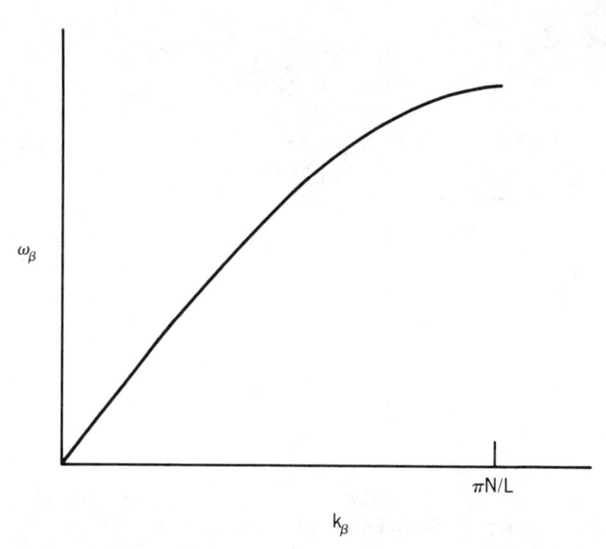

Figure 2.5. Dispersion curve for a sound wave in a one-dimensional lattice.

The values of β are not arbitrary. From Eqs. (2.31) and (2.34)

$$0 < \frac{2\beta}{N} < 1. \tag{2.36}$$

Therefore the maximum value of β is N/2 and the allowed values are

$$\beta = 1, 2, 3, \cdots, \frac{N}{2}. \tag{2.37}$$

Since there are two normal coordinates for each value of β (i.e., two oppositely propagating sound waves), then there are N normal coordinates, just as required for N atoms.

To show that the values of β are integers, it is necessary to impose boundary conditions on the chain. In any real problem, these conditions can be tricky and involve the surface physics of the medium in important ways. A convenient fiction that ignores these complications is to impose periodic boundary conditions, which work well for $N \gg 1$. One assumes that the chain is infinitely long, and that the basic chain of N atoms is repeated, i.e.,

$$r_\alpha = r_{\alpha+N} = r_{\alpha+2N}, \text{ etc.} \tag{2.38}$$

Thus from Eq. (2.24),

$$e^{i(\alpha+N)\beta\, 2\pi/N} = e^{i\alpha\beta\, 2\pi/N}, \tag{2.39}$$

which requires that

$$e^{i2\pi\beta} = 1.$$

This is true if

$$\beta = 1, 2, \cdots, \frac{N}{2}, \tag{2.40}$$

where we have used Eq. (2.36) for the upper limit. It is also clear from the preceding discussion that

$$\beta = 1 + \frac{N}{2}, 2 + \frac{N}{2}, \cdots, N \tag{2.41}$$

and

$$\beta = 1 + N, 2 + N, \cdots, \frac{3N}{2} \tag{2.42}$$

also satisfy all of the equations and give equivalent results for the normal coordinates and normal modes. The range defined by Eq. (2.40) is called the first Brillouin zone, by Eq. (2.41) is the second Brillouin zone, etc. For the light-sound interactions of interest, we restrict ourselves to the first Brillouin zone.

If one considers all three dimensions of the crystal, one finds that there are three acoustic modes, one longitudinal mode discussed already in this section, and two transverse modes. The transverse modes, which correspond to bending the chain of atoms, can be understood using methods similar to those developed here. These three modes correspond to the three translational degrees of freedom in three-dimensional space.

If there is more than one atom per unit cell, then, in addition to acoustic phonons, one has optical phonons. The distinction between acoustic and optical phonons is illustrated in Figure 2.6. In the ideal limit, optical phonons have motions within a unit cell, whereas acoustic modes move the unit cell as a whole. If the forces between the atoms in the unit cell are much stronger than the intercell binding forces (Figure 2.6c), the acoustic and optical phonons become decoupled, i.e., independent of one another. Then the optical phonon has a frequency ω_β that is independent of β and the intracell motions are not transferred from cell to cell. This describes N independent oscillators (for two atoms per unit cell). Such a circumstance is to be expected when the molecule is covalently bonded (e.g., solid organics). If the coupling strengths are all about equal (Figure 2.6d), the optical phonon is truly collective and is coupled to acoustic-type modes. There are still normal modes but their motions have both an acoustical and an optical phonon component. Even then, the dispersion in the quasi-optical phonon frequency is less than the quasi-acoustic phonon.

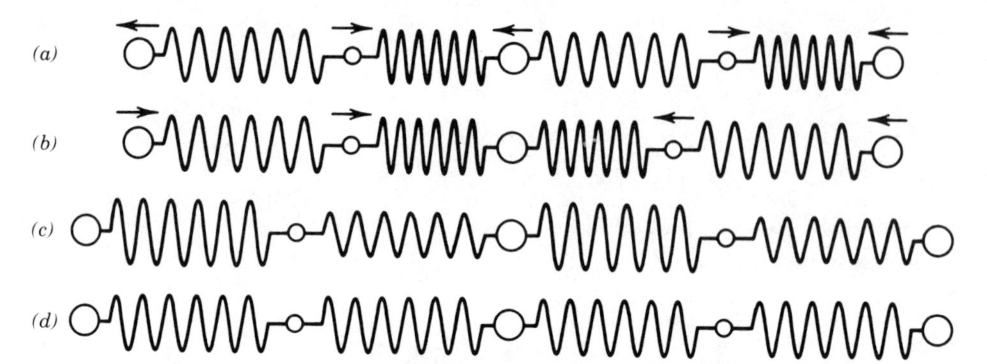

Figure 2.6. Lattice with two dissimilar atoms. Motion of a) optical phonons, b) acoustic phonons. c) Alternate springs different, properties of optical and acoustic phonon modes distinct. d) Similar springs, properties of modes mixed.

In this chapter we have reviewed the basic types of motion that occur in media. All motions can be expressed as a summation over the normal modes as described by normal coordinates. The motions of the normal modes are decoupled from one another. The modes considered here are those appropriate to insulators and consist of vibrations, rotations, sound waves, and electronic vibrations. In the next chapter we show how these couple to optical fields.

ADDITIONAL READING

Basic Mechanics

Goldstein, H. *Classical Mechanics*, 2nd Edition, Addison-Wesley, Reading, MA, 1980. Chapter 6 (1st Ed. Chapter 10).

Molecules

Flygare, W.H. *Molecular Structure and Dynamics*, Prentice Hall, Englewood Cliffs, NJ, 1978. Chapter 2.

Wilson, E.B., Decius, J.C., and Cross, P.C. *Molecular Vibration*, McGraw-Hill, New York, 1955. Chapter 2. This book has many useful figures depicting vibrational motion.

Herzberg, G. *Molecular Spectra and Molecular Structure*, Van Nostrand Reinhold, New York, 1945. Chapters 1,2.

Electrons as Vibrations

Fowles, G.R. *Introduction to Modern Optics*, Holt, Rinehart and Winston, New York, 1968. Chapter 7.

Sargent, M., Scully, M.O., and Lamb, W.E. Jr. *Laser Physics*, Addison-Wesley, London, 1974. Chapter 3.

Treating the Electron as a Normal Coordinate

Feynman, R.P., Leighton, R.B., and Sands, M. *The Feynman Lectures in Physics*, Addison-Wesley, Reading, MA, 1964. Volume I, pp. 31-38.

PROBLEMS

2.1. Define a general tensor

$$\mathbf{J} = \sum_\beta \mathbf{r}_\beta \mathbf{r}_\beta m_\beta$$

where $\mathbf{r}_\beta$ is the vector from the origin to a point mass m_β. $\mathbf{J}$ is, in general, dependent on the coordinate system. Show that there exists a coordinate system in which $\mathbf{J} = \mathbf{aa}M + \boldsymbol{I}$ where $\mathbf{r} = \mathbf{a} + \mathbf{r}'$,

$$M = \sum_\beta m_\beta,$$

$$\sum_\beta \mathbf{r}_\beta' \, m_\beta = 0$$

and

$$\boldsymbol{I} = \sum_\beta \mathbf{r}_\beta' \, \mathbf{r}_\beta' \, m_\beta.$$

Here $\mathbf{r}'$ is called the center of mass coordinate system, and $\boldsymbol{I}$ is diagonal in the same coordinate system in which the moment of inertia tensor is diagonal.

2.2. The normal modes of rotation are defined by the center of mass coordinate system in which $\boldsymbol{I}$ is diagonal. Using these facts, show that the rotational normal modes of the water molecule in Figure 3.3 are correctly drawn.

2.3. Verify Eqs. (2.4a), (2.4b), (2.5a), (2.5b) and (2.5c).

2.4 Show that, in a symmetric diatomic molecule, the cubic terms in Eq. (2.10) (i.e., those in Eqs. (2.12a) and (2.12b)) are zero.

2.5. Derive Eqs. (2.25) and (2.26) from Eqs. (2.23) and (2.24).

2.6. Consider the chain of molecules in Figure 2.4 for the case where the spring constants are all equal but the alternating masses are different. Derive the normal coordinates of this case by applying periodic boundary conditions. Hint: Proceed exactly as in the equal-mass case, but substitute the normal coordinates as

$$r_\alpha = \frac{1}{2} \sum_\beta q_\beta \, e^{2\pi i \alpha\beta/N} + cc, \quad \alpha = 1,3,5,\cdots$$

$$r_\alpha = \frac{\gamma}{2} \sum_\beta q_\beta \, e^{2\pi i\alpha\beta/N} + cc, \quad \alpha = 2,4,6,\cdots$$

where γ is a constant that is determined by requiring that the equation for q_β must be the same for all α.

2.7. (a) Consider the axial vibrations of the mass and spring assembly illustrated

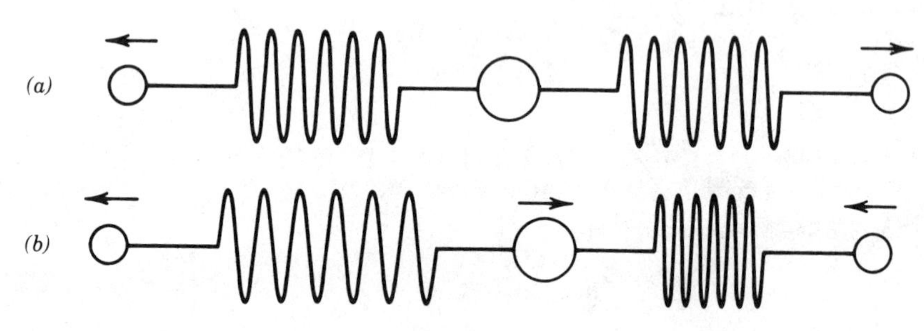

Figure 2.7. Symmetrical triatomic molecule: a) symmetric stretch and b) asymmetric stretch.

in Figure 2.7, where the outer masses are identical and the springs are identical. Show that the normal modes are the symmetrical stretch and the asymmetrical stretch. You need to define a center of mass system for this motion.

(b) Consider the bending motion of the system illustrated in Figure 2.7. show that there are two-degenerate bending vibrations. Show that, if the two-vibrations are 90° out of phase, they describe a rotation. This is called vibrational angular momentum.

(c) Assume that the mass of the central atom in Figure 2.7 is much less than the mass of the outer two. Estimate the frequencies of vibration of the two stretch modes.

3
The Interaction of Light with Matter

There are three basic elements involved in understanding the interaction of light with ordinary matter. First, light is an oscillating electromagnetic field that produces forces on charges via Eq. (1.9). Second, it is the electrons which are highly mobile and are therefore primarily responsible for the interaction. Third, the atoms and molecules are very small compared to the wavelength of the electromagnetic waves ($\lambda \simeq 1$ mm - 100 nm) of interest in this book. This means that, to a first approximation, the electrical forces and fields can be evaluated at the center of mass $\mathbf{r}_a$ of the a'th atom or molecule, i.e.,

$$\begin{aligned} F(\mathbf{r},t) &\simeq F(\mathbf{r}_a,t) \\ E(\mathbf{r},t) &\simeq E(\mathbf{r}_a,t). \end{aligned} \tag{3.1}$$

This is called the dipole approximation.

Our starting point is the interaction potential that describes the coupling of matter to an electromagnetic field. The leading term in this interaction is

$$V_{int} = -\int \mathbf{P}\cdot\mathbf{E}\,d\mathbf{r}, \tag{3.2}$$

where $\mathbf{P}$ is the polarization. The polarization is the sum over the dipoles of all of the molecules, which in turn can be expressed as a sum over all of the normal coordinates of the system. When the interaction potential is written as an integral, each point dipole is multiplied by a three dimensional delta function $\delta^{(3)}(\mathbf{r}-\mathbf{r}_a)$ (see Problem 3.10). Evaluating the integrals gives

$$V_{int} = -\sum_{\alpha} \mathbf{p}_{\alpha,a}\cdot E(\mathbf{r}_a). \tag{3.3}$$

Here $\mathbf{p}_\alpha$ is the dipole of the α'th normal coordinate of the a'th molecule. In writing down only the dipole term in Eq. (3.3) we implicitly assume that higher order effects involving magnetic dipoles, quadrapoles, octapoles, etc., can be neglected, which is usually the case. In Appendix B the magnetic dipole is discussed explicitly and exercises are given in which one can work out electric quadrapole interactions.

The motions of normal modes may or may not contribute to the dipole $\mathbf{p}_\alpha$. If the motion of the normal coordinate q_α is associated directly with an oscillating dipole moment (i.e., $q_\alpha \neq 0$ implies $\mathbf{p}_\alpha \neq 0$), then the mode is called dipole active. The motions associated with the normal mode may change a permanent dipole moment such as to give it an oscillating component. For example, a rotational normal mode which results in changes in the orientation of a permanent dipole moment (see Figure 2.3b) relative to the incident field is dipole active. In this molecule, the vibrational coordinate alters, sinusoidally, the permanent dipole moment, and hence it is also dipole active. Alternatively, the motion of q_α may cause an oscillating dipole in a molecule that does not otherwise have a dipole moment. These cases are referred to as induced dipoles.

If the motion of a normal coordinate q_β has no associated dipole (i.e., if $\mathbf{p} = 0$ when only q_β is nonzero) then the mode is said to be dipole inactive. Many dipole inactive modes are Raman active. Raman active modes modulate the dipole produced by a dipole active normal coordinate, i.e., $\mathbf{p}_\alpha$ is a function of q_β. In applications of Raman activity in the visible, it is nearly always the case that the dipole active coordinate is the electron motion. An incident electromagnetic field induces a polarization in each molecule via the molecular polarizability due to the electrons. This polarization, in turn, interacts with the incident field. If the vibrational or rotational normal modes modulate this polarization (and hence the interaction potential), these Raman active modes can be altered by the field, even if they are not dipole active. When the electron motion is the dipole active coordinate, then there is a strong tendency for dipole inactive rotations and vibrations to be strongly Raman active, and for dipole active rotations and vibrations to be weakly Raman active. A discussion of this tendency (which is not universal even for dipole active electron motions and need not apply at all if the dipole active coordinate is vibrational) involves quantum mechanics, and hence is strictly speaking beyond the scope of this book. Some discussion is given in Problem 3.3. Dipole and Raman activity are discussed in detail in this chapter.

3.1. DIPOLE ACTIVE MODES

We concentrate on dipole active modes because they lead to the strongest interaction with electromagnetic fields. We first inspect the electronic properties of the normal coordinates, i.e., the electronic normal modes. In Figure 3.1, the displacement of an electron on a spring is shown to induce a dipole moment. At rest, there may or may not be a permanent dipole moment. When the electron is displaced from its rest position, either a dipole is induced or a change in permanent dipole occurs. Therefore the electronic normal modes are dipole active. In order to avoid awkwardness in discussion, we stop making explicit reference to the case of changes in permanent dipole when discussing induced dipoles. Unless otherwise stated the two cases behave the same way.

Vibrational modes can also be dipole active. Consider, for example, a CO_2 molecule which is a linear molecule in its equilibrium configuration as illustrated in Figure 3.2a. The carbon, which has a small positive charge, is directly between the negatively charged oxygens, so that the average positions of the positive and negative charges coincide. There are 12 degrees of freedom, of which four are vibrational (the others are: three translational; three electronic; two rotational).

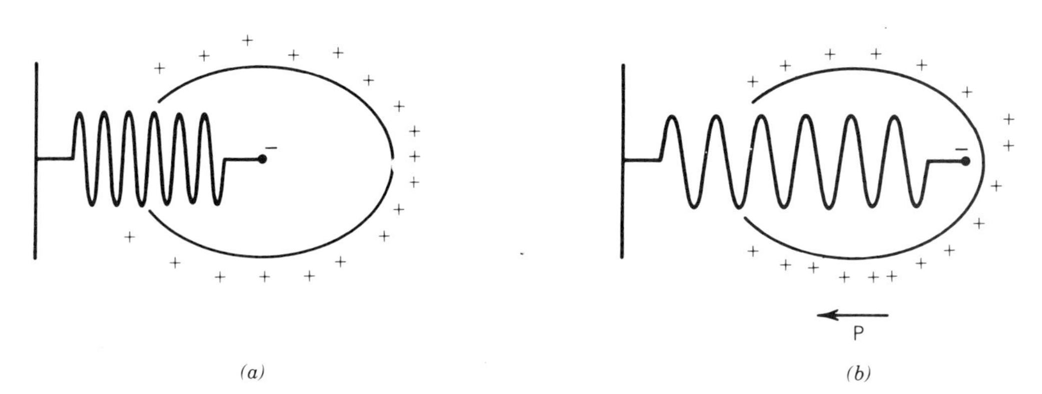

Figure 3.1. Induced dipole moment produced by the displacement of an electron from its rest position: a) electron at rest, and b) electron displaced producing a dipole labeled **p**.

One of the two degenerate bending motions of CO_2 is sketched in Figure 3.2, and the stretching motions are shown in Figure 2.7. Since the symmetric stretch, illustrated in Figure 2.7a, does not displace the centers of charge, no dipole is

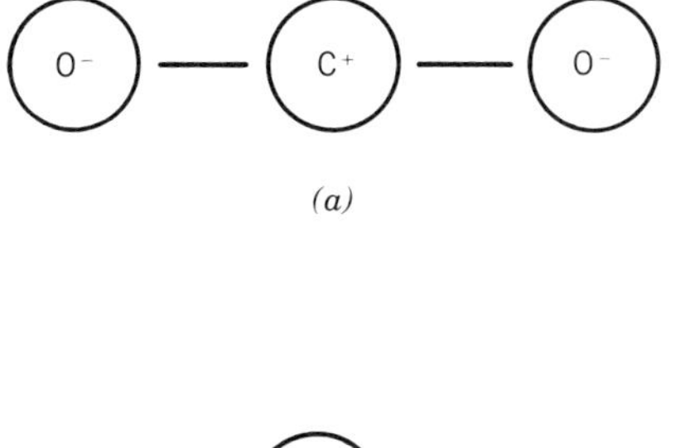

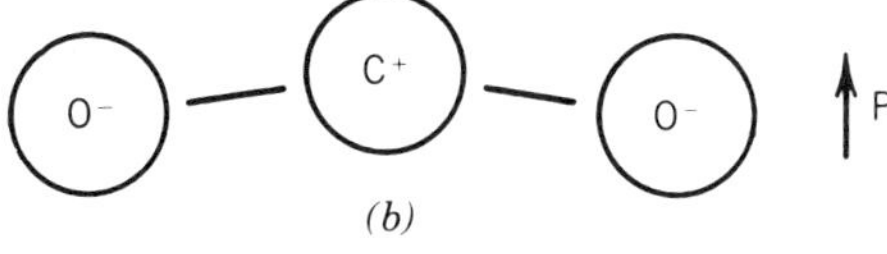

Figure 3.2. Dipole active bending mode of CO_2: a) CO_2 at rest, no dipole, and b) CO_2 bent with dipole **p**.

induced and the mode is not dipole active. On the other hand, the asymmetric stretch, illustrated in Figure 2.7b, does displace the centers of charge, thus inducing a dipole. Hence it is dipole active. For the bending modes, dipoles are induced when the molecules are bent and these normal modes are dipole active.

If the molecule has a permanent dipole moment, the same arguments given above determine whether a vibrational mode is dipole active or dipole inactive. That is, dipole active modes change the relative positions of the centers of negative and positive charge.

If a molecule has a permanent dipole moment, rotational modes may be dipole active. In Figure 3.3 we illustrate the dipole active and inactive modes of a

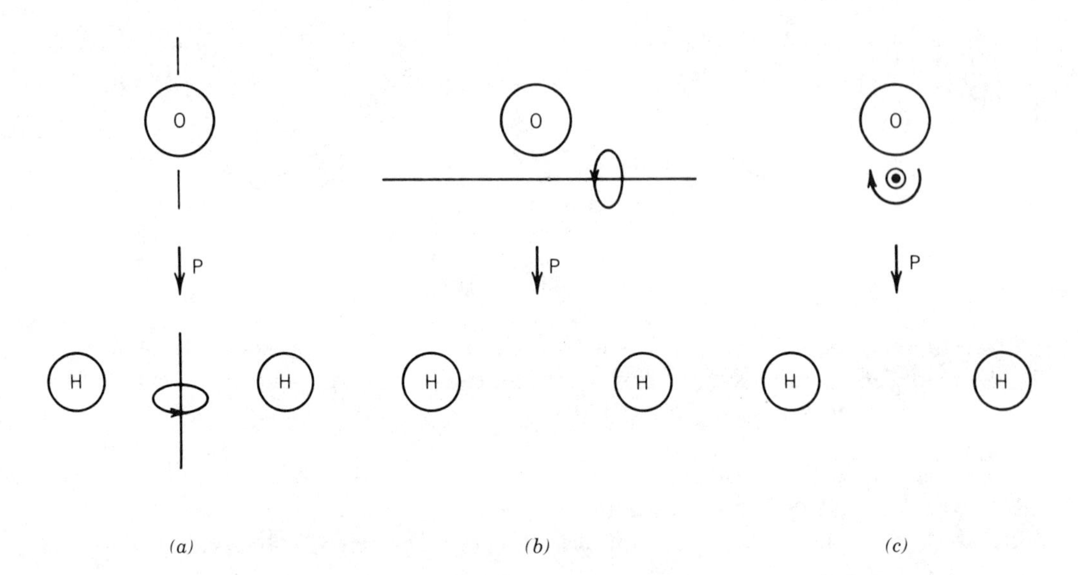

Figure 3.3. Dipole activity of rotational modes of water vapor; a) dipole inactive rotational mode, b) first dipole active rotational mode, c) second dipole active rotational mode. Permanent dipole of H_2O is labeled **p.** Dashed line is axis of rotation.

water vapor molecule which has a permanent dipole moment. If the rotation axis coincides with the dipole moment, rotation about that axis produces no change in the dipole moment relative to an electromagnetic field in space. Thus the mode is not dipole active. For the other two rotational modes, the orientation of the dipole relative to the field changes in time and hence the interaction potential (Eq. (3.3)) also changes. This is the case with the modes illustrated in Figure 3.3b and 3.3c. Therefore these two modes are dipole active.

3.2. DIPOLE ACTIVE MODES INTERACTING WITH ELECTROMAGNETIC FIELDS

For dipole active modes, the induced dipole is proportional to the normal coordinate. There is therefore a nonzero force (or torque) acting on the molecule because $\partial V_{int}/\partial q_\alpha$ is nonzero. For rotational normal modes there is a torque which tends to align the dipole along the field direction. For both vibrational and electronic normal modes, there are forces. The combination of forces and displacements leads to an energy exchange between the normal modes and an electromagnetic field. The goal of this section is to calculate the frequency distribution of the energy absorbed from the electromagnetic field.

We now proceed to evaluate the response of the normal coordinate to these forces. The dipole $\mathbf{p}_\alpha$ induced in the α'th normal coordinate is proportional to its amplitude of motion given by q_α, i.e.,

$$\mathbf{p}_\alpha = \boldsymbol{\ell}_\alpha \, q_\alpha \tag{3.4}$$

where $\boldsymbol{\ell}_\alpha$ is a vector which at all times points in the direction of the induced dipole. Note that $\boldsymbol{\ell}_\alpha$ is not necessarily parallel to any particular direction associated with the normal coordinate, e.g., in the case of rotational normal modes $\boldsymbol{\ell}_\alpha$ is usually normal to the axis of rotation. The magnitude of $\boldsymbol{\ell}_\alpha$ is a measure of the strength of the interaction between the normal coordinate and the field. In the usual case in which q_α has units of length, one can write $\boldsymbol{\ell}_\alpha$ as

$$\boldsymbol{\ell}_\alpha = e_\alpha \, \hat{\boldsymbol{\ell}}_\alpha, \tag{3.5}$$

where e_α is the effective charge separation produced per unit q_α (for generality we usually denote the amplitude of $\boldsymbol{\ell}_\alpha$ as ℓ_α). Thus the interaction potential (Eq. (3.3)) is written as

$$V_{int} = - \sum_\alpha q_\alpha \, \boldsymbol{\ell}_\alpha \cdot \mathbf{E} \tag{3.6}$$

and the force on the α'th normal coordinate is

$$F_\alpha = - \frac{\partial V_{int}}{\partial q_\alpha} = \boldsymbol{\ell}_\alpha \cdot \mathbf{E}. \tag{3.7}$$

Let us now include this force in Eq. (2.16) and introduce a phenomenological damping term linear in $\dot{q}_\alpha$ to give

$$\ddot{q}_\alpha + \Gamma_\alpha \dot{q}_\alpha + \omega_\alpha^2 q_\alpha = \frac{1}{M_\alpha} \boldsymbol{\ell}_\alpha \cdot \mathbf{E}. \tag{3.8}$$

Here Γ_α is the damping constant introduced to take into account collision effects and spontaneous radiation.

The solution for q_α is relatively straightforward to obtain in a gas. Assuming the incident field to be a plane wave of the form

$$\mathbf{E} = \frac{1}{2} \boldsymbol{\mathcal{E}} \, e^{i(\mathbf{k}\cdot\mathbf{r} - \omega t)} + cc, \tag{3.9}$$

the coordinate response can be written as

$$q_\alpha = \frac{1}{2} Q_\alpha \, e^{i(\mathbf{k}\cdot\mathbf{r} - \omega t)} + cc. \tag{3.10}$$

In a gas, the molecule is moving and, since the interaction must be evaluated at the instantaneous position of the molecule,

$$\mathbf{r}_a = \mathbf{r}_0 + \mathbf{v}(t - t_0) \tag{3.11}$$

where $\mathbf{r}_a$ is the position at time t of a molecule that was at position $\mathbf{r}_0$ at time t_0 (assumed to be $t_0 = 0$ for simplicity) moving with velocity $\mathbf{v}$. This translational motion gives rise to Doppler effects, and, for an ensemble (distribution) of velocities, to Doppler broadening of the spectral lines. Therefore the phasor term in Eq. (3.10) is given by

$$e^{i(\mathbf{k}\cdot\mathbf{r} - (\omega - \mathbf{k}\cdot\mathbf{v})t)} \tag{3.12}$$

There are subtleties involving the method of introducing and then eliminating $\mathbf{r}_a$ from the analysis that are discussed in Problem 3.10 and in Sargent, Scully and Lamb. A complete discussion of this issue involves too much detail to go into in this book. Substituting Eqs. (3.9) and (3.10) into (3.8) with phasor terms given by Eq. (3.12), we get

$$(-(\omega - \mathbf{k}\cdot\mathbf{v})^2 - i(\omega - \mathbf{k}\cdot\mathbf{v})\Gamma_\alpha + \omega_\alpha^2)Q_\alpha = \frac{\boldsymbol{\ell}_\alpha \cdot \boldsymbol{E}}{M_\alpha} \tag{3.13}$$

Following standard notation, we define

$$D_\alpha(\omega - \mathbf{k}\cdot\mathbf{v}) = \omega_\alpha^2 - (\omega - \mathbf{k}\cdot\mathbf{v})^2 - i\Gamma_\alpha(\omega - \mathbf{k}\cdot\mathbf{v}) \tag{3.14}$$

so that the solution for Q_α reads

$$Q_\alpha = -\frac{\boldsymbol{\ell}_\alpha \cdot \boldsymbol{E}}{M_\alpha D_\alpha(\omega - \mathbf{k}\cdot\mathbf{v})}. \tag{3.15}$$

The coupling of the electromagnetic field to the normal coordinates results in power absorbed from the field by the medium, and subsequently dissipated by the damping mechanism (via Γ_α). The mechanical energy, i.e., the energy stored in the normal mode, is

$$E_{mech} = \frac{1}{2}\sum_\alpha (M_\alpha \dot{q}_\alpha^2 + k_\alpha q_\alpha^2). \tag{3.16}$$

Taking the time derivative of this expression and using $k_\alpha = M_\alpha \omega_\alpha^2$ from Eq. (2.17) we obtain

$$\frac{d}{dt}E_{mech} = \sum_\alpha (M_\alpha \ddot{q}_\alpha \dot{q}_\alpha + M_\alpha \omega_\alpha^2 \dot{q}_\alpha q_\alpha). \tag{3.17}$$

Eliminating the $\ddot{q}_\alpha$ term via Eq. (3.8),

$$\frac{d}{dt}E_{mech} = -\sum_\alpha M_\alpha \Gamma_\alpha \dot{q}_\alpha^2 + \sum_\alpha \dot{q}_\alpha \boldsymbol{\ell}_\alpha \cdot \mathbf{E}. \tag{3.18}$$

The significant quantity is the time average of Eq. (3.18). It is left as an exercise to work this out to give

$$\frac{d}{dt}E_{mech} = -\frac{1}{2}\sum_{\alpha} M_{\alpha}(\omega - \mathbf{k}\cdot\mathbf{v})^2 \Gamma_{\alpha}|Q_{\alpha}|^2 + \frac{1}{2}\sum_{\alpha}(\omega - \mathbf{k}\cdot\mathbf{v})\,\mathrm{im}(Q_{\alpha}\boldsymbol{\ell}_{\alpha}\cdot\boldsymbol{E}^*). \quad (3.19)$$

The symbol "im" stands for the imaginary part of the argument. In Eq. (3.19) the first term on the right-hand side of the equality is the power dissipated by the medium (via damping of the normal modes), and the second is the power gained from the electromagnetic field by the medium. In the steady state case, these two terms are equal in magnitude but opposite in sign. The proof that $dE_{mech}/dt = 0$ in steady state is left as an exercise (Problem 3.1). For all calculations other than conservation of energy, we can use the fact that $\omega >> \mathbf{k}\cdot\mathbf{v}$ and write the following relation for the power absorbed by the normal coordinate under steady state conditions:

$$P_{abs} = \frac{1}{2}\,\omega^2 \sum_{\alpha} \Gamma_{\alpha} M_{\alpha} |Q_{\alpha}|^2. \quad (3.20)$$

In the next chapter, we show that this power is actually lost by the electromagnetic field, thus establishing that energy is conserved. Substituting Eq. (3.15) for Q_{α} gives

$$P_{abs} = \frac{1}{2}\,\omega^2 \sum_{\alpha} \frac{\Gamma_{\alpha}|\boldsymbol{\ell}_{\alpha}\cdot\boldsymbol{E}|^2}{M_{\alpha}|D_{\alpha}(\omega - \mathbf{k}\cdot\mathbf{v})|^2}, \quad (3.21)$$

which is the power absorbed by all of the normal modes (i.e., summation over α) by a single molecule.

3.2.1 Doppler broadening

In any real gas, intermolecular collisions occur, and there is a distribution of molecular velocities. Since the absorption is a maximum when $|D_{\alpha}(\omega - \mathbf{k}\cdot\mathbf{v})|^2$ is a minimum, there must also be a distribution of frequencies over which the absorption is a maximum. For light incident along the z axis, only velocity components along z, i.e., v_z, contribute to the broadening. For a Maxwellian velocity distribution (i.e., thermal equilibrium), the probability of finding molecules in the velocity interval $v_z \to v_z\, dv_z$ is

$$\rho(v_z)\, dv_z = \frac{1}{u\sqrt{\pi}}\, e^{-v_z^2/u^2}\, dv_z \quad (3.22)$$

where u is a temperature dependent parameter that determines the root mean squared velocity. Hence the power absorbed by N molecules is

$$P_{abs} = \frac{\omega^2}{2}\, N \sum_{\alpha} \frac{|\boldsymbol{\ell}_{\alpha}\cdot\boldsymbol{E}|^2\Gamma_{\alpha}}{M_{\alpha}} \int_{-\infty}^{\infty} \frac{\rho(v_z)}{|D_{\alpha}(\omega - kv_z)|^2}\, dv_z. \quad (3.23)$$

In most cases, both Γ_{α} and kv_z are small compared to ω and ω_{α}, and Eq. (3.23) can be simplified. In this limit

$$D_{\alpha}(\omega - kv_z) \simeq \omega_{\alpha}^2 - \omega^2 + 2kv_z\omega - i\Gamma_{\alpha}\omega.$$

Furthermore, since $\omega_\alpha \simeq \omega$,

$$D_\alpha(\omega - kv_z) \simeq 2\omega(\omega_\alpha - \omega) + 2kv_z\omega - i\,\frac{\Gamma_\alpha}{2}\,2\omega$$

and

$$|D_\alpha(\omega - kv_z)|^2 \simeq 4\omega^2[(\omega_\alpha - \omega + kv_z)^2 + (\Gamma_\alpha/2)^2]. \tag{3.24}$$

Finally, substituting this and Eq. (3.22) into Eq. (3.23) gives

$$P_{abs} = \frac{1}{8}\,\frac{N}{u\sqrt{\pi}}\,\sum_\alpha \frac{\Gamma_\alpha|\boldsymbol{\ell}_\alpha \cdot \boldsymbol{E}|^2}{M_\alpha} \int_{-\infty}^{\infty} \frac{e^{-v_z^2/u^2}\,dv_z}{(\omega_\alpha + kv_z - \omega)^2 + (\Gamma_\alpha/2)^2} \tag{3.25}$$

3.2.2 Absorption lines

For most spectroscopic measurements, the goal is to measure ω_α and hence identify the normal mode force constants. Usually, one works at low pressures to minimize Γ_α and make the spectral lines as sharp as possible. However, in the visible and infrared, $ku \gg \Gamma_\alpha$ and under these circumstances the absorption line is said to be Doppler, or inhomogeneously broadened. In this case, the denominator can be treated as a Dirac delta function. It is left as an exercise to work out the resulting expression. In the other limit, $\Gamma_\alpha \gg ku$, the Doppler distribution can be regarded as a delta function giving the absorption line

$$P_{abs} = \frac{1}{8}\,N\,\sum_\alpha \frac{|\boldsymbol{\ell}_\alpha \cdot \boldsymbol{E}|^2}{M_\alpha}\;\frac{\Gamma_\alpha}{(\omega_\alpha - \omega)^2 + (\Gamma_\alpha/2)^2} \tag{3.26}$$

This Lorentzian line shape corresponds to a homogeneously broadened line. The two cases are sketched in Figure 3.4. Therefore, in a gas, as frequency is tuned, a series of absorption lines are obtained, one for each of the dipole active normal modes. For most molecules, this means a series of absorptions in the far infrared due to rotational degrees of freedom, in the infrared due to coupling to vibrational modes, and in the visible and ultraviolet due to electronic degrees of freedom.

The situation in liquids and solids is more complicated. In a liquid, the local environment is different from molecule to molecule and hence there is a distribution in the frequencies ω_α. This results in a large broadening of the spectral lines from the electronic and vibrational degrees of freedom. In liquids, the rotation of molecules is limited to a diffusional motion, i.e., $\omega_\alpha \to 0$ and the corresponding absorption is centered at zero frequency. Similar considerations apply to an isotropic solid, in which local strains can lead to additional broadening of the absorption lines.

In crystalline media, absorptions occur primarily due to electronic and vibrational degrees of freedom. For certain soft molecular crystals, vibrational motion, i.e., rotational oscillations, occur in a unit cell and absorption due to these modes does appear in the infrared.

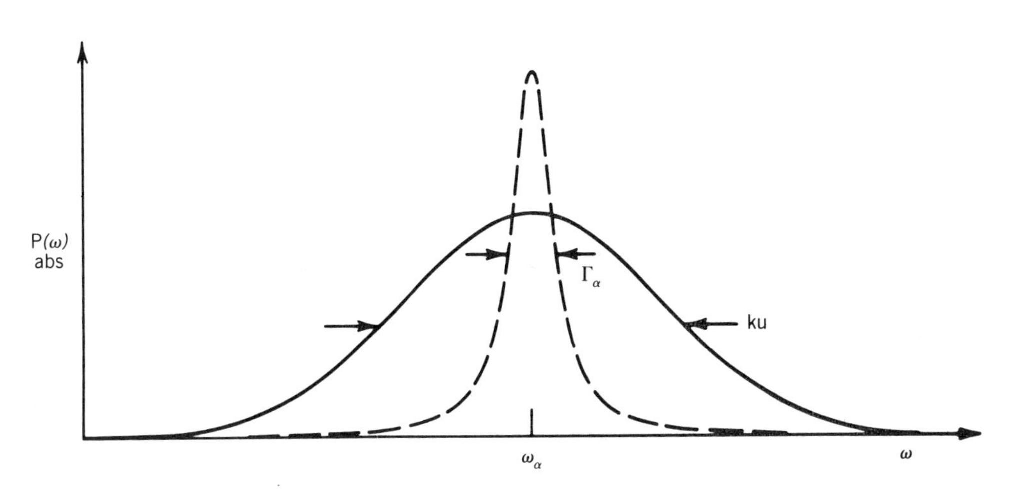

Figure 3.4. Absorption line; power absorbed as a function of frequency ω. Contribution from atoms with $\mathbf{k}\cdot\mathbf{v} = 0$ is the dashed line shown in the center.

3.3. RAMAN ACTIVE MODES

In a Raman active mode, the normal coordinate does not directly induce a dipole (as was the case for dipole active modes). An incident electric field produces a dipole in each molecule via that molecule's polarizability. If the induced dipole is modulated by a normal mode, that mode is Raman active. This is the basic mechanism by which Raman scattering occurs; hence the term "Raman active."

As an example, consider a symmetrical diatomic molecule oriented at an angle θ_a to an incident optical field as illustrated in Figure 3.5. The electrons can move much more freely along the axis of the diatomic than they can perpendicular to it. We assume, for simplicity, that we can neglect dipoles induced perpendicular to the molecular axis. The incident electric field causes a displacement of the centers of negative and positive charge, and a dipole is induced. The interaction potential between this induced dipole and the electric field is

$$V_{int} = - p_a E \cos\theta_a$$

where $\mathbf{p}_a$ is the dipole induced in the a'th molecule. Letting q_α be the electronic coordinate and ℓ_α the effective charge (which is a measure of the strength of the dipole), then

$$p_a = \sum_\alpha q_\alpha \ell_\alpha. \tag{3.27}$$

Therefore the interaction potential becomes

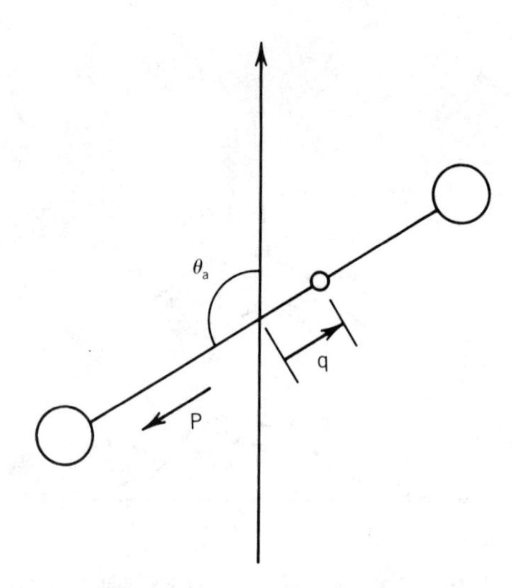

Figure 3.5. Dipole as a function of angle θ.

$$V_{int} = - \sum_{\alpha} \ell_{\alpha} q_{\alpha} E \cos\theta_a \tag{3.28}$$

If the molecule now rotates about one of its principal axes, there is a force on the molecule and the normal mode (rotation in this case) couples to the electromagnetic field. This normal mode is Raman active.

For Raman active modes, the interaction potential is not linear, but rather quadratic in the applied field. This occurs, of course, because $q_\alpha \sim E$ in Eq. (3.6). As a result, there is an ambiguity in the use of Eq. (3.6). This ambiguity may result in errors when q_a is written as an explicit function of θ_a, and we now consider this arrangement in more detail.

In the case in which one wishes to use an interaction potential written only in terms of the Raman active mode (which is nearly always the situation in complicated cases) one develops a phenomenological interaction potential as follows. Assume that $q_\alpha = A_\alpha(\theta_a)E$ where A_α is some function of θ_a. Then the change in the interaction potential as the electric field is increased from E to $E + \Delta E$ is

$$\Delta \mathcal{V}_{int} = - p_a \Delta E \cos\theta_a \tag{3.29}$$

$$\Delta \mathcal{V}_{int} = - \sum_{\alpha} \ell_{\alpha} A_{\alpha}(\theta_a) \cos\theta_a E\Delta E, \tag{3.30}$$

where, for the purpose of clarity, we denote the phenomenological interaction potential in italics. In later chapters this notation is dropped, since it is rarely the case in the literature that one finds explicit statements as to which form is used, and one must pick out the usage from the context of the problem. The phenomenological interaction potential is found by integrating the right-hand side of Eq. (3.29) for the electric field varying from 0 to some value E. This results in the expression

$$V_{int} = -\sum_{\alpha} \frac{1}{2} \ell_{\alpha} A_{\alpha}(\theta_a) \cos\theta_a E^2.$$

This potential can now be reexpressed in terms of q_{α} as

$$V_{int} = -\frac{1}{2} \sum_{\alpha} q_{\alpha}(\theta_a) \ell_{\alpha} \hat{\ell}_{\alpha} \cdot \mathbf{E}. \tag{3.31}$$

The factor of 1/2 in the interaction potential has been the source of some confusion in the literature, and it appears to be contradictory to have two different interaction potentials in the form of Eqs. (3.28) and (3.31). The key to understanding the difference is in recognizing that in Eq. (3.28) the coordinate q_{α} is not regarded as a function of θ_a, while in Eq. (3.31) it is a function of θ_a. Hence when finding the torque on the molecule, one does not take the derivative of q_{α} when using Eq. (3.28) and does take it when using Eq. (3.31). However one chooses to proceed (see Problems 3.5 and 3.6 for exercises) one obtains the same result. Moreover, Eq. (3.31) is the real interaction energy between the θ_a coordinate and the field. This comes about due to the fact that part of the interaction of θ_a with the field is in V_{int} and part is in the mechanical energy of q_{α}. Problem 3.7 is an exercise designed to clarify the way the energies work out.

Many vibrational modes are Raman active. For example, the symmetric stretch in Figure 2.7a is Raman active. An incident optical field displaces the centers of charge along the CO_2 axis and the induced dipole is modulated by the symmetric stretching vibration. On the other hand, the bending vibration (Figure 3.2b) is not Raman active. A dipole induced along the molecular axis is affected only in second order by vibrations perpendicular to that axis.

Collective modes are often Raman active. Raman activity can also occur in electron motions, but we do not discuss this case. Raman active interactions give rise to Brillouin scattering and Raman (rotational, vibrational, and electronic) scattering. They also are the basis of Raman-Nath scattering, the Debye-Sears effect, and all bulk acousto-optics. In the nonlinear regime, they cause stimulated scattering of all kinds, the ac and dc Kerr effect, two-photon absorption and resonant two-photon enhancement of other nonlinear effects (e.g., tripling of optical frequencies and real-time holography). It is the basis of all Raman spectroscopy including coherent anti-Stokes Raman scattering (CARS) and the Raman-induced Kerr effect spectroscopy (RIKES). It is the dominant cause of self-focusing and self-defocusing and hence is involved in associated phenomena of optical bistability and real-time holography (optical phase conjugation, or real-time adaptive optics). These applications are dealt with in detail in the second volume of this book. Since these effects are so pervasive in modern optics, we now cover the basic physics of the interaction in detail.

3.4. ELECTROMAGNETIC FIELD - RAMAN ACTIVE MODE INTERACTIONS

We now discuss how Raman active modes interact with an incident electromagnetic field. In dealing with Raman active modes, there are two strategies that one can follow. One can write down the equations for all of the normal coordinates first, using Newton's laws in the form of Eq. (3.8) with forces derived from the interaction potential in Eq. (3.28). This gives rise to coupled equations which are often difficult to solve in complex cases. This procedure is always correct, but it is often cumbersome and is rarely done in practice. It is much simpler to use the phenomenological potential.

We therefore use the more customary phenomenological approach, even though in the cases discussed here, it is not all that difficult to carry out the alternative (see Problems 3.5 and 3.6). Let us use the polarizability of the molecule coming from the electronic degrees of freedom. As discussed in Section 3.3, the induced dipole is itself a function of the incident field. That is, in the interaction potential (Eq. (3.31))

$$V_{int} = -\frac{1}{2}\sum_{\alpha} q_{\alpha}\, \boldsymbol{\ell}_{\alpha}\cdot \mathbf{E}, \tag{3.32}$$

q_{α} is a function of the applied field $\mathbf{E}$, and it is an explicit function of the normal coordinate q_{β} in which we are ultimately interested. Using the solution for q_{α} given by Eqs. (3.10) and (3.15),

$$V_{int} = \frac{1}{4}\sum_{\alpha}\left(\frac{\boldsymbol{\ell}_{\alpha}\cdot \mathbf{E}}{M_{\alpha}D_{\alpha}(\omega-\mathbf{k}\cdot\mathbf{v})}\, e^{i(\mathbf{k}\cdot\mathbf{r}-\omega t)} + cc\right)\boldsymbol{\ell}_{\alpha}\cdot \mathbf{E}. \tag{3.33}$$

We now define the molecular polarizability tensor $\boldsymbol{\alpha}$ as

$$\boldsymbol{\alpha} = \sum_{\alpha} \boldsymbol{\alpha}_{\alpha} \tag{3.34}$$

where α runs over the normal coordinates, and

$$\boldsymbol{\alpha}_{\alpha} = \frac{\boldsymbol{\ell}_{\alpha}\boldsymbol{\ell}_{\alpha}}{M_{\alpha}D_{\alpha}(\omega-\mathbf{k}\cdot\mathbf{v})}. \tag{3.35}$$

In all practical applications involving Raman active modes, the interaction is very far from the resonance of the dipole active coordinate, and the resonance denominator $D_{\alpha}(\omega-\mathbf{k}\cdot\mathbf{v})$ is real and can be taken to be independent of $\mathbf{v}$. (If $\omega \ll \omega_{\alpha}$, D_{α} can be approximated by ω_{α}^{2} insofar as the computation of the Raman active coordinate is concerned.) In the following we make use of this simplification that allows us to remove $\boldsymbol{\alpha}$ from within the parentheses of Eq. (3.33). The induced dipole is then

$$\mathbf{p} = \boldsymbol{\alpha}\cdot \mathbf{E} \tag{3.36}$$

and the corresponding interaction potential is

$$V_{int} = -\frac{1}{2}\sum_{\alpha} (\boldsymbol{\alpha}_{\alpha}\cdot \mathbf{E})\cdot \mathbf{E}. \tag{3.37}$$

The nature of this interaction is different from the dipole active case. Since the potential is proportional to E^2, rather than E, the interaction involves two photons, rather than one. The process in which Raman active modes absorb light is called two-photon absorption. It scales as the intensity squared and thus requires the high power available from lasers to be observed.

Our formulation of the motions in terms of normal coordinates means that we are concerned with coordinates fixed to the molecule. As a result, the polarizability tensors defined by Eqs. (3.34) and (3.35) are diagonal in form, i.e., the axes correspond to the normal coordinates for the three electronic degrees of freedom. Therefore, the projection of the fields implied by Eq. (3.37) is into the molecule frame of reference. This projection is changed with the orientation of a molecule relative to the incident field which makes some rotational modes Raman active. For example, let $\alpha_{\parallel}$ represent the polarizability along the axis of a linear diatomic molecule oriented at an angle θ_β to the incident field as illustrated in Figure 3.5. We assume for simplicity that we can neglect the influence of the polarizability perpendicular to the axis. Therefore from Eq. (3.37) we obtain

$$V_{int} = -\frac{1}{2}\,\alpha_{\parallel}\cos^2\theta_\beta\,E^2. \tag{3.38}$$

The restoring torque is

$$\tau_\beta = -\,\alpha_{\parallel}\sin\theta_\beta\cos\theta_\beta\,E^2. \tag{3.39}$$

For rotational degrees of freedom $\theta_\beta = \omega_\beta t$, where ω_β is the angular frequency of rotation, and the relevant equation for the normal coordinates is

$$\ddot{\theta}_\beta + \Gamma_\beta\,\dot{\theta}_\beta = -\,\frac{\alpha_{\parallel}}{I_\beta}\sin\theta_\beta\cos\theta_\beta\,E^2 \tag{3.40}$$

where I_β is the moment of inertia. One can now solve for θ_β and then follow the sequence of Eqs. (3.15) to (3.21) to obtain the absorbed power. The case of rotational absorption is left as problems for the reader. Note that there are no frequencies of rotation appearing in this formula or in the formula for the case of a permanent dipole. This means that the classical rotational absorption spectrum lacks the discrete absorption lines that one obtains from a proper quantum mechanical solution.

3.5. DOPPLER-FREE TWO-PHOTON ABSORPTION

This section is an extended homework exercise where all of the steps are indicated and many are worked through. The point is to show that we are basically finished with the conceptual foundations of classical optical physics and are now in a position to apply them to problems. Two-photon absorption is really a part of nonlinear optics, but we are in a position to deal with it now. One of the interesting features of two-photon spectroscopy is that it can yield spectra that are not broadened by molecular velocity distributions, in which case the absorption is said to be Doppler-free. Since analysis of rotations tends to be messy, we consider the case of Raman active vibrations. These vibrations affect the polarizability of the molecule as discussed in the previous section. Write the normal vibrational coordinate as q_β. Then from Eq. (3.37) we obtain

$$\ddot{q}_\beta + \Gamma_\beta \dot{q}_\beta + \omega_\beta^2 q_\beta = \frac{1}{2M_\beta} \sum_\alpha \left(\frac{\partial \boldsymbol{\alpha}_\alpha}{\partial q_\beta} \cdot \mathbf{E}\right) \cdot \mathbf{E}. \tag{3.41}$$

It is the tensor elements $\partial \boldsymbol{\alpha}_\alpha / \partial q_\beta$ which determine whether a vibration is Raman active or not. We shall consider the case of two incident oppositely-directed fields propagating along the z axis, i.e.,

$$\mathbf{E} = \frac{1}{2} \hat{\mathbf{e}}(E_+ e^{i(\mathbf{k}\cdot\mathbf{r} - \omega t)} + E_- e^{i(-\mathbf{k}\cdot\mathbf{r} - \omega t)} + \text{cc}) \tag{3.42}$$

where $\hat{\mathbf{e}}$ is a unit vector in the direction of $\mathbf{E}$. The resulting diadic product of the fields is

$$\begin{aligned} \mathbf{EE} = \frac{1}{4} \hat{\mathbf{e}}\hat{\mathbf{e}}\{&|E_+|^2 + |E_-|^2 + E_+^2 e^{i(2\mathbf{k}\cdot\mathbf{r} - 2\omega t)} \\ &+ E_-^2 e^{i(-2\mathbf{k}\cdot\mathbf{r} - 2\omega t)} + 2 E_+ E_- e^{-2i\omega t} \\ &+ 2 E_+ E_- e^{2i\mathbf{k}\cdot\mathbf{r}} + \text{cc}\} \end{aligned} \tag{3.43}$$

Since only terms which oscillate in time can be resonant with ω_α, the dc terms are ignored. There are three ac components characterized by different spatial dependencies which give rise to three separate components of q_α, i.e.,

$$\begin{aligned} q_\beta = \frac{1}{2} \{&Q_{\beta+} e^{i(2\mathbf{k}\cdot\mathbf{r} - 2\omega t)} \\ &+ Q_{\beta-} e^{i(-2\mathbf{k}\cdot\mathbf{r} - 2\omega t)} + Q_{\beta 0} e^{-2i\omega t} + \text{cc}\}. \end{aligned} \tag{3.44}$$

Solving Eq. (3.41) with Eqs. (3.42) through (3.44) gives

$$Q_{\beta+} = \frac{1}{4M_\beta} \frac{1}{D_\beta(2\omega - 2\mathbf{k}\cdot\mathbf{v})} \sum_\alpha \left(\frac{\partial \boldsymbol{\alpha}_\alpha}{\partial q_\beta} \cdot \hat{\mathbf{e}}\right) \cdot \hat{\mathbf{e}} \quad E_+^2 \tag{3.45a}$$

$$Q_{\beta-} = \frac{1}{4M_\beta} \frac{1}{D_\beta(2\omega + 2\mathbf{k}\cdot\mathbf{v})} \sum_\alpha \left(\frac{\partial \boldsymbol{\alpha}_\alpha}{\partial q_\beta} \cdot \hat{\mathbf{e}}\right) \cdot \hat{\mathbf{e}} \quad E_-^2 \tag{3.45b}$$

and

$$Q_{\beta 0} = \frac{1}{2M_\beta} \frac{1}{D_\beta(2\omega)} \sum_\alpha \left(\frac{\partial \boldsymbol{\alpha}_\alpha}{\partial q_\beta} \cdot \hat{\mathbf{e}}\right) \cdot \hat{\mathbf{e}}\, E_- E_+. \tag{3.45c}$$

We note that the third term does not contain the velocity $\mathbf{v}$ and hence the associated absorption is not Doppler broadened.

The power absorbed is now calculated in exactly the same way as for dipole active modes. The pertinent equation is (3.20), which requires that $|Q_\beta|^2$ be evaluated. In this case

$$Q_\beta = Q_{\beta+} e^{2ikz} + Q_{\beta 0} + Q_{\beta-} e^{-2ikz}, \tag{3.46}$$

and

$$|Q_\beta|^2 = (|Q_{\beta+}|^2 + |Q_{\beta-}|^2 + |Q_{\beta 0}|^2)$$
$$+ [Q_{\beta+}Q_{\beta-}e^{i4kz} + Q_{\beta+}Q_{\beta 0}e^{i2kz} + Q_{\beta-}Q_{\beta 0}e^{-2ikz} + cc]. \tag{3.47}$$

The terms in round brackets have no spatial periodicity. All of the other terms correspond to spatially periodic gratings which average out approximately to zero for incident fields that overlap many wavelengths along the propagation axis z. Since this is the usual case, even for subpicosecond pulses we retain only the static terms and obtain

$$P_{abs} = 2\omega^2 \sum_\beta \Gamma_\beta M_\beta [|Q_{\beta+}|^2 + |Q_{\beta-}|^2 + |Q_{\beta 0}|^2] \tag{3.48}$$

The apparent discrepancy of a factor of four between Eqs. (3.20) and (3.48) arises because the ω in Eq. (3.20) is replaced by 2ω in (3.48): this can be verified by carrying out the calculation in detail.

One now makes the same assumptions in evaluating Eq. (3.48) via Eqs. (3.45) as for the dipole active case. For a dilute gas consisting of N molecules whose velocities obey a Maxwellian distribution, we obtain

$$P_{abs} = \frac{N}{32} \sum_{\alpha,\beta} \frac{\Gamma_\beta}{M_\beta} |\hat{\mathbf{e}} \cdot \frac{\partial \boldsymbol{\alpha}_\alpha}{\partial q_\beta} \cdot \hat{\mathbf{e}}|^2 \{|E_+|^4 \int_{-\infty}^{\infty} \frac{\rho(v_z)dv_z}{(2\omega - 2\mathbf{v}\cdot\mathbf{k} - \omega_\beta)^2 + (\Gamma_\beta/2)^2}$$
$$+ |E_-|^4 \int_{-\infty}^{\infty} \frac{\rho(v_z)dv_z}{(2\omega + 2\mathbf{v}\cdot\mathbf{k} - \omega_\beta)^2 + (\Gamma_\beta/2)^2}$$
$$+ |E_+|^2|E_-|^2 \frac{4}{(2\omega - \omega_\beta)^2 + (\Gamma_\beta/2)^2}\}. \tag{3.49}$$

Simplifying in the limit $|E_+|^2 = |E_-|^2 = |E|^2$, and noting that $\rho(v_z) = \rho(-v_z)$, we finally get

$$P_{abs} = \frac{N}{8} \sum_{\alpha,\beta} \frac{\Gamma_\beta}{M_\beta} |\hat{\mathbf{e}} \cdot \frac{\partial \boldsymbol{\alpha}_\alpha}{\partial q_\beta} \cdot \hat{\mathbf{e}}|^2 |E|^4$$
$$\times \{\frac{1}{(2\omega - \omega_\beta)^2 + (\Gamma_\beta/2)^2} + \int_0^{\infty} \frac{\rho(v_z)dv_z}{(2\omega - 2\mathbf{v}\cdot\mathbf{k} - \omega_\beta)^2 + (\Gamma_\beta/2)^2}\}. \tag{3.50}$$

The absorption associated with Eq. (3.50) is sketched in Figure 3.6. The broad peak is a Doppler broadened line and the sharp peak in the middle is resonant for $\omega = \omega_\beta/2$. This narrow peak is independent of **v** and provides a means of locating a spectral line maximum without the degradation in resolution caused by Doppler broadening. This technique is called two-photon Doppler-free spectroscopy.

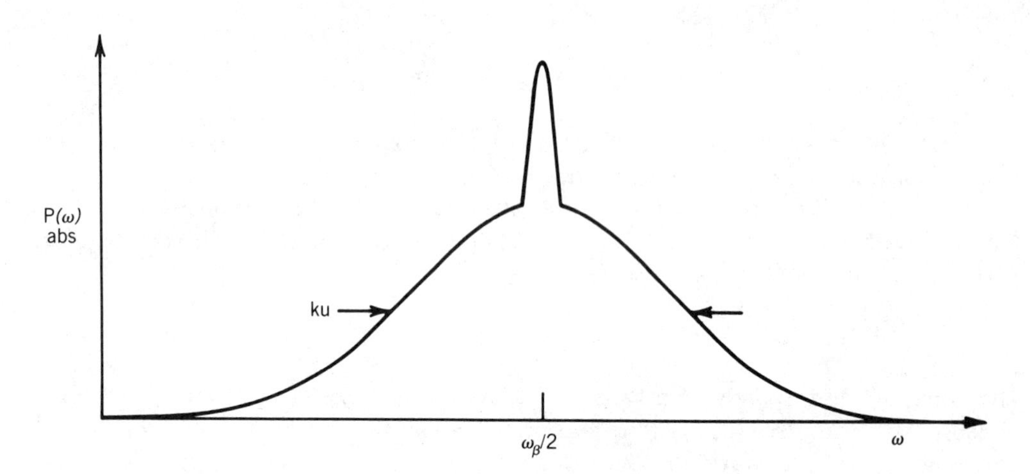

Figure 3.6. Two-photon resonance.

ADDITIONAL READING

Very little discussion is available on the classical aspect of the interaction of light with matter, but the same interaction occurs in quantum mechanics where the discussion is extensive.

Quantum mechanical discussion

Sargent, M., Scully, M.O., and Lamb, W.E. Jr. *Laser Physics*, Addison-Wesley, Reading, MA, 1974. Chapter 2. This reference also discusses classical absorption models (pp. 34-40) and the Doppler effect (Chapter 10), see also pp. 102-104 for technique to obtain Eq. (3.12) systematically.

Classical Absorption

Feynman, R.P., Leighton, R.B., and Sands, M. *The Feynman Lectures in Physics*, Addison-Wesley, Reading, MA, 1964. Volume I, pp. 31-38.

Work done in creating a dipole

Jackson, J.D. *Classical Electrodynamics*, 2nd Edition, Wiley, New York, 1975 158-162 (1st Ed., 1962, pp. 123-127).

PROBLEMS

3.1. Derive Eq. (3.18). In Eq. (3.19), use Eq. (3.15) to show that the $dE_{mech}/dt = 0$ in steady state.

3.2. (a) Discuss the dipole activity of the vibrational modes of CO_2 shown in Figures 3.2 and 2.7.

(b) How many vibrational modes does water have? Are any of them dipole active? Explain your result.

3.3. In many cases, a normal mode is either dipole active or Raman active, but not both. Consider the vibration of the homonuclear diatomic of Figure 2.2c, the vibrational modes of CO_2, and the water vapor vibrations developed in Problem 3.2b. Give a brief and intuitive discussion of why this rule is valid or invalid for the various cases.
Hint; the difficult part of this problem is intuiting the underlying electronic structures that give large Raman effects. Try to envision the charge cloud that surrounds the atoms, and recognize that Raman activity is highest when the motion of the normal coordinate distorts the charge cloud. In cases in which there are strong and highly asymmetric forces on the charge cloud due to the nuclei plus core electrons, the cloud tends to shift rather than distort.

3.4. (a) Compute the interaction potential of a linear diatomic molecule which has a permanent dipole (e.g., Figure 2.3b) with an optical field.

(b) In the absence of external forces the equation for a linear rotator is $I\ddot{\theta}_\alpha = 0$ where θ_α is an angle defined about an inertia axis perpendicular to the molecular axis and I is the moment of inertia of the molecule. The energy is $E_{mech} = 1/2\ I\dot{\theta}_\alpha^2$. Introduce a damping that takes the molecule to the state of zero rotational energy, i.e., $\dot{\theta}_\alpha = 0$, in the absence of interaction with light. Write the equation of motion for θ_α. Write the expressions for the absorbed and dissipated power.

(c) Let the incident optical field be circularly polarized, and let the molecule stay in the plane of the electric fields at all times. Write the solution for the rotation in the limit of small and large incident optical frequency ω. Sketch the nature of the absorption band. Note that there are no spectral lines. This is a case in which quantum mechanics is needed to give an accurate idea of the spectrum.

(d) In real gases, the damping does not take the molecule to $\dot{\theta}_\alpha = 0$, instead they have nonzero angular velocities due to thermal excitation, which is collisional and hence is part of the damping process. Based on extrapolation from (c), sketch the absorption spectrum for weak incident fields for a thermally excited gas (i.e., a gas with a distribution of initial angular frequencies ω_i).

3.5. Consider the simple model of a diatomic molecule illustrated in Figure 3.5. Write the kinetic and potential energies as

$$T = \frac{1}{2} M_1 \dot{q}_1^2 + \frac{1}{2} I \dot{\theta}^2$$

$$V = \frac{1}{2} k_1 q_1^2$$

and write the interaction potential as

$$V_{int} = - p_1 \cdot E = - q_1 \ell_1 E \cos\theta .$$

(a) Write down the equations of motion for q_1 and θ first, introducing a phenomenological damping term into both equations. Next solve for q_1. In this solution, assume that θ can be treated as a constant. The assumption that θ is constant is one version of the Born-Oppenheimer approximation, which is based on the fact that the motion of electrons is rapid compared to the nuclei. The approximation itself involves taking the nuclei to be stationary in the computation of the electron motion. The validity of the phenomenological potential is contingent on the validity of the Born-Oppenheimer approximation. Substitute the result for q_1 into the equation of motion for θ.

(b) Substitute the solution for q_1 obtained in (a) into the phenomenological interaction potential

$$\mathcal{V}_{int} = -\frac{1}{2} q_1(\theta)\, \ell_1\, E \cos\theta$$

so that q_1 does not appear explicitly in $\mathcal{V}_{int}$ (i.e., $\mathcal{V}_{int}$ should be expressed in terms of θ only, and should not contain the symbols q_1 or Q_1). Develop the equation of motion for θ using this potential (this is just a repeat of the development in Section 3.4). Show that the methods in parts (a) and (b) of this problem give the same equations of motion for θ.

(c) Using the equation of motion, show that the strongest interaction occurs for the condition $\omega = \omega_\alpha$, where ω is the optical frequency and ω_α is the rotational frequency (do not solve the equation, just set $\theta = \omega_\alpha t$ and examine the time dependence of the force). This resonance appears to contradict the idea that this is a two-photon absorption. However, note that the charge cloud of the homonuclear diatomic has half the period of the molecular rotation (i.e., the frequency of rotation of the charge cloud is $2\omega_\alpha$, and the resonance is $2\omega = 2\omega_\alpha$). In the case of a molecule with a permanent dipole moment, the frequency of rotation of the dipole is the same as rotational frequency of the molecule. In this case the resonance reads $\omega = \omega_\alpha$.

3.6. (a) Repeat the exercise in Problem 3.5a and 3.5b, this time using the Raman active vibrational coordinate denoted q_2. For simplicity assume that the axis of the molecule is oriented parallel to the electric field vector ($\cos\theta = 1$). The potential and kinetic energies, the interaction potential, and phenomenological interaction potential read

$$T = \frac{1}{2} M_1 \dot{q}_1^2 + \frac{1}{2} M_2 \dot{q}_2^2$$

$$V = \frac{1}{2} k_1 q_1^2 + \frac{1}{2} k_2 q_2^2$$

$$V_{int} = -\mathbf{p}_1 \cdot \mathbf{E} = -q_1\, \ell_1 (1+\gamma q_2) E \tag{3.51}$$

$$\mathcal{V}_{int} = -\frac{1}{2} q_1(q_2)\, \ell_1 (1+\gamma q_2) E, \tag{3.52}$$

where γ is a coefficient.

(b) Try to intuit the limits of validity of the Born-Oppenheimer approximation in terms of inequalities involving ω_1, ω_2, ω, Γ_1, and Γ_2.

3.7. Repeat Problem 3.6 in the case in which the optical frequency ω is very different from the resonance frequency ω_1, but is still much larger than ω_2. In this case it is not necessary to include phenomenological damping, so leave out the damping. Develop the formula for q_1 using the Born-Oppenheimer approximation. Evaluate explicitly the mechanical energy for mode 1, i.e.,

$$E_{mech} = \frac{1}{2} M_1 \dot{q}_1^2 + \frac{1}{2} k_1 q_1^2.$$

Add this energy to the interaction potential in Eq. (3.51). Express the result in terms of q_2 only (i.e., q_1 and its amplitude should not appear explicitly in the formula). Consider only the time varying term in this sum and show that it is the same as the phenomenological potential in Eq. (3.52).
Hint: this problem is somewhat easier if you do not use the complex amplitude formulation. Instead use the field in the form $\mathbf{E} = \boldsymbol{\mathcal{E}} \sin\omega t$

3.8. Use the solution of Problem 3.6 to construct the polarizability tensor $\boldsymbol{\alpha}$, which is to be written as an explicit function of q_2. Use this expression to evaluate the right-hand side of Eq. (3.41), in which $\beta = 2$ and the subscript α is restricted to the single mode $\alpha = 1$.

3.9. In Eq. (3.25) treat the Lorentzian denominator as a Dirac delta function. Show that in the limit $ku \gg \Gamma_\alpha$,

$$P_{abs} = \frac{\pi}{4} \frac{N}{u\sqrt{\pi}} e^{-(\omega-\omega_\alpha)^2/(ku)^2} \sum_\alpha \frac{\Gamma_\alpha |\boldsymbol{\ell}_\alpha \cdot \boldsymbol{\mathcal{E}}|^2}{M_\alpha}. \tag{3.53}$$

3.10. The Doppler effect is often introduced into the formalism by writing

$$\frac{d}{dt} = \frac{\partial}{\partial t} + \frac{d\mathbf{r}}{dt} \cdot \nabla = \frac{\partial}{\partial t} + \mathbf{v} \cdot \nabla.$$

(a) Use this form of the derivative in Eq. (3.8) and use $\mathbf{r}$ in the form of Eq. (3.12) to obtain Eq. (3.13). Show that the term of $\mathbf{r}_a$ drops out of the expression for Q_α.

(b) The term $\mathbf{r}_a$ reappears when one constructs a macroscopic polarization, which is defined formally in Eq. (5.1). Let $\mathbf{P}_{\alpha,a}$ denote the contribution of the α'th normal mode of atom a to the total polarization field, then

$$\mathbf{P}_{\alpha,a} = \boldsymbol{\ell}_{\alpha,a} q_{\alpha,a} \delta^{(3)}(\mathbf{r} - \mathbf{r}_a),$$

where $\mathbf{r}_a$ is the instantaneous position of the atom given in Eq. (3.12) and $\delta^{(3)}(\mathbf{r}) = \delta(x)\delta(y)\delta(z)$ is the three-dimensional delta function. Multiply $\mathbf{P}_{\alpha,a}$ by a probability density of the form $\rho(\mathbf{r}_a,\mathbf{v})$ which is the probability of finding an atom at $\mathbf{r}_a$ with velocity $\mathbf{v}$. Integrate this over all positions and velocities. Show that $\mathbf{r}_a$ drops out of the final answer. Show that if you take $\rho(\mathbf{r}_a,\mathbf{v})$ to describe a Maxwellian velocity distribution and a constant density, you get the same result obtained in the text. (i.e., the result obtained by ignoring $\mathbf{r}_a$ from the beginning and multiplying the single atom result by Eq. (3.22)).

3.11. Repeat the exercise in Problems 3.7 and 3.8 but allow the angle θ_a in Figure 3.5 to be general.

(a) Compute, in general, the term

$$\left(\frac{\partial \boldsymbol{\alpha}_\alpha}{\partial q_\beta} \cdot \hat{\mathbf{e}}\right) \cdot \hat{\mathbf{e}}.$$

(b) Use this result to compute P_{abs} for two-photon absorption. In computing the slowly varying terms, assume that there is only the term E_+ in Eq. (3.42) and neglect Doppler broadening. Remember that the coordinate q_β goes as 2ω (see, e.g., Eq. (3.44)).

(c) Use the solution in Eq. (3.45b) and verify that $dE_{mech}/dt = 0$ in steady state.

(d) Using the results of parts (a) - (c), and assuming $\omega \ll \omega_\alpha$ write the power absorbed as

$$P_{abs} = \frac{\Gamma_B \omega^2 \gamma^2 \ell_\alpha^2}{\varepsilon_0^2 c^2 M_\beta M_\alpha^2 \omega_\alpha^4} \frac{1}{|D(2\omega)|^2} \mathcal{S}^2. \tag{3.54}$$

4

Electrodynamics in Dilute Media

In the preceding chapters we have set the stage for analyzing the propagation of an electromagnetic wave through a medium composed of dipole active molecules. The fields radiated by an electric dipole are written down in Chapter 1. The basic field-dipole interaction is discussed in Chapter 3, as is the energy absorbed from an electromagnetic field by damped normal modes. In this chapter we treat the interaction from the viewpoint of the incident and transmitted fields. In particular, we consider interference effects between the incident and reradiated fields, define absorption and scattering cross sections, and show how the velocity of light is modified by the action of the dipoles, giving rise to the refractive index.

4.1. ABSORPTION OF RADIATION

As a precursor to a detailed discussion of absorption, it is useful to examine what we mean by this term. From Eq. (3.20), power absorbed from an electromagnetic field is directly proportional to Γ_α, the normal mode decay rate. In any medium, there are a large number of mechanisms which can contribute to Γ_α, reradiation by the induced dipoles, local heating, coupling to other local or collective modes, etc. By absorption we mean the total energy removed from the incident beam via all of these physical mechanisms. (This does not include, for example, stray scattering due to impurities, inhomogeneities, etc.) Occasionally in the literature one finds this version of absorption referred to as extinction, and absorption taken to mean energy converted into local heating. (This is prevalent in the laser damage literature.)

The goal of this section is to show that the absorption of radiation from the incident field (assumed to be a plane wave) is obtained from interference effects between that field and the fields reradiated by the induced dipoles. The pertinent geometry is illustrated in Figure 4.1. For convenience, the incident field $\mathbf{E}_I$ is assumed to propagate along the z axis with the electric field polarized along the x axis, i.e.,

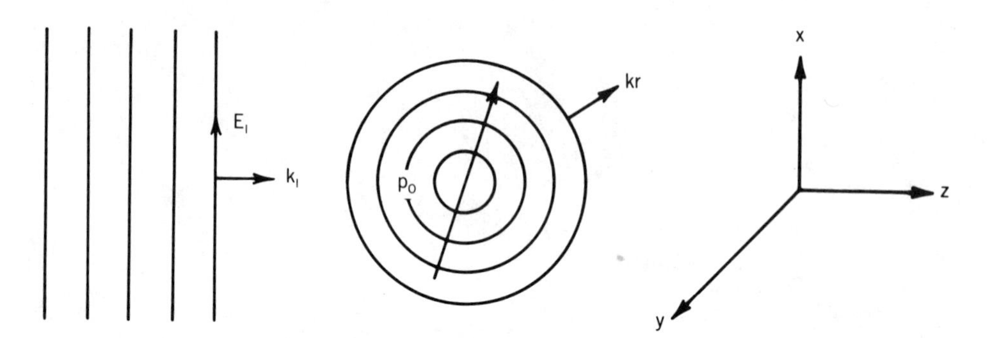

Figure 4.1. Left, incident plane wave interfering with the field of a radiating dipole. Right, coordinate system for the analysis.

$$\mathbf{E}_I = \frac{1}{2}\, \mathcal{E}_I\, \hat{\mathbf{x}}\, e^{i(k_I z - \omega t)} + \text{cc}. \tag{4.1}$$

For the field radiated by the induced dipole in the far field ($kr \gg 1$) we write (see Eq. (1.48))

$$\mathbf{E}_S = \frac{1}{2}\, \frac{k^2}{4\pi\varepsilon_0}\, \mathbf{O}(\hat{\mathbf{r}})\cdot\mathbf{p}_0\, \frac{e^{i(kr-\omega t)}}{r} + \text{cc}, \tag{4.2}$$

where $\hat{\mathbf{r}}$ is the unit vector to the observation point and, in keeping with the notation of Chapter 3, $\mathbf{p}_0 = \hat{\ell} p_0$.

The total electromagnetic field is the sum of Eqs. (4.1) and (4.2). That is

$$\begin{aligned} \mathbf{E} &= \mathbf{E}_I + \mathbf{E}_S \\ \mathbf{B} &= \mathbf{B}_I + \mathbf{B}_S, \end{aligned} \tag{4.3}$$

where $\mathbf{B}$ is obtained from Eq. (1.4), and the Poynting vector is given by

$$\mathbf{S} = \frac{1}{\mu_0}\,(\mathbf{E}_I \times \mathbf{B}_I + \mathbf{E}_I \times \mathbf{B}_S + \mathbf{E}_S \times \mathbf{B}_I + \mathbf{E}_S \times \mathbf{B}_S). \tag{4.4}$$

The first and last terms are clearly the incident and scattered fluxes $\mathbf{S}_I$ and $\mathbf{S}_S$, respectively. The cross terms describe the interference effects we wish to evaluate. They give rise to absorption of light from the incident beam. We therefore call these terms $\mathbf{S}_{abs}$ where

$$\mathbf{S}_{abs} = \frac{1}{\mu_0} (\mathbf{E}_I \times \mathbf{B}_S + \mathbf{E}_S \times \mathbf{B}_I), \tag{4.5}$$

and show by subsequent calculation that the results agree with Eq. (3.20) for the energy absorbed by the normal modes.

The first step is to eliminate the terms which oscillate as $\exp(\pm 2i\omega t)$. These terms oscillate too fast to be detected and average to zero leaving

$$\mathbf{S}_{abs} = \frac{k^2 c}{16\pi} [\hat{\mathbf{x}} \times (\hat{\mathbf{r}} \times \hat{\boldsymbol{\ell}}) + (\mathbf{O}(\hat{\mathbf{r}}) \cdot \hat{\boldsymbol{\ell}}) \times \hat{\mathbf{y}}] \left(\frac{\mathcal{E}_I p_0^*}{r} e^{i(k_I z - kr)} + cc\right), \tag{4.6}$$

where $\mathbf{S}_{abs}$ is now the time-averaged absorbed flux. The terms in the square brackets are formidable vector products of the fields. The round brackets contain terms which vary rapidly in space, except along the direction of the incident light where $k_I z - kr \simeq 0$.

The next step is to understand the directional characteristics of $\mathbf{S}_{abs}$. To do this, we consider the flux passing through a sphere centered at the radiating dipole which permits us to inspect the quantity $\mathbf{S}_{abs} \cdot \hat{\mathbf{r}}$. Utilizing the spherical symmetry of the problem, we reexpress Eq. (4.6) in terms of spherical coordinates, r, θ, and ϕ, where θ is the angle $\hat{\mathbf{r}}$ makes with the z axis (incident light propagation direction; note that this angle is called the scattering angle and is different from the one used in Eq. (1.50)), and ϕ is the angle from the x axis in the x-y plane. Noting that $\cos\theta = \hat{\mathbf{r}} \cdot \hat{\mathbf{z}}$ and $k_I = k$, then we obtain

$$\mathbf{S}_{abs} \cdot \hat{\mathbf{r}} = \frac{k^2 c}{16\pi} f(\theta,\phi) \left(\frac{\mathcal{E}_I p_0^*}{r} e^{-ikr(1-\cos\theta)} + cc\right), \tag{4.7}$$

where

$$f(\theta,\phi) = \hat{\mathbf{r}} \cdot [\hat{\mathbf{x}} \times (\hat{\mathbf{r}} \times \hat{\boldsymbol{\ell}}) + (\mathbf{O}(\hat{\mathbf{r}}) \cdot \hat{\boldsymbol{\ell}}) \times \hat{\mathbf{y}}]. \tag{4.8}$$

There are two distinct types of variation with angle θ in Eq. (4.7). The function $f(\theta,\phi)$ contains terms like $\sin\theta$, $\cos\theta$, $\sin\theta\cos\theta$, etc., all of which vary slowly with angle θ. On the other hand, the phasor $\exp(-ikr(1-\cos\theta))$ oscillates rapidly with θ for the case in which $kr \gg 1$, except near $\theta \simeq 0$ where $(1-\cos\theta) \simeq 0$. This behavior is sketched in Figure 4.2a. As indicated in Figure 4.2b, $\mathbf{S}_{abs} \cdot \hat{\mathbf{r}}$ fails to average to zero spatially only near the z axis, that is, downstream from the radiating dipole. The flux $\mathbf{S}_{abs}$ is drawn as a negative quantity indicating absorption. Because $\mathbf{S}_I$ and $\mathbf{S}_{abs}$ are appreciable in size and opposite in sign near the z axis, the total flux $(\mathbf{S}_I + \mathbf{S}_{abs})$ is small near the axis. This is the shadow of the oscillating dipole, and the radiation pattern is the same as that of a small, opaque disk located at the position of the dipole. This is the origin of the concept of an optical cross section denoted σ. We can take any induced dipole (remember that p_0 is proportional to E_I in practice) and treat it optically as if it were an opaque disk of area σ.

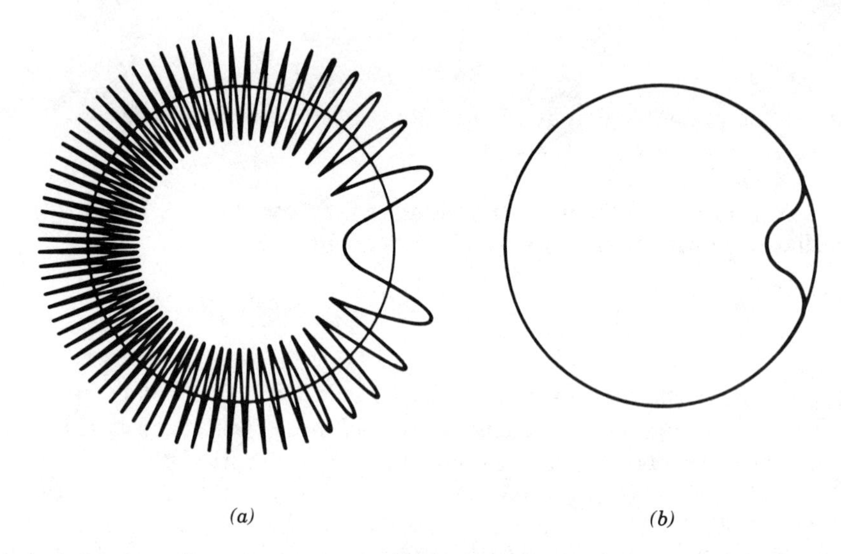

Figure 4.2. Interference pattern observed around a sphere surrounding the dipole. The circle intersects the graph of the flux at points where the flux is zero. Fluxes inside the circle are negative, outside the circle are positive. (a) Detailed structure of the interference. (b) Result of average over a small area.

We now proceed to evaluate the cross section σ. The total absorbed power is obtained by integrating the flux over the area of the sphere of Figure 4.2. Despite the fact that the contribution to this integral is limited to $\theta \simeq 0$, it is convenient to formally treat the full spherical surface. The appropriate integral is (with $\mu = \cos\theta$),

$$P_{abs} = - \frac{k^2 c}{16\pi} \int \mathcal{E}_I \, p_0^* \, \frac{e^{-ikr(1-\mu)}}{r} f(\mu,\phi) \, r^2 d\mu d\phi + cc. \tag{4.9}$$

Note that the absorbed power as defined in Section 3.2.1 is an explicitly positive quantity, and we anticipate from the construction in Figure 4.2 that this power is being removed from the field. We have accordingly introduced a minus sign into Eq. (4.9), and we next show that with this definition, the formulae for P_{abs} in this and the previous chapter are equivalent.

There are some tricks which can be used to evaluate Eq. (4.9) without actually carrying out any integrals. In the far field, the parameter P_{abs} must be independent of r, the distance to the observation point. Thus the nonphasor part of Eq. (4.9) should be manipulated to yield a series of terms in r^{-m} with the leading term characterized by $m = 0$. Terms with $m > 0$ become negligible as $r \to \infty$. Evaluating the angular part of Eq. (4.9) by parts gives

$$\int_{-1}^{1} r\, f(\mu,\varphi)\, e^{ikr\mu}\, d\mu = f(\mu,\varphi)\, r\, \frac{e^{ikr\mu}}{ikr}\Bigg|_{-1}^{1} - \frac{\partial f(\mu,\phi)}{\partial \mu}\, r\, \frac{e^{ikr\mu}}{(ikr)^2}\Bigg|_{-1}^{1} + \int_{-1}^{1} \frac{\partial^2 f(\mu,\varphi)}{\partial \mu^2}\, r\, \frac{e^{ikr\mu}}{(ikr)^3}\, d\mu$$

i.e.,

$$\int_{-1}^{1} rf(\mu,\varphi)\, e^{ikr\mu}\, d\mu = -\frac{i}{k}\, f(\mu,\varphi)\, e^{ikr\mu}\Bigg|_{-1}^{1} + O(r^{-m}) \tag{4.10}$$

where $O(r^{-m})$ refers to an arbitrary term of order r^{-m}. Clearly the first term falls off as r^0 and is the surviving term as $r \to \infty$.

To evaluate the first term in Eq. (4.10), note that $\mu = 1, -1$ corresponds to $\theta = 0, \pi$ respectively, and the vector products are

$$f(\theta=0,\varphi) = 2\, \hat{\mathbf{x}}\cdot\hat{\boldsymbol{\ell}}; \quad f(\theta=\pi,\varphi) = 0;$$

then

$$\frac{i}{k}\, f(\mu,\phi) e^{ikr\mu}\Bigg|_{-1}^{1} = -\frac{2i}{k}\, \hat{\mathbf{x}}\cdot\hat{\boldsymbol{\ell}}\, e^{ikr} \tag{4.11}$$

The integral over φ gives 2π, and the result for P_{abs} reads

$$P_{abs} = +\frac{kc}{4}\, [i\, E_I\, p_0^*\, \hat{\mathbf{x}}\cdot\hat{\boldsymbol{\ell}} + cc] \tag{4.12}$$

As one might expect, the important contribution is from the component of the dipole parallel to the incident field. Rewriting Eq. (4.12) (using im(a) = - im(a*)) as

$$P_{abs} = \frac{\omega}{2}\, \mathrm{im}(E_I^*\, p_0\, \hat{\mathbf{x}}\cdot\hat{\boldsymbol{\ell}}) \tag{4.13}$$

and noting that $E_I\hat{\mathbf{x}} = E_I\hat{\mathbf{e}}$ and that $p_0 = Q_\alpha \ell_\alpha$, we find that Eq. (4.13) is identical to the second term in Eq. (3.19), as expected. This result verifies that the energy lost from the electromagnetic field is absorbed (and dissipated by) the normal modes of the molecules.

The results of Eq. (4.13) can now be expressed in terms of an absorption cross section, i.e.,

$$P_{abs} = \sigma_{abs}\, S_I. \tag{4.14}$$

The quantity $\sigma_{abs}\, S_I$ is the amount of power blocked by an opaque object of cross sectional area σ_{abs}, which explains the terminology. Because the cross section is related to power, the cumulative influence of many cross sections is additive.

Hence there is considerable flexibility in its use. Depending on the context, the cross section can be associated with single molecules or single normal coordinates, or it can refer to statistical averages over the orientations and velocities of the molecules which constitute the medium, i.e.,

$$\sigma_{abs}\, S_I = \frac{\omega}{2} \langle \mathrm{im}(E_I\, \hat{\mathbf{x}} \cdot \hat{\boldsymbol{\ell}}\, p_0^*) \rangle$$

where the pointed brackets signify the statistical average. The averaged version is the customary one, and we use it here.

Having now established the equivalence of the power lost by the electromagnetic field, and the power absorbed by the normal modes, we use Eq. (3.21) to define σ_{abs}, i.e.,

$$\sigma_{abs}\, S_I = \frac{\omega^2}{2} \sum_{\alpha} \frac{\Gamma_\alpha}{M_\alpha} \left\langle \frac{|\boldsymbol{\ell}_\alpha \cdot \hat{\mathbf{x}}|^2}{|D_\alpha(\omega - \mathbf{k} \cdot \mathbf{v})|^2} \right\rangle |E_I|^2. \tag{4.15}$$

Since

$$S_I = \frac{1}{2} \left(\frac{\varepsilon_0}{\mu_0}\right)^{1/2} |E_I|^2,$$

we obtain the result

$$\sigma_{abs} = \frac{\omega^2}{\varepsilon_0 c} \sum_{\alpha} \frac{\Gamma_\alpha}{M_\alpha} \left\langle \frac{|\boldsymbol{\ell}_\alpha \cdot \hat{\mathbf{x}}|^2}{|D_\alpha(\omega - \mathbf{k} \cdot \mathbf{v})|^2} \right\rangle \tag{4.16}$$

Beer's law of absorption follows directly from this last result. Consider a volume of unit area and thickness dz filled with molecules with a number density N. The total number of atoms in the volume is Ndz, and total effective area blocked is $\sigma_{abs} N dz$. As a result

$$S_I - S_{out} = (\sigma_{abs}\, N\, dz)\, S_I,$$

and writing $dS = S_{out} - S_I$ and dropping the subscript I,

$$\frac{dS}{dz} = -\gamma S, \tag{4.17}$$

which is Beer's law with an absorption coefficient

$$\gamma = \sigma_{abs} N. \tag{4.18}$$

In its traditional form, Beer's law is written as the solution to Eq. (4.17), namely

$$S = S_0\, e^{-\gamma z}, \tag{4.19}$$

where S_0 is the flux at $z = 0$. (In this form, the loss of energy at the boundary is included in the value of S_0.)

4.2. SCATTERING CROSS SECTION

A fraction of the energy absorbed from the incident beam is subsequently reradiated by the induced oscillating dipole. The scattered power, P_{scatt}, the last term in Eq. (4.4), is the integral over a surface centered around the dipole. This integral is reviewed in Chapter 1, i.e., given by Eq. (1.51). For the present discussion we specialize to the case where only one normal coordinate (mode) dominates the scattering for the frequency interval of interest. We continue to label the coordinate as q_α, but drop the summation. The scattered power is given by Eq. (1.51). Since $\mathbf{p}_0 = \boldsymbol{\ell}_\alpha Q_\alpha$ and Q_α is given by Eq. (3.15),

$$\frac{1}{4\pi\varepsilon_0}\frac{1}{3}ck^4|\mathbf{p}_0|^2.$$

Since $\mathbf{p}_0 = \boldsymbol{\ell}_\alpha Q_\alpha$ and Q_α is given by Eq. (3.15),

$$P_{scatt} = \frac{\omega^4}{12\pi\varepsilon_0 c^3}\frac{|\boldsymbol{\ell}_\alpha|^2}{M_\alpha^2}\left\langle\frac{|\boldsymbol{\ell}_\alpha\cdot\hat{\mathbf{x}}|^2}{|D_\alpha(\omega-\mathbf{k}\cdot\mathbf{v})|^2}\right\rangle|E_I|^2, \tag{4.20}$$

where the brackets again signify a statistical average. Writing

$$P_{scatt} = \sigma_{scatt}S_I, \tag{4.21}$$

the cross section σ_{scatt} is given by

$$\sigma_{scatt} = \frac{\omega^4}{6\pi\varepsilon_0^2 c^4}\frac{|\boldsymbol{\ell}_\alpha|^2}{M_\alpha^2}\left\langle\frac{|\boldsymbol{\ell}_\alpha\cdot\hat{\mathbf{x}}|^2}{|D_\alpha(\omega-\mathbf{k}\cdot\mathbf{v})|^2}\right\rangle. \tag{4.22}$$

This area (σ_{scatt}) gives the amount of incident power blocked by a single molecule due to reradiation by the dipole induced in the α'th normal mode.

If scattering is the only mechanism for dissipating energy, the absorption and scattering cross sections should be equal. From Eqs. (4.16) and (4.22), they do not even have the same frequency dependence, i.e., ω^2 versus ω^4. This is a generic defect in trying to use a classical approach to solve the basically quantum mechanical problem of spontaneous emission. This defect is not too serious if one stays in the neighborhood of a resonance, i.e., $\omega \simeq \omega_\alpha$. In that case, one can equate the scattered to the absorption cross sections, provided that the classical decay rate Γ_α takes on the particular form

$$\Gamma_{rad,\alpha} = \frac{1}{4\pi\varepsilon_0}\frac{2}{3}\frac{\omega_\alpha^2|\boldsymbol{\ell}_\alpha|^2}{M_\alpha c^3}. \tag{4.23}$$

When $|\boldsymbol{\ell}_\alpha|$ is just the charge on an electron, this becomes the radiative decay rate of the Lorentz electron. It can also be related to the quantum mechanical decay rate

$$\frac{1}{4\pi\varepsilon_0}\frac{4\omega_\alpha^3|\mu_\alpha|^2}{3c^3\hbar},$$

where μ_α is the dipole matrix element (in MKS) for the α'th transition and $\hbar$ is Planck's constant divided by 2π. These relations allow an interpretation of phenomenological coefficients in terms of basic quantum-mechanical quantities.

Two terms are used in the literature to describe scattering. When the scattering is far from resonance, it is called Rayleigh scattering. It is the ω^4 dependence of the scattered power which makes the sky appear blue, i.e., sunlight is scattered more efficiently toward the blue rather than the red region of the visible spectrum. Near a resonance, the scattering is usually called resonance fluorescence.

Also there is a variety of terms used for absorption and scattering, primarily in the high power laser field. It is quite common to define absorption as the optical power that is dissipated in the form of collisional relaxation leading to heat, as opposed to scattering which does not lead to local heating. In this formulation, the absorption cross section is viewed as the difference between the quantities σ_{abs} and σ_{scatt} developed above, and the sum is called the cross section for extinction.

4.3. DISPERSION IN THE REFRACTIVE INDEX

In addition to absorbing and scattering light, an ensemble of induced dipoles also affects the velocity of light as it passes through the medium. This is an effect that cannot be obtained from a single molecule, but is, rather, a many-molecule effect. The field at any point in a medium is a sum of the incident field and all of the fields reradiated by the induced dipoles. If the sum of the dipolar fields is not in phase with the incident field, the phase of the total field (and hence the phase velocity) is affected. This is the phenomenon responsible for the refractive index.

Let us consider a volume of molecules of density N which is infinite in extent in the transverse (i.e., x and y) dimensions, as illustrated in Figure 4.3. By symmetry, the fields radiated by the induced dipoles, when summed, must produce plane waves propagating in the z direction. In recognition of this fact, we consider only the projections of the dipole fields onto this plane wave.

4.3.1 Projections

At this point we digress temporarily to discuss how to calculate the projection of one field onto another. Let us assume that one can find solutions for all of the possible electromagnetic plane waves which can propagate in a given volume V. We are interested in the complete sets which are orthogonal to one another. In that case, any arbitrary field can be expanded as a linear combination of the electromagnetic normal modes, i.e.,

$$\mathbf{E}(\mathbf{r},t) = \frac{1}{2} \sum_n \mathcal{E}_n e^{i(\mathbf{k}_n \cdot \mathbf{r} - \omega t)} + cc, \tag{4.24}$$

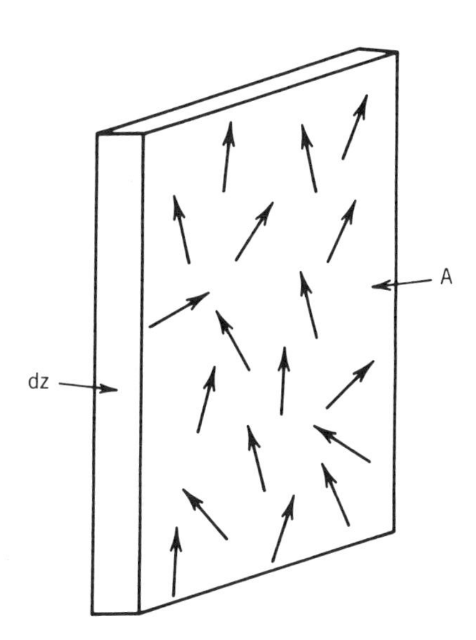

Figure 4.3. Illustration of sheet of dipoles of density N, thickness dz and area A, where we are ultimately interested in the limit $A \to \infty$.

where $\mathcal{E}_n$ and $\mathbf{k}_n$ are the amplitude and wave vector, respectively, of the n'th normal mode of the optical field. Mode orthogonality can be expressed mathematically by requiring that the energy carried by a field, which is the sum of many normal modes, is just the sum of the energies carried by the individual modes, i.e.,

$$\sum_{nm} \int_S [(\mathbf{E}_m + \mathbf{E}_n) \times (\mathbf{H}_m + \mathbf{H}_n)] \cdot d\mathbf{S} = 2 \sum_n \int_S (\mathbf{E}_n \times \mathbf{H}_n) \cdot d\mathbf{S},$$

where S denotes the surface of integration. This implies that

$$\int_S (\mathbf{E}_m \times \mathbf{H}_n) \cdot d\mathbf{S} = \int_S e^{i(\mathbf{k}_m - \mathbf{k}_n)\cdot\mathbf{r}} \frac{1}{c} [\mathcal{E}_m \times (\mathbf{k}_n \times \mathcal{E}_n^*)] \cdot d\mathbf{S} + cc = 0.$$

For this equation to be satisfied in general

$$\int_S e^{i(\mathbf{k}_m - \mathbf{k}_n)\cdot\mathbf{r}} \, dS = \delta_{mn} A, \tag{4.25}$$

where A is the area of the plane surface. Let us denote by n = p the component of the field traveling along the z axis, i.e., the component with $\mathbf{k}_p = \mathbf{k}_l$. To evaluate $\mathcal{E}_p$, which is actually the sum of two degenerate normal modes with orthogonal polarizations, one forms the integral equation

$$E_p = \frac{2}{A} \int_S \mathbf{E}(\mathbf{r},t)\, e^{-i\mathbf{k}_p \cdot \mathbf{r}}\, dS. \tag{4.26}$$

Substituting Eq. (4.24) into Eq. (4.26), we get

$$E_p = \frac{1}{A} \sum_n \int_S E_n\, e^{i[(\mathbf{k}_n - \mathbf{k}_p)\cdot \mathbf{r} - \omega t]}\, dS, \tag{4.27}$$

where we note that the phasor containing $(\mathbf{k}_n + \mathbf{k}_p)\cdot\mathbf{r}$ invariably vanishes in the integration, and so it is not written in the formula. From the orthogonality condition (4.25) we obtain the identity

$$E_p = E_p$$

as required. Thus E_p is the amplitude of the wave traveling parallel to the z axis as a result of the projection process in Eq. (4.26).

Since the vector potential A (see Chapter 1) is always parallel to the induced dipole, we use it to carry out the projection operation. The vector potential of the resulting plane wave, i.e., the wave traveling along the z axis, is

$$\mathbf{A}_{R,T} = \frac{1}{2} A_{R,T}\, e^{i(kz - \omega t)} + cc, \tag{4.28}$$

where the symbol R refers to the field radiated by the dipoles. We use the symbol T to denote the radiated field from all the dipoles, which is the sum of the fields from the individual dipoles. When referring to the contribution of one dipole, labeled a, to the total, the subscript reads R,a. The vector potential due to a single dipole is

$$\mathbf{A}_a = \frac{1}{2} \frac{-i\omega\mu_0}{4\pi} \mathbf{p}_0 \frac{e^{i(kr' - \omega t)}}{r'} + cc, \tag{4.29}$$

where r' is the distance from the a'th dipole to the field point $\mathbf{r}$. The projection amplitude for this single molecule ($A_{R,a}$) is given by Eq. (4.29) as

$$A_{R,a} = \frac{-i\omega\mu_0}{4\pi} p_0 \frac{1}{A} \int_S \frac{e^{i(kr' - kz)}}{r'}\, dS. \tag{4.30}$$

The integral in Eq. (4.30) has received considerable discussion in literature that discusses the contribution of a sheet of dipoles to the refractive index. Difficulties arise because the simplest way to do the integral in closed form is to locate the surface, S, symmetrically with respect to the z axis. Under these circumstances, there are interference phenomena that cause ambiguities in the answer. These disappear if S is not symmetrical about the z axis (in effect we note that only one dipole in the sheet is symmetrically located with respect to S), but the integrals can no longer be carried out in closed form.

We set up this problem in a special way that allows us to do the integral, which we denote by I (the constant term exp(−ikz) is factored out of I) to a reasonable approximation, and that explicitly show that the fringing fields vanish.

Consider the coordinate system in Figure 4.4. Let S be a disk whose center is offset by an amount y_0 from the z axis (x=0, y=0) and which is directly

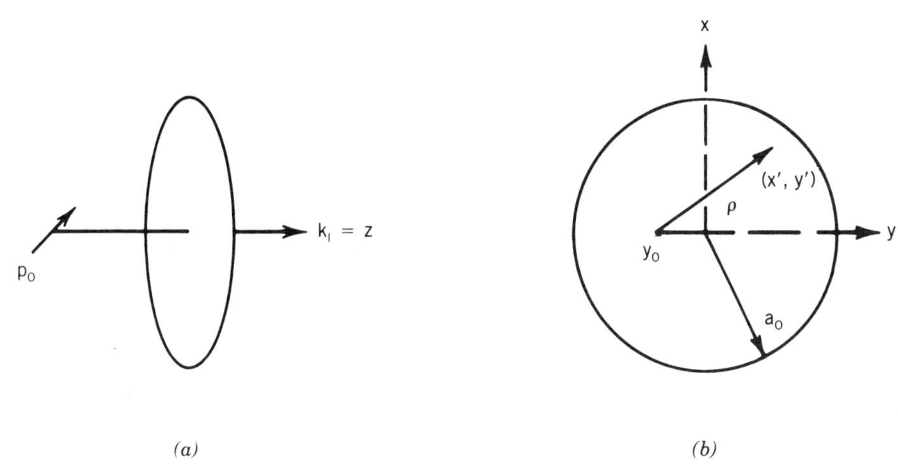

Figure 4.4. Coordinate system for computing projection of a dipole field onto a plane wave. a) Side view. b) Front view.

downstream from the dipole. Using the coordinates defined in Figure 4.4, the integral reads

$$I = \int_0^{2\pi} d\phi \int_0^{\rho_u(\phi)} \rho d\rho \, \frac{e^{ik(\rho^2+z^2)^{1/2}}}{(\rho^2+z^2)^{1/2}}, \tag{4.31}$$

where $\rho_u(\phi)$ is the upper bound on ρ, which depends explicitly on ϕ through the relation

$$\rho_u(\phi) = (a_0^2 + y_0^2 - 2a_0 y_0 \cos\phi)^{1/2},$$

where a_0 is the radius of the circle. One can change the variable of integration to $\zeta = (\rho^2+z^2)^{1/2}$ and, along with the relation $\rho d\rho = \zeta d\zeta$, Eq. (4.31) becomes

$$I = \int_0^{2\pi} d\phi \int_z^{(z^2+\rho_u^2(\phi))^{1/2}} e^{ik\zeta} \, d\zeta.$$

Carrying out the integral over ζ gives

$$I = -\frac{i}{k} \int_0^{2\pi} d\phi \, \{ e^{ik(z^2+\rho_u^2(\phi))^{1/2}} - e^{ikz} \}. \tag{4.32}$$

The aim of the sheet of dipoles argument, which in our notation involves integrating this expression over all $\mathbf{y}_0$, is to show that the contribution from the first term in Eq. (4.32) is negligible compared to the second. To do this properly involves considerable algebra that serves no useful purpose here, and can be found in the appropriate literature. Instead we introduce some approximations to evaluate the integral. We now expand

$$(z^2+\rho_u^2(\phi))^{1/2} \simeq (z^2+a_0^2+y_0^2)^{1/2} - \frac{a_0y_0\cos\phi}{(z^2+a_0^2+y_0^2)^{1/2}}$$
$$= r_0 - \frac{a_0y_0\cos\phi}{r_0},$$

where $r_0 = (z^2+a_0^2+y_0^2)^{1/2}$. The validity of this expansion requires that $r_0 \gg 2a_0y_0\cos\phi$. Equation (4.32) now becomes

$$I = -\frac{i}{k}\int_0^{2\pi} d\phi\, (e^{ikr_0} e^{-i\, ka_0y_0\cos\phi/r_0} - e^{ikz}),$$

which gives

$$I = +\frac{i}{k}\, 2\pi\, e^{ikz} - \frac{i\pi}{k}\, e^{ikr_0}\, J_0(a_0y_0k/r_0). \tag{4.33}$$

Here J_0 is the zero'th order Bessel function. For $a_0 \gg z$ and $y_0 \gg \lambda$, the argument of the Bessel function is very large and the value of the Bessel function becomes vanishingly small. Thus, in the integration over all y_0, the first term in Eq. (4.32) is nonzero for a negligibly small region $y_0 < \lambda$ and contains the contribution due to diffraction from the edge of the medium where $y_0 \simeq a_0$, which is negligible in the limit of infinite a_0 and y_0. Thus we retain only the first term in Eq. (4.33). Finally, substituting this result back into Eq. (4.30), and noting that the factor exp(ikz) cancels the factor exp(-ikz) in Eq. (4.30), we obtain

$$A_{R,a} = \frac{\mu_0 c}{2A}\, p_0. \tag{4.34}$$

Note that the amplitude does not have a factor of i, as does the vector potential amplitude from a single dipole, (Eq. (1.44)).

The total dipole-induced field at a field point is obtained by summing over all of the molecules, which is now easy. This sum simply involves averaging $\mathbf{p}_0$ over all of the possible molecular orientations, and multiplying by the total number of participating molecules. For a slab of thickness dz, $N = \mathcal{N}Adz$ and

$$A_{R,T} = \frac{\mu_0 c}{2}\, \mathcal{N} dz\, \langle p_0 \rangle, \tag{4.35}$$

which is the amplitude of the vector potential projected onto waves traveling along the z axis.

4.3.2 Refractive Index

We are now in a position to derive an expression for the refractive index using the previous results. We begin by using standard relations to evaluate the magnetic and electric fields. Bearing in mind that the field is propagating along the z axis, from Eqs. (4.28), (4.35), and $\mathbf{B} = \nabla \times \mathbf{A}$, we have

$$\mathbf{B}_{R,T} = \frac{1}{2}\,\frac{i\mu_0\omega}{2}\,\mathcal{N}dz\,\hat{\mathbf{z}} \times \langle \mathbf{p}_0 \rangle\, e^{i(kz-\omega t)} + \text{cc}. \tag{4.36}$$

From Eq. (1.60)

$$\mathbf{E}_{R,T} = \frac{1}{2}\,\frac{i\mu_0 c^2 k}{2}\,\mathcal{N}dz\,\mathbf{O}(\hat{\mathbf{z}})\cdot\langle \mathbf{p}_0 \rangle\, e^{i(kz-\omega t)} + \text{cc}, \tag{4.37}$$

which is the electric field created by the induced dipoles in an infinite slab of thickness dz.

What remains is to carry out the statistical average over orientations required in Eq. (4.37). For a plane wave traveling along the z axis there are two possible polarizations for the field, one polarized along the x axis (parallel to $\mathbf{E}_I$, denoted with the subscript $\parallel$) and one polarized along the y axis (perpendicular to $\mathbf{E}_I$, denoted $\perp$). Thus

$$\mathbf{E}_{R,T} = \frac{1}{2}\,\frac{i\mu_0 c^2 k\mathcal{N}dz}{2}\, e^{i(kz-\omega t)}\,[\langle \hat{\mathbf{x}}\cdot\mathbf{p}_0\rangle_{\parallel}\,\hat{\mathbf{x}} + \langle \hat{\mathbf{y}}\cdot\mathbf{p}_0\rangle_{\perp}\,\hat{\mathbf{y}}] + \text{cc}. \tag{4.38}$$

The dipoles are all induced by the incident field, i.e.,

$$\mathbf{p}_0 = \boldsymbol{\alpha}\cdot E_I;$$

therefore,

$$\langle \hat{\mathbf{x}}\cdot\mathbf{p}_0\rangle_{\parallel} = \langle \hat{\mathbf{x}}\cdot\boldsymbol{\alpha}\cdot\hat{\mathbf{x}}\rangle_{\parallel}\, E_I \tag{4.39a}$$

and

$$\langle \hat{\mathbf{y}}\cdot\mathbf{p}_0\rangle_{\perp} = \langle \hat{\mathbf{y}}\cdot\boldsymbol{\alpha}\cdot\hat{\mathbf{x}}\rangle_{\perp}\, E_I. \tag{4.39b}$$

Because the molecular orientation is random, the net contribution along the y axis averages to zero in nearly all cases (see Chapter 6 for exceptions, but note that one needs to deal with multipole fields to discuss these exceptions at the microscopic level), so that

$$\langle \hat{\mathbf{y}}\cdot\mathbf{p}_0\rangle_{\perp} = \langle \hat{\mathbf{y}}\cdot\boldsymbol{\alpha}\cdot\hat{\mathbf{x}}\rangle_{\perp}\, E_I = 0.$$

The result of Eq. (4.39a) depends on the details of the problem and is left in general form.

The total field is the sum of the incident and radiated fields. That is

$$E_T = E_I + E_{R,T},$$

which, when dotted into $\hat{\mathbf{x}}$, gives

$$E_T = E_I + \frac{ikN}{2\varepsilon_0}\, dz \, \langle \hat{\mathbf{x}} \cdot \boldsymbol{\alpha} \cdot \hat{\mathbf{x}} \rangle \, E_I. \tag{4.40}$$

For small dz, the second term is small compared to the first. Noting that exp(ix) $\simeq 1 + ix$ for $|x| \ll 1$, then Eq. (4.40) can be written as

$$E_T = E_I \, e^{i\frac{kN}{2\varepsilon_0} dz \, \langle \hat{\mathbf{x}} \cdot \boldsymbol{\alpha} \cdot \hat{\mathbf{x}} \rangle}, \tag{4.41}$$

which indicates that the radiated field affects the phase of the total field. Taking into account that $\boldsymbol{\alpha}$ is in general a complex quantity (Eqs. (3.14) and (3.35)),

$$E_T = E_I \, e^{-(\gamma/2)dz - i(\frac{\partial\phi}{\partial z})dz}, \tag{4.42}$$

where ϕ is the optical phase. The first exponent corresponds to absorption, and is consistent with the power absorption coefficient defined in Eq. (4.17). It is left as an exercise to show that these two results for γ are equivalent. Note that the absorption coefficient is defined as the decay rate of the flux (see Eq. (4.18)), and that the flux goes as the amplitude squared, so therefore the field decays as $\gamma/2$.

It is the second term in the argument of the exponential in Eq. (4.42) which gives the refractive index, and its dispersion. Here

$$\frac{\partial\phi}{\partial z} = -\frac{kN}{2\varepsilon_0} \, \mathrm{re}(\langle \hat{\mathbf{x}} \cdot \boldsymbol{\alpha} \cdot \hat{\mathbf{x}} \rangle).$$

The total phase shift at the field point is obtained by summing the phase shift from individual slabs, i.e., integrating over z. This gives

$$\phi = -\frac{kNz}{2\varepsilon_0} \, \mathrm{re}(\langle \hat{\mathbf{x}} \cdot \boldsymbol{\alpha} \cdot \hat{\mathbf{x}} \rangle). \tag{4.43}$$

Thus the total field can be written as

$$E = \frac{1}{2} E_I \, e^{i(k_T z - \omega t) - \gamma z/2} + cc, \tag{4.44}$$

where

$$k_T = k\{1 - \frac{N}{2\varepsilon_0} \, \mathrm{re}(\langle \hat{\mathbf{x}} \cdot \boldsymbol{\alpha} \cdot \hat{\mathbf{x}} \rangle)\}. \tag{4.45}$$

Since the velocity of the wave is given by $v = \omega/k_T$,

$$\frac{1}{v} = \frac{1}{c} \{1 - \frac{N}{2\varepsilon_0} \, \mathrm{re}(\langle \hat{\mathbf{x}} \cdot \boldsymbol{\alpha} \cdot \hat{\mathbf{x}} \rangle)\}. \tag{4.46}$$

We therefore define the refractive index as n = c/v or

$$n = 1 - \frac{N}{2\varepsilon_0}\,\mathrm{re}(\langle \hat{\mathbf{x}}\cdot\boldsymbol{\alpha}\cdot\hat{\mathbf{x}}\rangle). \tag{4.47}$$

Using Eq. (3.35) for the polarizability of the normal modes gives

$$n = 1 + \frac{N}{2\varepsilon_0}\,\mathrm{re}\sum_\beta \left\langle \frac{|\hat{\mathbf{x}}\cdot\boldsymbol{\ell}_\beta|^2}{M_\beta D_\beta(\omega-\mathbf{k}\cdot\mathbf{v})}\right\rangle. \tag{4.48}$$

Finally, substituting for the distribution of molecular velocities in Eq. (3.22) gives

$$n = 1 + \frac{N}{2\varepsilon_0}\sum_\beta \left\langle \frac{|\hat{\mathbf{x}}\cdot\boldsymbol{\ell}_\beta|^2}{m_\beta}\int dv_z\, \frac{\rho(v_z)(\omega_\beta^2-(\omega-kv_z)^2)}{(\omega_\beta^2-(\omega-kv_z)^2)^2+\Gamma_\beta^2(\omega-kv_z)^2}\right\rangle, \tag{4.49}$$

which is the final result.

Clearly, from Eq. (4.49) the refractive index becomes strongly frequency dependent when the incident frequency ω is approximately equal to one of the normal mode resonance frequencies. This variation is sketched for one normal mode (i.e., restrict the sum in Eq. (4.49) to a single mode $\beta = \alpha$) in Figure 4.5. For $\omega_\alpha > \omega$, one has normal dispersion in which the slope of the cudispersion curve is positive and the light moves at a velocity less than c. For $\omega > \omega_\alpha$, the velocity can be greater than c. The portions of the dispersion curve with negative slope have anomalous dispersion. The condition v > c is not a violation of the fact that energy cannot be transported at velocities greater than c. Energy is transported at the group velocity (see Problem 10.14), while the refractive index determines the phase velocity v.

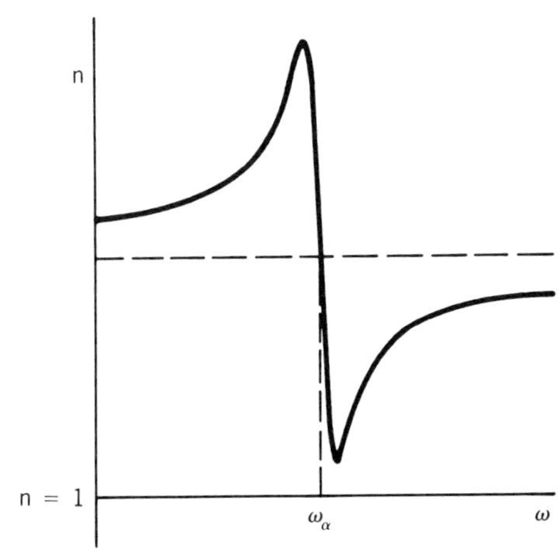

Figure 4.5. Contribution to the index of refraction from a single normal coordinate near the resonance at ω_α.

4.4 OPTICAL SPECTRA

Both the absorption and the refractive index undergo characteristic behavior when dipole active modes are near the resonance $\omega \simeq \omega_\alpha$. These characteristic features are measured and are referred to as optical spectra. In Figure 4.6 we show the complete optical spectrum, i.e., the index and absorption, of a gas

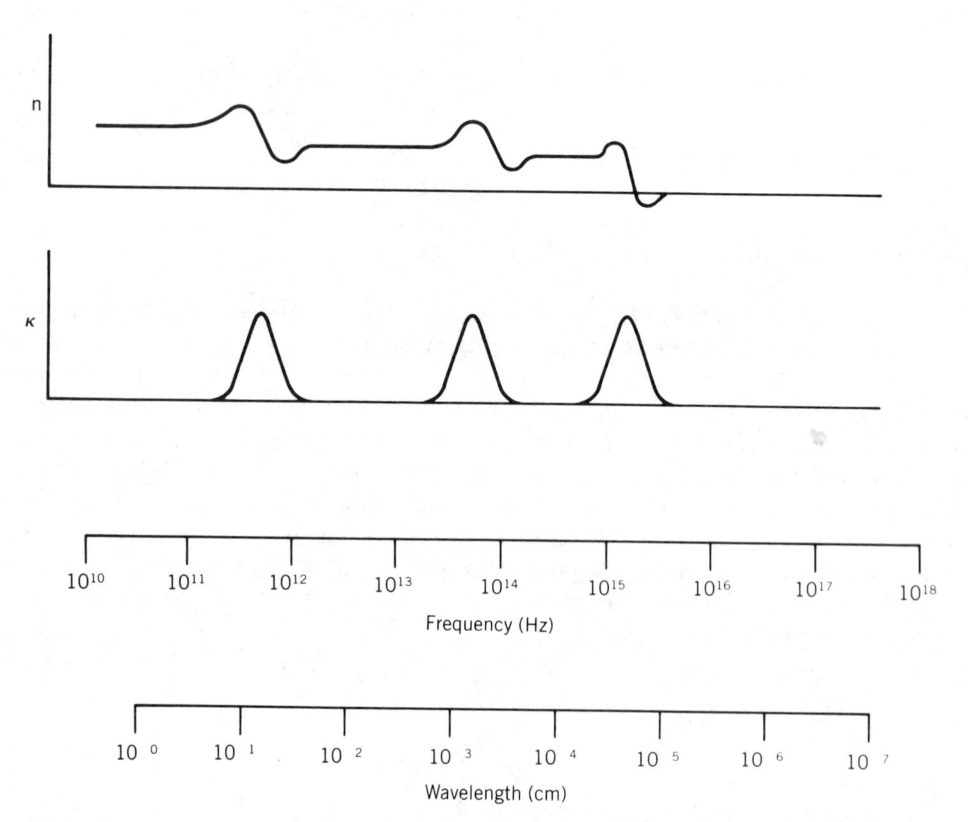

Figure 4.6. Real part $n(\omega)$ and imaginary part $\kappa(\omega) = \gamma(\omega)c/\omega$ of the complex index of refraction for a molecule with a permanent dipole moment.

containing molecules with a permanent dipole moment. In a gas, normal modes due to molecular rotation show spectral features that appear in the far infrared. Vibrational modes give features in the near infrared, and electronic transitions dominate the spectral features in the visible and ultraviolet.

If the absorption lines and dispersion are examined in more detail, they are found to be more complicated than we have indicated. This occurs because in reality there is some coupling between the normal modes. For example, when a diatomic (or triatomic, etc.) molecule undergoes vibrational motion, the resulting displacements in the atomic masses cause changes in the moments of inertia. This leads to vibrational-rotational coupling, which is not of concern in this book.

There are standard derivations with which one can now make contact. For electronic degrees of freedom,

$$(n+i\kappa)^2 = 1 + \frac{Ne^2}{m\varepsilon_0}\sum_\alpha \frac{f_\alpha}{(\omega_\alpha^2-\omega^2)-i\omega\Gamma} \tag{4.50}$$

where m refers to the electron mass, f_α is called the oscillator strength for the transition associated with the resonance frequency ω_α, and $n+i\kappa$ is the complex index of refraction given by kc/ω (see Chapter 5 for a more complete discussion of these terms). Comparing with Eq. (4.49), the oscillator strength is given by

$$f_\alpha = \frac{|\ell_\alpha|^2}{e^2}, \tag{4.51}$$

which is a useful relationship since f_α is tabulated for many transitions.

4.5 LIMITS OF VALIDITY

The formulas derived in this chapter are limited in their validity to dilute media such as gases. The wave equations of electrodynamics contain second derivatives, as discussed in Chapter 1. The first order integration techniques used here treat a wave equation that involves only first derivatives. We have, in effect, used the slowly varying amplitude and phase approximation, which is discussed in Chapter 10. The validity of this approximation is limited to small induced polarizations, but we need some further development to explain what a small polarization is. In the context of the present chapter, a small polarization requires that $1 \gg \gamma\lambda$ and that $|n-1| \ll 1$. The former requirement is widely met, but the latter is not typically true for solids and liquids.

ADDITIONAL READING

Phase shift and dipole sheet argument

Feynman, R.P., Leighton, R.B., and Sands, M. *The Feynmann Lectures in Physics*, Addison-Wesley, Reading, MA, 1964. pp. 31-2 to 31-8, and 30-10.

Sargent, M., Scully, M.O., and Lamb, W.E. Jr. *Laser Physics*, Addison-Wesley, Reading, MA, 1974. pp. 358-360.

Scattered and absorbed power

Feynman, Volume 1, Chapter 32.

Strong, J.M. *Radiation and Optics*, McGraw-Hill, New York, 1963. Chapter 12.

Technique for calculating absorption

Cocke W.J. Stimulated emission and absorption in classical systems, *Physical Review A* Volume 17, pp. 1713-1720.

PROBLEMS

4.1. (a) Calculate the scattering cross section from Eq. (4.22) in the limits $\omega \ll \omega_\alpha$ and $\omega \gg \omega_\alpha$ (set $\mathbf{k}\cdot\mathbf{v} = 0$, and take $\Gamma_\alpha \ll \omega_\alpha$).

(b) Instead of introducing a phenomenological decay in Eq. (3.8), introduce a radiation-reaction force term proportional to $\dddot{q}$. Show how this acts as a decay term.

(c) Equate the absorbed power from a single mode (i.e., Eqs. (4.14) and (4.16)) with the scattered power (i.e., Eqs. (4.20) and (4.22)). Obtain the frequency-dependent radiative decay rate.

$$\Gamma_{rad,\alpha} = \frac{1}{2\pi\varepsilon_0}\frac{\omega^2|\ell_\alpha|^2}{3c^3 M_\alpha}$$

4.2. (a) Assume a field and polarization of the form

$$E = \frac{1}{2}\hat{x}\,E_o\,e^{i[(\frac{n\omega}{c} + i\frac{\gamma}{2})z - \omega t]} + cc$$

$$P = \frac{1}{2}\hat{x}\,P_o\,e^{i[(\frac{n\omega}{c} + i\frac{\gamma}{2})z - \omega t]} + cc.$$

Write Eq. (1.15) with $\nabla\cdot\mathbf{E} = 0$, which is appropriate for the isotropic media discussed here. For $n \simeq 1$ (i.e., assume $n^2 - 1 \simeq 2(n - 1)$) and for small γ (i.e., drop terms in γ^2) show that

$$\gamma = \frac{\omega}{c\varepsilon_0}\frac{\mathrm{im}(P_o E_o^*)}{|E_o|^2}$$

$$n = 1 + \frac{1}{2\varepsilon_0}\frac{\mathrm{re}(P_o E_o^*)}{|E_o|^2}.$$

(b) Define the polarization as

$$\mathbf{P} = N\langle\mathbf{p}\rangle. \tag{4.52}$$

Use Eqs. (3.34) and (3.36) to develop this polarization. Show that a derivation along this line gives the same answers as found in Eqs. (4.16) - (4.18) and Eq. (4.48).

4.3. (a) Show that the cross section in Eq. (4.16) can be re-expressed near line center using Eqs. (4.23) and (3.22) as (note only one α is used and ℓ_α is taken to be parallel to $\mathbf{E}_I$)

$$\sigma_\alpha(\omega) \simeq \frac{3\pi}{2}\frac{c^2}{\omega_\alpha^2}\Gamma_\alpha\Gamma_{rad,\alpha}\int\frac{dv_z}{u\sqrt{\pi}}\frac{e^{-v_z^2/u^2}}{(\omega_\alpha-(\omega-kv_z))^2 + (\Gamma_\alpha/2)^2}$$

(b) Show the following:

$$\sigma_\alpha(\omega=\omega_\alpha) = \frac{3\lambda^2}{2\pi}\frac{\Gamma_{rad,\alpha}}{\Gamma_\alpha}, \qquad \Gamma_\alpha >> ku$$

$$\sigma_\alpha(\omega=\omega_\alpha) = \frac{3\lambda^2}{2\pi}\frac{\Gamma_{rad,\alpha}}{ku}\frac{\sqrt{\pi}}{2}, \qquad \Gamma_\alpha << ku$$

Note that if radiative decay is the dominant process in the homogeneous limit, then $\sigma = 3\lambda^2/4\pi$. These formulae are the same as obtained quantum mechanically. Note also that these formulae are independent of electromagnetic units.

4.4. Show that the absorption coefficient derived in Eqs. (4.15) to (4.17) is the same as the coefficient derived by starting with Eq. (4.42) as the definition of γ, and following the subsequent analysis in the text to relate γ to im$\langle\hat{\mathbf{x}}\cdot\boldsymbol{\alpha}\cdot\hat{\mathbf{x}}\rangle$.

4.5. Based on Figure 4.6 sketch the absorption coefficients and index of refraction for the molecules HF, N_2, CO_2, He.

4.6. (a) The relationship between the field energy density u, where

$$u = \frac{1}{2}(\mathbf{E}\cdot\mathbf{D} + \mathbf{B}\cdot\mathbf{H}),$$

the Poynting vector, and the current reads

$$\frac{\partial u}{\partial t} + \nabla\cdot\mathbf{S} = -\,\mathbf{J}\cdot\mathbf{E}. \tag{4.53}$$

Consider the case where the field amplitudes are time independent so that $\partial\bar{u}/\partial t = 0$, where $\bar{u}$ is the time average energy. Use Eq. (1.14) to show that time-independent parts of Eq. (4.53) give

$$\nabla\cdot\bar{\mathbf{S}} = -\,\frac{\omega}{2}\,\mathrm{im}(P\cdot E^*). \tag{4.54}$$

(b) Use Eqs. (4.52), (3.35) and (1.30) and set $\mathbf{k}\cdot\mathbf{v} = 0$ to obtain

$$\nabla\cdot\bar{\mathbf{S}} = -\,\omega N\,\frac{\mathrm{im}\langle\alpha\rangle}{\varepsilon_0 c}\,\bar{S}. \tag{4.55}$$

(c) Use Eqs. (3.34) and (3.35) (still for $\mathbf{k}\cdot\mathbf{v} = 0$) to obtain the relationships in Eqs. (4.16) - (4.18).

(d) Specialize Eq. (4.53) to a single atom at $\mathbf{r}=(0,0,0)$ using

$$\mathbf{P} = \sum_\alpha \mathbf{p}_\alpha\,\delta^{(3)}(\mathbf{r}), \tag{4.56}$$

(see Problem 3.10) and Eqs. (3.4) and (3.10) to show

$$\nabla \cdot \bar{S} = -\frac{\omega}{2} \sum_{\alpha} \mathrm{im}(Q_\alpha \hat{\boldsymbol{\ell}}_\alpha \cdot \boldsymbol{E}^*)\ \delta(r)$$

(e) Integrate this equation over a volume containing the molecule. Then use Gauss' theorem and Eqs. (3.17) and (3.20) to show that in steady state

$${}_S\!\int \bar{S} \cdot d\sigma = P_{abs}.$$

This is the simplest method for showing conservation of energy in the matter-field interaction.

4.7. Repeat the exercise in Problem 4.6d and 4.6e for the case in which the matter system developed in Eq. (3.8) and Eqs. (3.16) - (3.18) is undamped. In this case assume that Q_α varies slowly in time (see Problem 10.19), i.e., that $\dot{Q}_\alpha < \omega Q_\alpha$. Write the slowly varying part of E_{mech} and show that

$${}_S\!\int \bar{S} \cdot d\sigma = \frac{dE_{mech}}{dt}.$$

5 Macroscopic Electrodynamics

In Chapter 4 we discussed the propagation of electromagnetic radiation through a dilute medium. In this chapter we show how to generalize the result to dense media. The major results of Chapter 4 are written as statistical averages over dipoles induced in normal modes. The computation of averages over random molecular orientations is conceptually straightforward. An example is worked out in Section 5.5, and there are numerous examples in later chapters. There is also a spatial averaging procedure that reduces the matter system to the continuum needed for consistency with macroscopic electrodynamics. This average leads to consequences discussed in Section 5.2.

The macroscopic polarization is defined in terms of normal coordinates as

$$\mathbf{P} = \sum_{a} \sum_{\alpha} \langle \mathbf{p}_{\alpha,a} \rangle \tag{5.1}$$

where α identifies the normal mode and the summation over a runs over all molecules. The effects of fluctuations are deferred to Chapter 12 which deals with scattering. We assume that the electromagnetic wavelengths are much larger than the typical molecular dimensions. The spatial average implicit in Eq. (5.1) homogenizes the polarization over spatial scales of the order of an optical wavelength giving a continuous polarization field, rather than one that is singular at each molecule.

In addition to electric dipole effects which lead to the macroscopic polarization **P**, there may also be a macroscopic magnetization **M** resulting from magnetic dipoles induced in individual molecules. This is usually a weak effect which is discussed in Chapter 6, Optical Activity.

5.1 WAVE PROPAGATION IN DENSE MATTER

As the result of averaging over individual dipoles, the optical properties of matter can be described by constants which can depend on both frequency and wave vector. Let us consider a plane-wave field $\mathbf{E}(\mathbf{r},t)$ of the form given by Eq. (1.20), with a corresponding polarization field

$$\mathbf{P}(\mathbf{r},t) = \frac{1}{2}\,\boldsymbol{P}(k,\omega)\,e^{i(\mathbf{k}\cdot\mathbf{r}-\omega t)} + cc. \tag{5.2}$$

We refer to the $\mathbf{E}(\mathbf{r},t)$ and $\mathbf{P}(\mathbf{r},t)$ fields as Maxwell fields when we need to distinguish them from the local fields discussed in Section 5.2. Eqs. (1.20) and (5.2) are trial solutions of the wave equations, which eventually determine the relation between $\mathbf{k}$ and ω. We assume that, for the spatially homogeneous media of interest here, the electric and polarization fields are related through a susceptibility $\boldsymbol{\chi}$ such that

$$\boldsymbol{P}(k,\omega) = \varepsilon_0\,\boldsymbol{\chi}(k,\omega)\cdot\boldsymbol{E}(k,\omega). \tag{5.3}$$

Note that $\boldsymbol{\chi}$ can be derived from molecular quantities and is in general a function of both $\mathbf{k}$ and ω. The dependence on $\mathbf{k}$, called spatial dispersion, occurs in limited number of cases. For example, in semiconductors, electron-hole pairs can be formed which lead to wave vector-dependent susceptibilities. Another example is the effects of quantum pressure in an electron gas, i.e., a metal. Since these are special cases, we drop the dependence on $\mathbf{k}$ and write the susceptibility as $\boldsymbol{\chi}(\omega)$. Since

$$\mathbf{D} = \varepsilon_0\,\mathbf{E} + \mathbf{P}$$

then, from Eq. (5.3), we can write

$$\boldsymbol{D} = \boldsymbol{\varepsilon}\cdot\boldsymbol{E} \tag{5.4}$$

where the dielectric tensor $\boldsymbol{\varepsilon}$ is given by

$$\boldsymbol{\varepsilon} = \varepsilon_0\,(\mathbf{I} + \boldsymbol{\chi}). \tag{5.5}$$

The form of the dielectric tensor is determined by the symmetries of the medium. In gases, liquids, amorphous solids and several crystal classes, the dielectric tensor can be written as

$$\boldsymbol{\varepsilon} = \varepsilon\,\mathbf{I}, \tag{5.6}$$

where $\mathbf{I}$ is the unit tensor. For crystalline solids with fewer symmetries the dielectric tensor takes on more complicated forms. The resulting complexities are discussed in Chapter 7.

In the present section we deal only with optically isotropic media, i.e, media with a dielectric tensor in the form of Eq. (5.6). For this case

$$\nabla\cdot\mathbf{E} = \frac{1}{2}\,\mathbf{k}\cdot\boldsymbol{E}(k,\omega)\,e^{i(\mathbf{k}\cdot\mathbf{r}-\omega t)} + cc \tag{5.7}$$

and from the relation

$$\mathbf{k} \cdot \boldsymbol{E}(\mathrm{k},\omega) = \frac{1}{\varepsilon(\omega)} \mathrm{k} \cdot \boldsymbol{D}(\mathrm{k},\omega) \tag{5.8}$$

Eq. (1.25) gives

$$\mathbf{k} \cdot \boldsymbol{E}(\mathrm{k},\omega) = 0. \tag{5.9}$$

In isotropic media we have

$$\nabla \cdot \mathbf{E}(z,t) = 0. \tag{5.10}$$

We now derive the dispersion relation, i.e., the relation between $|k|$ and ω obtained from the wave equation. We go through a somewhat tedious derivation of an expression which can be written down almost by inspection in order to emphasize the logic. In later chapters in which the algebra is by no means simple, it is the logic that is the key to identifying the plane wave solutions and obtaining dispersion relations. From Eq. (1.11), with $\mathbf{M} = \mathbf{J} = 0$ and reducing the curl curl to $-\nabla^2$ because of our assumption of an isotropic medium we obtain

$$-\nabla^2\mathbf{E} + \mu_0 \frac{\partial^2\mathbf{D}}{\partial t^2} = 0.$$

Remember that everything covered in this section is based on the unjustified assumption that it is possible to write the fields in the plane-wave form of Eq. (1.20). We need to verify the validity of this assumption. Substituting Eq. (5.6) and (1.20) gives

$$-\nabla^2\mathbf{E} + \frac{\varepsilon}{\varepsilon_0 c^2} \frac{\partial^2\mathbf{E}}{\partial t^2} = 0, \tag{5.11}$$

which gives

$$(k^2 - \frac{\varepsilon}{\varepsilon_0 c^2} \omega^2)\, \boldsymbol{E}(\mathrm{k},\omega) = 0. \tag{5.12}$$

This equation defines a relationship between k^2 and ω^2 that must be true if $\boldsymbol{E}$ is to be nonzero. In other words, the plane wave field in Eq. (5.12) exists only if the dispersion relation,

$$k^2 - \frac{\varepsilon(\omega)}{\varepsilon_0 c^2} \omega^2 = 0, \tag{5.13}$$

is true. Otherwise Eq. (5.12) requires that the field amplitude be zero. This logical interrelationship between the existence of plane wave fields and dispersion relations is the key to discovering plane-wave solutions in cases in which it is not as simple as it is here.

Since ε is in general a complex number, so also is the wave vector

$$k = \frac{\omega}{c}\left(\frac{\varepsilon}{\varepsilon_0}\right)^{1/2}. \tag{5.14}$$

Expressing the real and imaginary parts of k in terms of the real and imaginary parts of ε, i.e., $\varepsilon = \varepsilon_R + i\varepsilon_I$, gives

$$\mathrm{re}(k) = \frac{1}{(2\varepsilon_0)^{1/2}}\frac{\omega}{c}\left([\varepsilon_R^2 + \varepsilon_I^2]^{1/2} + \varepsilon_R\right)^{1/2} \tag{5.15}$$

$$\mathrm{im}(k) = \frac{1}{(2\varepsilon_0)^{1/2}}\frac{\omega}{c}\left([\varepsilon_R^2 + \varepsilon_I^2]^{1/2} - \varepsilon_R\right)^{1/2}. \tag{5.16}$$

The optical properties of matter are often written in terms of the complex refractive index denoted $n + i\kappa$. These quantities are obtained from k through the relationship

$$n + i\kappa = \frac{kc}{\omega}. \tag{5.17}$$

For most cases of interest in this book, the absorption is small, so that $\varepsilon_R >> \varepsilon_I$, re(k) >> im(k), we write the formulae for n and κ explicitly for that case

$$n = \left(\frac{\varepsilon_R}{\varepsilon_0}\right)^{1/2} \tag{5.18}$$

$$\kappa = \frac{\varepsilon_I}{2(\varepsilon_R\varepsilon_0)^{1/2}}. \tag{5.19}$$

From Eq. (5.5), in the case of isotropic media we have

$$n = (1 + \chi_R)^{1/2}, \tag{5.20}$$

where χ_R is the real part of the susceptibility. Note that this formula is quite different from Eq. (4.48). In dilute media the index is proportional to the density through the approximate relationship

$$n \simeq 1 + \frac{\chi_R}{2}. \tag{5.21}$$

It is left as an exercise to show that Eqs. (5.21) and (4.49) are equivalent. However, we see that for any n that is appreciably larger than one, which is the case in all solids and liquids, it is not in general possible to use relationships derived from dilute media. For the next four chapters, we elaborate on the procedures sketched in this section to deal with dense media. In Chapter 10, we discuss the methods by which we can exploit the substantial simplifications made possible when there are weak optical interactions occurring in dense media, while avoiding the substantial errors that can result when the weak-interaction analysis is used inappropriately.

5.2 LOCAL FIELDS

The response of local molecules is not given correctly by the macroscopic Maxwell field **E**, i.e, the field computed using Eq. (1.15) with a polarization given by Eq. (5.1). The correction to the Maxwell field required to describe the local field is called the local field correction. In this book, local field effects are not of great importance, and hence the present discussion is limited to an outline of the basic concepts.

The local field arises as a consequence of the passage from microscopic electrodynamics to macroscopic electrodynamics. There are no **H** and **D** fields in microscopic electrodynamics. Instead optical media are always viewed as being a vacuum filled with real charges. It is necessary to compute the fields at each point in space as radiated from each individual charge. Some simplification occurs from treating charge pairs as dipoles, but it is still necessary to compute the field at any point as an explicit sum of the fields from each dipole. The formula used to describe the field at the a'th molecule is written

$$\mathbf{E}_a = \mathbf{E}_I + \sum_b{}' \mathbf{E}_{a,b} + \mathbf{E}_a. \tag{5.22}$$

The field $\mathbf{E}_I$ is the field incident on the medium from the outside, the source field $\Sigma' \mathbf{E}_{a,b}$ is the sum of the fields of all molecules other than a (signified by the prime on the sum, and each term in the sum is given by Eq. (1.47)), and $\mathbf{E}_a$ is the self field of the a'th molecule acting on itself. The self field is infinite, and no calculations can be performed until it is eliminated. The process of elimination gives the $\dddot{q}$ term in Problem 4.1. This in turn gives the radiative decay which is introduced phenomenologically in Chapter 3, and leads to scattering as discussed in Section 4.2. The elimination procedure, a matter of current research, does not have any further bearing on the material in this book.

Once the self field is eliminated from the problem, the force on the a'th molecule is computed from the local field, which is the sum of the remaining terms

$$\mathbf{E}_{loc,a} = \mathbf{E}_I + \sum_b{}' \mathbf{E}_{a,b}. \tag{5.23}$$

In the passage from microscopic to macroscopic electrodynamics, the source field is computed from the polarization which is summed over all atoms. Thus the Maxwell field **E** takes the form

$$\mathbf{E} = \mathbf{E}_I + \sum_b \langle \mathbf{E}_{a,b} \rangle.$$

Here the brackets around the field indicate that it is not the field of the point dipole, but rather the field that has been averaged over a neighborhood surrounding the position of the a'th molecule, which is one of the averages involved in defining the macroscopic polarization in Eq. (5.1). There are two potential sources of error in using the macroscopic Maxwell field, i.e., the field derived from the macroscopic polarization, to compute forces. First, in a crystal, it may be a poor approximation to average over the position of the a'th molecule. This gives rise to a structure factor s in the formulas given below, but further discussion of this aspect of the correction is beyond the scope of this book, and we continue the discussion as if s were zero. An error that is always present is that the Maxwell field reintroduces the self field. In particular,

$$\mathbf{E} = \mathbf{E}_I + \sum_b{}' < \mathbf{E}_{a,b} > + < \mathbf{E}_a >. \tag{5.24}$$

The averaged field $<\mathbf{E}_a>$ is not infinite, but it doesn't belong in the force at all; it was eliminated from the calculation when radiative damping was included. Therefore the local field is related to the Maxwell field by

$$\mathbf{E}_{loc,a} = \mathbf{E} - < \mathbf{E}_a >. \tag{5.25}$$

The term $<\mathbf{E}_a>$ is called the local field correction. Its computation can be found in Van Kranendonk and Sipe, who show that

$$\mathbf{E}_{loc} = \mathbf{E}_I + \frac{\zeta \mathbf{P}}{\varepsilon_0}, \tag{5.26}$$

where

$$\zeta = (\frac{1}{3} + s). \tag{5.27}$$

The interaction potential V_{int} at the molecule site can be written as $\mathbf{p} \cdot \mathbf{E}_{loc}$, in which case the potential contains the mutual interaction between dipoles. It is frequently called the dipole-dipole interaction. We do not use this interaction potential here. Instead, assuming that we know E_{loc}, the induced polarization is given by (for one molecule)

$$\mathbf{p} = \alpha \mathbf{E}_{loc} \tag{5.28}$$

where the polarizability is assumed to be a scalar coming from an average over all possible orientations. The macroscopic polarization is therefore

$$\mathbf{P} = N \alpha \mathbf{E}_{loc}, \tag{5.29}$$

where N is the number of molecules per unit volume. Substituting for E_{loc} from Eq. (5.26),

$$\mathbf{P} = \frac{N\alpha}{1 - \zeta N\alpha/\varepsilon_0} \mathbf{E}_I \tag{5.30}$$

or, eliminating **P**,

$$\mathbf{E}_{loc} = \frac{1}{1 - \zeta N\alpha/\varepsilon_0} \mathbf{E}_I. \tag{5.31}$$

Comparing Eqs. (5.3) and (5.30)

$$\chi = \frac{1}{\varepsilon_0} \frac{N\alpha}{1 - \zeta N\alpha/\varepsilon_0}. \tag{5.32}$$

Alternative forms for this result are the Clausius-Mossotti equation (stated for $s = 0$).

$$\alpha = \frac{3\varepsilon_0}{N} \frac{\varepsilon - \varepsilon_0}{\varepsilon + 2\varepsilon_0} \tag{5.33}$$

and the Lorentz-Lorenz equation

$$\alpha = \frac{3\varepsilon_0}{N} \frac{n^2 - 1}{(n^2 + 2)}. \tag{5.34}$$

Equations (5.32) - (5.34) are various ways of scaling the susceptibility with density. Since quantities such as α are expected to change with density as well, both through the decay coefficients Γ_α and resonance frequencies ω_α of the normal modes, these scaling laws are, at best, approximate.

Eliminating α from Eq. (5.31) via (5.34) gives

$$\mathbf{E}_{loc} = \frac{n^2 + 2}{3} \mathbf{E}_I, \tag{5.35}$$

for which we still have $s = 0$. Thus for all cases in which n is not nearly unity, i.e., for all cases in which a dense-medium rather than dilute-medium analysis is needed, the difference between the local and Maxwell fields is large. For $n \simeq 2$, $\mathbf{E}_{loc} \simeq 2\mathbf{E}$. For semiconductors in which n can be as large as 3 or 4, the local field effects are even more important.

The local field effects discussed here do not by themselves involve any additional exchange of energy. The absorption and other optical interactions are determined by the dielectric constant, which includes local field effects in its value. The local field effect is incorporated into all of the measured coefficients used to describe electromagnetic phenomena in condensed matter (see e.g., Problem 5.5). It is not consistent to take a formula using measured coefficients and reintroduce the correction. Hence we drop all reference to the local field and regard the correction as having been included in the phenomenological coefficients from the beginning.

5.3 SELLMEIER FORMULA

It is frequently important to be able to interpolate the refractive index at wavelengths intermediate to those at which accurate measurements have been made. A fair analytical approximation to this wavelength variation is given by the Sellmeier formula.

We start by returning to the basic definition for the polarizability:

$$\boldsymbol{\alpha} = \sum_{\beta} \frac{\boldsymbol{\ell}_{\beta} \boldsymbol{\ell}_{\beta}}{M_{\beta} D_{\beta}(\omega)} \tag{5.36}$$

where we have assumed dense matter by dropping the dependence on velocity in D_β. At this point, we may also assume that local field effects have been included in ℓ_β. Thus, in Eq. (5.3)

$$\boldsymbol{\chi} = \frac{N}{\varepsilon_0} \boldsymbol{\alpha}. \tag{5.37}$$

Noting that $\chi_R = n^2 - 1$,

$$n^2 - 1 = \frac{N}{\varepsilon_0} \sum_{\beta} \text{re} \left(\frac{|\ell_\beta|^2}{M_\beta D_\beta(\omega)}\right). \tag{5.38}$$

In the visible, usually $\omega_\beta >> \omega$ and $\omega_\beta >> \Gamma_\beta$. Hence in this regime, writing all frequencies in terms of the wavelength of light in vacuum denoted λ_v,

$$n^2 - 1 = \sum_{\beta} A_\beta \frac{\lambda_v^2}{(\lambda_v^2 - B_\beta)} \tag{5.39}$$

which is the Sellmeier relation. Here

$$A_\beta = \frac{N |\ell_b|^2}{M_\beta \omega_\beta^2} \tag{5.40}$$

and

$$B_\beta = \left(\frac{2\pi c}{\omega_\beta}\right)^2.$$

Usually, only one value of β is assumed, and the coefficients are determined by fitting the curve to the index at two different wavelengths. The coefficients are tabulated in many standard optics texts. Sellmeier curves can be used whenever accuracies of the order of 0.01 are acceptable, provided one stays away from regions of high absorption. Such accuracies are satisfactory for many applications, but are rarely good enough for nonlinear optics, especially for phase-matching applications (see Chapter 10) where accuracies of 0.001 are needed.

5.4 POYNTING VECTOR

We now write down the relationships among the various field amplitudes for future reference. The magnetic fields are perpendicular to both **E** and **k**. The D vector amplitude is given by Eq. (5.4). The magnetic field amplitude H is given by Maxwell's equations as

$$H(k,\omega) = n(\omega)\, c\varepsilon_0\, E(k,\omega). \tag{5.41}$$

The amplitude of the Poynting vector **S** reads

$$S = \frac{n(\omega)\varepsilon_0 c}{2}\, |E(k,\omega)|^2. \tag{5.42}$$

5.5 DIPOLE AVERAGES IN DENSE MATTER

The key quantity in describing the propagation of radiation through matter is the polarizability α, which gives rise to dispersion in the refractive index as well as absorption. In Chapter 4 we assumed a gas of molecules with no permanent dipole moments and random molecular orientation. In this section we include molecules with permanent dipole moments, and we discuss what happens to the dipole averages in dense matter such as a liquid.

Three kinds of normal modes contribute to the polarizability: electronic, vibrational and rotational. As the density of the medium increases, the rotational motion becomes hindered. In the limit of a solid, these degrees of freedom vanish and there is no rotational contribution to the polarizability. Both electronic and vibrational modes survive as the density increases. In dense media, local intermolecular forces can shift the characteristic resonance frequencies of vibrational and electronic normal coordinates. These forces can also affect the damping Γ via coupling to lattice degrees of freedom such as acoustic phonons. In general, the changes in the vibrational and electronic degrees of freedom are relatively small as the medium becomes more dense. It is the changes in the rotational modes that need to be discussed in some detail.

5.5.1 Gas Molecules with Permanent Dipole Moments

As a preliminary exercise to discussing an example of absorption in liquids we consider a dense gas made up of molecules with permanent dipole moments $\mathbf{p}_0$. When an electric field is applied, there is a torque on the dipole (molecule) which tries to align the dipole parallel to the incident **E** field. As the molecule rotates, it undergoes collisions with other molecules which re-randomize its orientation relative to the field. In cases in which the gas is so dense that the collision rate exceeds the rotational frequency, it is no longer useful to describe the medium in terms of the detailed motion of the normal modes. Instead, a statistical treatment is adopted from the beginning. Since the collision frequency depends on the temperature, the net orientation of the molecules into the field direction depends strongly on temperature.

When a field is applied, any permanent dipole has the interaction energy

$$V_{int} = -p_0\, E \cos\theta \tag{5.43}$$

where θ is the angle between the dipole $\mathbf{p}_0$ and the electric field **E** (see Figure 3.5). The corresponding contribution to the net induced dipole is $p_0\cos\theta$. From statistical mechanics, the probability $P(\theta)$ of finding the dipole in the angular range θ to $\theta + d\theta$ is

$$P(\theta) = e^{-p_0 E\cos\theta/KT}\, 2\pi \sin\theta\, d\theta \qquad (5.44)$$

where 2!sinθdθ is the solid angle subtended by a ring of angular "thickness" dθ, T denotes temperature, and K is the Boltzmann constant. Therefore the average contribution of the dipole to the net induced dipole moment for a random molecular orientation (in the absence of the field) is

$$\langle p_0\cos\theta\rangle = \frac{\int p_0\cos\theta\, e^{-p_0 E\cos\theta/KT}\sin\theta\, d\theta}{\int e^{-p_0\cos\theta\, E/KT}\sin\theta\, d\theta} = p_0\, L(p_0E/KT), \qquad (5.45)$$

where L(x) is the Langevin function, coth(x) - 1/x. In practice, $p_0E \ll KT$ and, in this limit, $L(x) \rightarrow x/3$. For our case,

$$\langle p_0\cos\theta \rangle = \frac{p_0^2 E}{3KT} \qquad (5.46)$$

and therefore the contribution to the polarizability is

$$\alpha = \frac{p_0^2}{3KT}\, \mathbf{I}. \qquad (5.47)$$

The contribution due to the induced dipoles which comes in via the dot product between the incident field and polarizability tensor (due to the vibrational and electronic degrees of freedom) is not included in this equation.

5.5.2 Molecular Diffusion Effects in Dense Gases and Liquids

The formulation in the preceding section was for an applied field. In dense gases and liquids, the molecule is not free to rotate through large angles directly, but rather undergoes a diffusional motion due to intermolecular collisions. There is a characteristic relaxation time τ associated with this motion.

The relaxation time can be estimated from the mechanism which inhibits reorientation. For example, in a liquid it is intermolecular collisions which occur due to the Brownian motion of the molecules - this is usually characterized by the viscosity η. The mean square value of the re-orientation angle θ achieved in a molecule of dimension ζ after a time t via Brownian motion is

$$\langle \theta^2 \rangle = \frac{KT}{4\pi\zeta^3\eta}\, t \equiv \frac{t}{\tau}. \qquad (5.48)$$

Thus

$$\tau = \frac{4\pi\eta\zeta^3}{KT} \qquad (5.49)$$

is the characteristic time of Brownian motion. For a water molecule at room temperature ($\zeta = 2.3 \times 10^{-10}$ m, $\eta = 0.001$) we find $\tau \simeq 4 \times 10^{-11}$ seconds. Since this time is shorter than any of the characteristic times it takes for a free molecule to rotate through an angle of 2π via its rotational normal modes, it follows that the molecule cannot rotate at any of its normal mode frequencies. Instead, dispersion in the refractive index associated with molecular reorientation takes place at frequencies $\omega \simeq 1/\tau$.

In a diffusional process, it is the macroscopic polarizability which is described by the diffusion process. That is, it makes no sense to talk about individual dipoles since we are dealing with a statistical process such as Brownian motion. In a diffusion process, the rate of decay of a quantity is proportional to the instantaneous value of that quantity. That is, if a macroscopic polarization is induced in the medium (for example by an applied DC field) and then allowed to relax (by turning off the field), the pertinent equation is

$$\frac{d\mathbf{P}}{dt} = -\frac{\mathbf{P}}{\tau} \tag{5.50}$$

where **P** is the macroscopic polarizability. Of course the dipoles in each molecule do not disappear, but rather the distribution thermalizes. In the case of an applied electromagnetic field **E**,

$$\mathbf{P} + \tau\frac{d\mathbf{P}}{dt} = \frac{\mathcal{N}p_0^2}{3KT}\mathbf{E}, \tag{5.51}$$

where the right-hand side is the driving polarization source created by the applied field. It is left as an exercise to derive the absorption and index implied by Eq. (5.51). The frequency dependence of the refractive index and the absorption for this case are shown in Figure 5.1. The absorption peaks at $\omega \simeq 1/\tau$. Both the dispersion and absorption are different in form to the free molecule case shown in Figures 3.4 and 4.5.

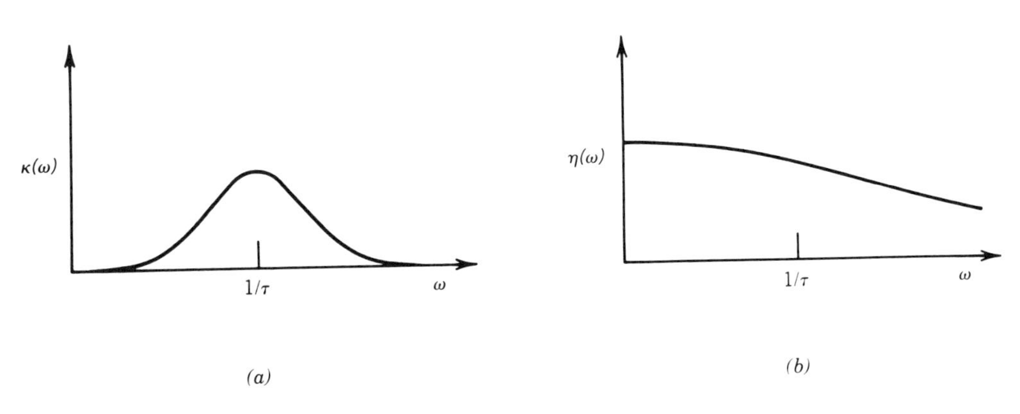

Figure 5.1. Real (n) and imaginary (k) parts of the index of refraction due to frustrated rotations of a liquid made up of molecules with permanent dipole moments.

We close this section with a few introductory comments on solids, which are discussed in detail in Chapter 7. In gases and liquids, the statistical average over the induced or permanent dipoles always leaves the net radiated field (traveling along the incident field direction) parallel to the incident field. That is, no change in polarization occurs, as follows from the equation

$$\langle \hat{\mathbf{y}} \cdot \hat{\mathbf{p}}_0 \rangle_\perp = \langle \hat{\mathbf{y}} \cdot \boldsymbol{\alpha} \cdot \hat{\mathbf{x}} \rangle_\perp = 0$$

taken from Chapter 4. In a crystalline solid, the molecules are on regular sites, and their orientations are all parallel to one another. In this case, the dipole radiated field inside the medium is not necessarily parallel to the **E** field. As we show in Chapter 7, this is a consequence of light not propagating along one of the crystal symmetry directions, and it gives rise to the many interesting phenomena of crystal optics.

ADDITIONAL READING

Electromagnetic theory

Jackson, J.D. *Classical Electrodynamics*, 2nd Edition, Wiley, New York, 1975 (or 1st Ed. 1962).

Local field correction

Jackson, 1975, pp. 152-162 (1962, pp. 116-119).
Strong, J.M. *Radiation and Optics*, McGraw-Hill, New York, 1963. Chapter 15.
Van Kranendonk, J. and Sipe, J.E. in *Progress in Optics*, E. Wolf ed., North Holland, Amsterdam, 1977. Volume XV, p. 245.

PROBLEMS

5.1. (a) Prove the following relations for plane wave fields in isotropic media assuming small κ (neglect terms in κ^2)

$$n^2 = 1 + \frac{1}{\varepsilon_0} \frac{\mathrm{re}\langle \boldsymbol{P} \cdot \boldsymbol{E}^* \rangle}{|E|^2} \tag{5.52}$$

$$\kappa = \frac{1}{2n\varepsilon_0} \frac{\mathrm{im}\langle \boldsymbol{P} \cdot \boldsymbol{E}^* \rangle}{|E|^2}. \tag{5.53}$$

(b) Suppose that in using these formulae it was discovered that $\boldsymbol{P}$ was not parallel to $\boldsymbol{E}$. Would the formulae be valid? Give reasons.

(c) Consider a polarization of the form

$$\mathbf{P}_T = \mathbf{P} + \mathbf{P}'$$

where **P** is the polarization used in part (a) and **P'** is small (see end of Chapter 4 for definition of small). Let P be such that it gives only a real part of the index (i.e., $\kappa = 0$ in part (a)). Show that the real and imaginary parts of the index given by P_T (denoted n' and κ') are given by

$$n' \simeq n + \frac{\mathrm{re}\langle P'E^*\rangle}{2n\varepsilon_0 |E|^2} \tag{5.54}$$

$$\kappa' \simeq \frac{\mathrm{im}\langle P'E^*\rangle}{2n'\varepsilon_0 |E|^2}. \tag{5.55}$$

5.2. Using Eq. (5.52), show that Eq. (5.51) predicts

$$n^2 = 1 + \left(\frac{1}{\varepsilon_0}\right)\left(\frac{1}{3KT}\right)\frac{Np_0^2}{1 + \omega^2\tau^2}$$

$$\kappa = \left(\frac{1}{2n\varepsilon_0}\right)\left(\frac{1}{3KT}\right)\frac{\omega\tau Np_0^2}{1 + \omega^2\tau^2}.$$

5.3. Derive Eqs. (5.38) - (5.40), from Eq. (5.52). In order to answer the next problem it is useful to retain the unapproximated form of the answer.

5.4. Consider Eq. (4.49) in the homogeneously broadened limit (treat $\rho(v_z)$ as a Dirac delta function). Compare it with the results of Problem 5.3. Show that these are related in a manner consistent with the approximation in Eq. (5.21).

5.5. Use the local field in Eq. (5.26) in place of the Maxwell field in Eq. (3.8). Assume a homogeneous medium (no averaging over velocity in Eq. (4.52)) and Eq. (3.4) to construct an equation for q_α.

(a) Show that this results in a shift in the resonance from ω_α to ω_α' where

$$\omega_\alpha' = \omega_\alpha \left(1 - \frac{1}{3}\frac{N|\ell_\alpha|^2}{\varepsilon_0 M_\alpha \omega_\alpha^2}\right)^{1/2}.$$

(b) Use Eqs. (3.34) and (3.35) in the case of a single normal coordinate. Obtain Eq. (5.32) by evaluating the resulting polarizability far off resonance.

5.6. Show that Eq. (4.13) matches the second term on the right-hand side of Eq. (3.19) (with $\mathbf{v} = 0$) when the electric field amplitude in Eq. (3.15) is replaced by the amplitude of the local field of Eq. (5.26).

6
Optical Activity

In this chapter we deal with the subject of optical activity. Optical activity is caused by electric quadrapoles and magnetic dipoles. It is one of the few exceptional cases in which one cannot ignore interactions other than those involving electric dipoles. We discuss this effect here because it is an inherently interesting electromagnetic phenomenon and because this is a way of introducing the concept of an eigenvector in a relatively simple context. For simplicity, we confine our attention to the case of transparent media.

Magnetic forces arise from currents in the media. Since it is the normal coordinates which describe the relative motion of charges, one associates currents with the velocities $\dot{q}_\alpha/c$. The magnitude of the velocity is of the order of $\omega q_\alpha/c$, or $2\pi q_\alpha/\lambda$. Since the range of the normal coordinates in a molecule is typically a few angstroms, whereas the electromagnetic wavelengths of interest are larger than a few thousand angstroms, the magnetic forces are weak. In cases when the electric dipole forces are zero, these magnetic forces become important. This is not the case in optical activity where it is the effect of magnetic dipoles accumulated over thousands of wavelengths which produces a significant perturbation on the electric dipole effects.

6.1 EFFECT OF MAGNETIZATION ON POLARIZATION

In all electromagnetic interactions, there is a magnetization **M** (which comes from q_α) that is usually about one thousand times smaller than **P**. Most of the time, **M** is orthogonal to **P** since they both arise from the same charge motions. From Eq. (1.13) with $\mathbf{J} = 0$ and $\nabla \cdot \mathbf{E} = 0$ (which we show later is appropriate to this case)

$$- \nabla^2 \mathbf{E} + \frac{1}{c^2} \frac{\partial^2 \mathbf{E}}{\partial t^2} = - \mu_0 \frac{\partial^2 \mathbf{P}}{\partial t^2} - \mu_0 \frac{\partial}{\partial t} \nabla \times \mathbf{M}. \tag{6.1}$$

Assuming that the molecules have induced electric dipoles **p** and magnetic dipoles **m**, the macroscopic quantities **P** and **M** are given by

$$\mathbf{P} = N \langle \mathbf{p} \rangle \tag{6.2}$$

and

$$\mathbf{M} = N \langle \mathbf{m} \rangle, \tag{6.3}$$

where N is the molecular number density.

The effect of the magnetic dipoles on the propagation of light in a material depends on whether **M** is perpendicular or parallel to **P**. For linear motion of charges, the electric and magnetic dipoles are orthogonal to one another and hence **M** and **P** are orthogonal. From Eq. (6.1), $\nabla \times \mathbf{M}$ is parallel to the polarization **P** and in this case ($\mathbf{P} \perp \mathbf{M}$), $\nabla \times \mathbf{M}$ can at most add a small correction term to **P**. The net result is a small modification of the dielectric constant. This situation is illustrated in Figure 6.1a.

Figure 6.1. Illustration of the influence of the magnetic term on the polarization: (a) **M** perpendicular to **P**, no optical activity; (b) **M** parallel to **P**, optically active.

In some special cases, **M** is parallel to **P** as shown in Figure 6.1b. As a result $\nabla \times \mathbf{M}$ is orthogonal to **P**. The net effect is that there is a small correction to **P** which tends to rotate it (and the electric field **E**). This small polarization is given by $P_{eff} = \hat{\mathbf{k}} \times \mathbf{M}/c$.

This means that linearly polarized light continually changes its polarization direction. Hence linear polarization is not an eigenvector in such materials. (An eigenvector must preserve both its sense and direction of polarization as it propagates.) Instead the eigenvectors consist of circularly polarized waves. This is shown later.

6.2 PHYSICAL BASIS OF OPTICAL ACTIVITY

Let us now consider what kind of material systems exhibit this type of behavior. The situation in which we are interested is the case of liquids containing molecules which are handed, which means that the molecules have distinguishable mirror images. These occur quite commonly in carbon compounds due to the tetrahedral character of saturated carbon bonds.

An example is shown in Figure 6.2 in which four different halides are bound to a carbon atom. The molecule on the right is the mirror image of the one on the

Figure 6.2. Carbon bonded tetrahedrally to four different halogens. Molecule on left cannot be superimposed on molecule to right. These molecules are optically active.

left. (The mirror reflection is taken in the plane containing F, C and I.) These two molecules cannot be superimposed and are different. The two forms are called optical isomers, and they are labeled by l- or d-, depending on whether the electric field **E** is rotated to the left or right respectively. More natural examples are found in naturally occurring sugars, with sucrose being a classic example.

In many models of optical activity the electrons are treated as if they moved along the coil of a spring, as illustrated in Figure 6.3. The electrons are taken to be situated between the carbon atoms, and the electrons move back and forth within the chain in response to an applied electric field that points in the $\hat{\mathbf{v}}$ direction. Because of the helical structure of the molecule, the net motion of the electrons is also helical. An important property of this helix is its symmetry on rotation about the $\hat{\mathbf{u}}$ or $\hat{\mathbf{w}}$ axis (Figure 6.3). Rotation by 180° about both of these axes leaves the molecule invariant. Hence, if **m** is parallel to **p**, it will not vanish on statistical averaging.

The optical activity arises when the electromagnetic field **E** is parallel to the $\hat{\mathbf{v}}$ axis. Consider the spiral viewed along the $\hat{\mathbf{v}}$ axis, as shown in Figure 6.4a. When the electric field points along the $+\hat{\mathbf{v}}$ axis of Figure 6.3, the electrons move along the $-\hat{\mathbf{v}}$ axis. This corresponds to a current circulating in a clockwise fashion

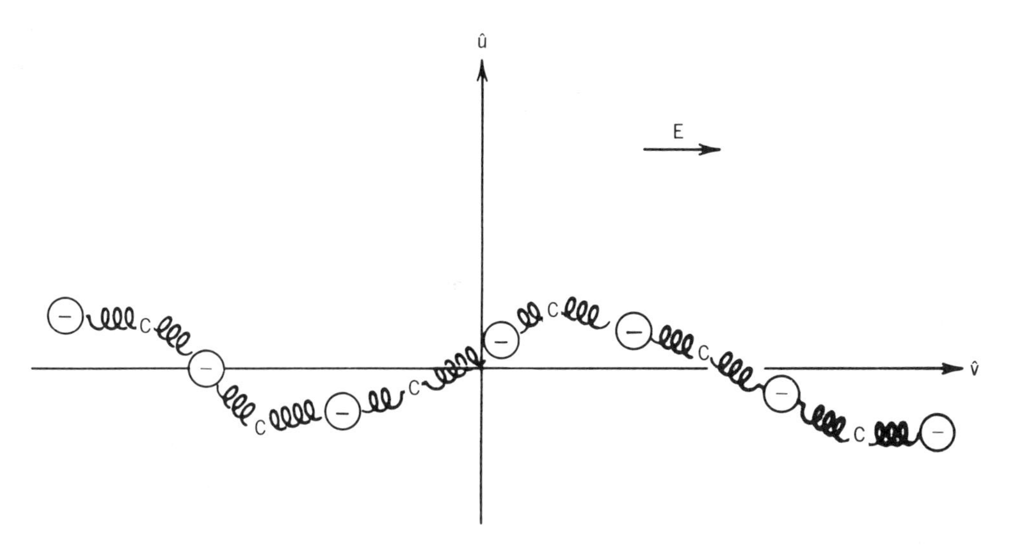

Figure 6.3. Central carbon chain of an unsaturated linear hydrocarbon. Carbons are configured as a large coiled spring. Gives optical activity if **E** is as shown.

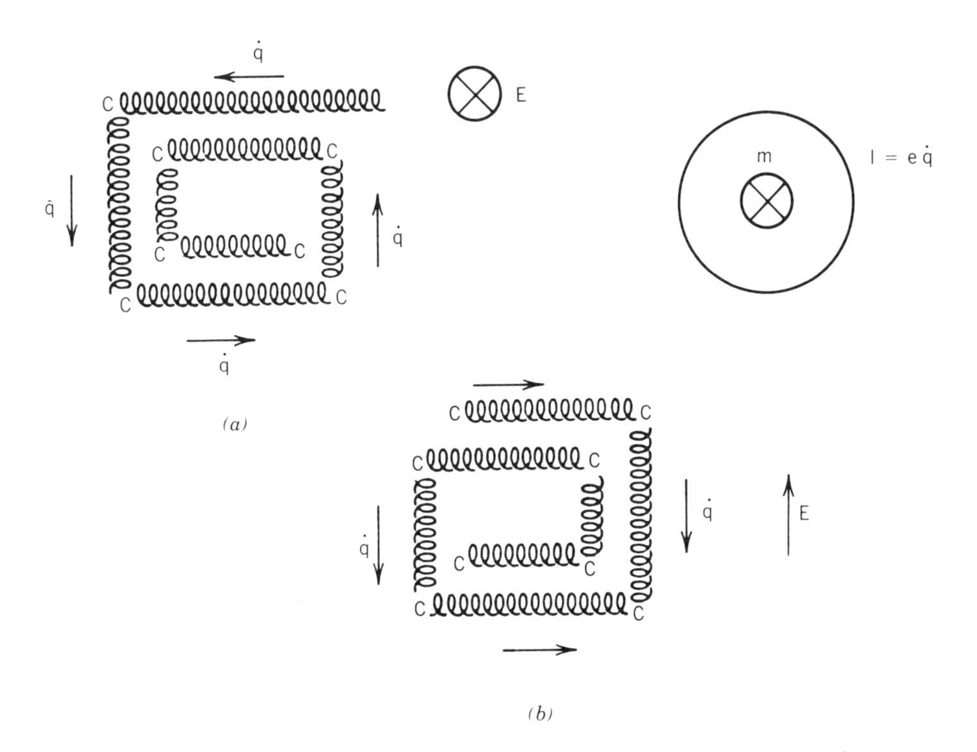

Figure 6.4. (a) Response of normal coordinates when E is as shown in Figure 6.3. Arrows labeled $\dot{q}$ show direction of electron velocity which lead to the current loop sketched on the right. The vectors **m** and **E** are parallel. Since P is parallel to E, this case corresponds to the optically active case in Figure 6.1b. (b) **E** orthogonal to $\hat{v}$; no current and no optical activity.

when viewed along the $+\hat{\mathbf{v}}$ axis. An induced magnetic dipole occurs along the $-\hat{\mathbf{v}}$ axis, parallel to the incident field and the induced electric dipole. A half cycle later, the current is reversed, and **m** points into the plane. The net result is a magnetic dipole **m** which systematically is parallel to **E**. This is the condition required for optical activity.

No optical activity is obtained if the incident electromagnetic field is perpendicular to the $\hat{\mathbf{v}}$ (spiral) axis. This case is illustrated in Figure 6.4b. The velocity components parallel to the $\hat{\mathbf{v}}$ axis cancel and only velocity components parallel to the incident field **E** remain. As discussed in Appendix B, the value of the magnetic dipole perpendicular to the electric dipole depends explicitly on the coordinate system in which it is computed, and hence it is not, in general, well defined. It corresponds to the case illustrated in Figure 6.1a. Since it gives rise to no interesting optical consequences, we take it to be zero.

6.3 ELECTROMAGNETIC PROPAGATION IN OPTICALLY ACTIVE MEDIA

The magnetization is related to the fields through phenomenological relations similar to those used previously for the electric field. First

$$M(\omega,\mathbf{k}) = \boldsymbol{\chi}_m(\omega) \cdot H(\omega,k) \tag{6.4}$$

where $\boldsymbol{\chi}_m(\omega)$ is the magnetic susceptibility and M and H are the vector amplitudes. The magnetic analog to the dielectric constant is the permittivity, which is usually written as $\boldsymbol{\mu}(\omega)$. The macroscopic field relations are

$$B(\omega,\mathbf{k}) = \boldsymbol{\mu}(\omega) \cdot H(\omega,\mathbf{k}) \tag{6.5}$$

and

$$\boldsymbol{\mu}(\omega) = \mu_0 (\mathbf{I} + \boldsymbol{\chi}_m(\omega)) \tag{6.6}$$

Since the optically active component of M is parallel to E, we can write the χ_m as a scalar when relating these quantities. It must be left in tensor form when relating the magnetic fields to each other, since M is perpendicular to H and B. From Eqs. (6.4) and (5.41) we obtain

$$M(\omega,k) = \frac{n\,\chi_m}{c\mu_0} E(\dot{\omega},k). \tag{6.7}$$

Where χ_m is the coefficient giving the strength of the magnetic effect.

We now examine the wave equation, and its solutions in the presence of optical activity. Note that a liquid is optically isotropic insofar as **P** is concerned, so that one can incorporate its effect into a dielectric constant. We write Eq. (6.1) as

$$-\nabla^2\mathbf{E} + \frac{n^2}{c^2}\frac{\partial^2\mathbf{E}}{\partial t^2} = -\frac{n\chi_m}{c}\frac{\partial}{\partial t}(\nabla\times\mathbf{E}). \tag{6.8}$$

This is the basic wave equation of optical activity. Since the medium has no preferred direction, we can, without loss of generality, take $\mathbf{k} = k\hat{\mathbf{z}}$ to point in the $\hat{\mathbf{z}}$ direction. However, the only real practical value of this discussion is to make clear why one cannot use optical activity to make optical isolators the way one can with magneto-optics (see Problem 6.2 and contrast the results with Section 9.3). For this purpose we consider, simultaneously, waves propagating in the $\pm\hat{\mathbf{z}}$ direction, which have wave vectors $\pm k$. Substituting a plane wave as a trial solution of $\mathbf{E}$ in Eq. (6.8),

$$\left((k^2 - \frac{n^2\omega^2}{c^2})\,\mathbf{I}\cdot \pm \frac{kn\omega\,\chi_m}{c}\,\hat{\mathbf{z}}\times\right)\,\boldsymbol{E} = 0. \tag{6.9}$$

In component form, this yields two equations

$$(k^2 - \frac{n^2\omega^2}{c^2})\,E_x \mp \frac{kn\omega\,\chi_m}{c}\,E_y = 0 \tag{6.10a}$$

and

$$(k^2 - \frac{n^2\omega^2}{c^2})\,E_y \pm \frac{kn\omega\,\chi_m}{c}\,E_x = 0 \tag{6.10b}$$

Equations (6.10) determine both the allowed values of $|k|$ and the field eigenvectors. First note that the only allowed solutions for which $\boldsymbol{E} \neq 0$ require some linear combination of E_x and E_y. This combination can be found by setting the determinant of the coefficients to zero. This yields the dispersion relation

$$\begin{vmatrix} (k^2 - \frac{n^2\omega^2}{c^2}) & -\frac{kn\omega\,\chi_m}{c} \\ \frac{kn\omega\,\chi_m}{c} & (k^2 - \frac{n^2\omega^2}{c^2}) \end{vmatrix} = 0 \tag{6.11}$$

where "$|\ |$" denotes the determinant. Evaluating the determinant explicitly gives

$$(k^2 - \frac{n^2\omega^2}{c^2})^2 + (\frac{kn\omega\,\chi_m}{c})^2 = 0. \tag{6.12}$$

From this we obtain

$$k^2 - \frac{n^2\omega^2}{c^2} = \pm\, i\,\frac{kn\omega\,\chi_m}{c}. \tag{6.13}$$

As we show below, χ_m is imaginary, and Eq. (6.13) describes two wave vectors of different magnitudes. The tricky aspect of this problem is to associate the correct wave vector $\pm k\hat{\mathbf{z}}$ with the correct field amplitude, and then to associate the result with right and left circularly polarized waves. In so doing, remember that the $\mp$ in Eq. (6.10a) and the $\pm$ in Eq. (6.10b) refer to the direction of propagation, and the $\pm$ in Eq. (6.13) refers to the magnitude of the wave vector. Substituting Eq. (6.13) into Eqs. (6.10a) gives

$$E_y = \mp i E_x \tag{6.14}$$

for the wave propagating in the +k direction and

$$E_y = \pm i E_x \tag{6.15}$$

for the wave propagating in the -k direction. It is left as an exercise to show that one obtains identically the same result with a substitution into Eq. (6.10b). In Eqs. (6.13), (6.14) and (6.15) one chooses either the set of upper signs or the set of lower signs to describe a plane-wave eigenvector. Therefore the eigenvectors can be written as

$$\boldsymbol{E}_\pm = \frac{1}{\sqrt{2}} (\hat{x} \pm i\hat{y}) E, \tag{6.16}$$

which describes two circularly polarized waves. It is left as an exercise to show that the choice $E_\pm$ describes right and left circularly polarized fields for propagation in the +z direction, and left and right circularly polarized waves for propagation in the -z direction. The resulting associations of wave vector directions, magnitudes, field amplitudes and senses of polarization are summarized in Table 6.1

Table 6.1. Association of Field Amplitudes and Wave Vectors with Sense of Polarization

Amplitude	Wave vector	Sense of Circular Polarization
E_+	$+k_+$	right hand
E_-	$+k_-$	left hand
E_-	$-k_+$	right hand
E_+	$-k_-$	left hand

We now evaluate the wave vectors in Eq. (6.13). Bearing in mind that $\chi_m \ll 1$, this solution reads

$$\frac{k_\pm}{\omega} = \frac{n}{c} \left(1 \pm i \frac{\chi_m}{2}\right). \tag{6.17}$$

We now show that χ_m is imaginary. Since

$$p \propto q_\alpha$$

and

$$m \propto \dot{q}_\alpha = i\omega q_\alpha \tag{6.18}$$

the induced electric and magnetic dipoles are $\pi/2$ out of phase with one another. Since the electric susceptibility is dominant, and is real for transparent media, χ_m can be expected to be imaginary. Writing $\chi_m = i\ \mathrm{im}(\chi_m)$, this gives

$$\frac{1}{v_\pm} = \frac{k_\pm}{\omega} = \frac{n}{c}\left(1 \mp \frac{\mathrm{im}(\chi_m)}{2}\right), \tag{6.19}$$

where the velocities $v_\pm$ are associated with the $E_\pm$ of Eq. (6.16) through the results stated in Table 6.1.

Now let us prove what we anticipated at the beginning of this section: the action of the magnetization causes the polarization direction to rotate. The incident field at the boundary of the medium is taken to be linearly polarized in the x direction. This sense of polarization is not an eigenvector of the optically active medium. Hence we must write the solution as a linear combination of the eigenvectors which are circularly polarized waves in this case. This sum is then matched to the incident field via the boundary conditions. We therefore write the total field as

$$E_T = \frac{1}{2}\{\boldsymbol{E}_+ e^{i(k_+z-\omega t)} + \boldsymbol{E}_- e^{i(k_-z-\omega t)} + cc\}. \tag{6.20}$$

At the point $z = 0$ we have

$$\hat{\mathbf{x}}\, E_{in} = \boldsymbol{E}_+ + \boldsymbol{E}_-. \tag{6.21}$$

From Eq. (6.16),

$$E_+ = E_- = \frac{E_{in}}{\sqrt{2}} \tag{6.22}$$

with the result that

$$E_T = \frac{1}{2}\frac{E_{in}}{\sqrt{2}}\left[(\hat{\mathbf{x}}+i\hat{\mathbf{y}})\, e^{i(k_+z-\omega t)} + (\hat{\mathbf{x}}-i\hat{\mathbf{y}})\, e^{i(k_-z-\omega t)} + cc\right]. \tag{6.23}$$

We now define

$$\bar{k} = \frac{k_+ + k_-}{2}; \qquad \Delta_\pm = k_\pm - \bar{k} \tag{6.24}$$

and then write this solution as

$$\mathbf{E}_T = \frac{1}{2}\frac{E_{in}}{\sqrt{2}}\left([(\hat{\mathbf{x}}+i\hat{\mathbf{y}})\, e^{i\Delta_+ z} + (\hat{\mathbf{x}}-i\hat{\mathbf{y}})\, e^{i\Delta_- z}]\, e^{i(\bar{k}z-\omega t)} + cc\right). \quad (6.25)$$

It is then straightforward to use Eq. (6.24) plus (6.19) to show that if we define

$$\Delta_\pm = \mp\,\Delta; \qquad \Delta = \frac{im(n\omega\, \chi_m)}{2c}, \quad (6.26)$$

then Eq. (6.24) simplifies to give

$$\mathbf{E}_T = \frac{1}{2}\,\{\hat{\mathbf{x}}\cos\Delta z + \hat{\mathbf{y}}\sin\Delta z\}\,\{E_{in}\, e^{i(\bar{k}z-\omega t)} + cc\}. \quad (6.27)$$

This shows explicitly the rotation of the direction of polarization at the rate Δ as z increases. This rotation is what is meant by optical activity or optical rotation. The parameter Δ, the rotary power coefficient, is usually a function of frequency, both via $n(\omega)$ and $\chi_m(\omega)$. This is called rotary dispersion.

We used the full formalism of the dispersion relation to obtain this relationship mainly as an exercise in seeing how it is done. We simplified the discussion somewhat by guessing that $\nabla\cdot\mathbf{E}$ was zero. The fact that we were able to find a solution with $\mathbf{E}\cdot\mathbf{k} = 0$ showed that this assumption was valid. In the next section we go through the exercise of finding eigenvectors in a crystal. There we approach the eigenvectors in the opposite way. We guess what they are based on intuition from the physics, and confirm by substitution that they have well-defined **k** vectors.

ADDITIONAL READING

Rotary Power Coefficients

Fowles, G.R. *Introduction to Modern Optics*, Holt, Rinehart and Winston, New York, 1968. p. 185.

Jenkins, F.A. and White, H.E. *Fundamentals of Optics*, McGraw-Hill, New York, 1957. p. 574 (see also the problem set on p. 587).

CRC Handbook for Chemistry and Physics, 53rd edition, Weast, R.C. editor, Chemical Rubber Company, Cleveland, 1972-73, pp. E-233 E-234.

Optical activity

Feynman, R.P., Leighton, R.B., and Sands, M. *The Feynmann Lectures in Physics*, Addison-Wesley, Reading, MA, 1964. pp. 33.6-33.7.

PROBLEMS

6.1. Verify Table 6.1.

6.2. Place an optically active liquid in a cell in front of a mirror as shown in Figure 6.5. Consider a linearly polarized incident wave. Show that the reflected wave is always linearly polarized, and that it is polarized in the same direction as the incident field.

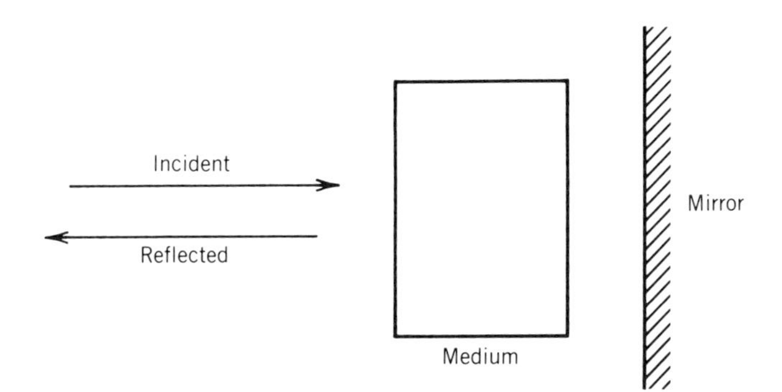

Figure 6.5. Illustration of medium configuration for Problem 6.2. The reflected field has the same polarization as the incident field, even when the medium is optically active.

6.3. Solve Eq. (6.8) by guessing that the eigenvector is a circularly polarized wave in the first place. The consistency of this choice is proved by deriving the dispersion relations for the fields $E_\pm$.

6.4. (a) Optical activity is often introduced by means of a Hermitian dielectric tensor. We write it for propagation in the $\hat{z}$ direction, i.e.,

$$\varepsilon_{eff} = \begin{pmatrix} \varepsilon & i\gamma & 0 \\ -i\gamma & \varepsilon & 0 \\ 0 & 0 & \varepsilon \end{pmatrix} .$$

Develop the wave equation for this form, and show that it gives optical activity.

(b) What happens to this tensor if $-\hat{z}$ is taken to be the propagation direction.

6.5. (a) Show that the signs in Eq. (6.9) follow from Eq. (6.8) for the two opposite directions of propagation.

(b) Show that if Eq. (6.13) is true, then Eq. (6.10a) is precisely the same as Eq. (6.10b), i.e., only one component of Eq. (6.9) is an independent equation.

6.6. Compute the rotary dispersion $\Delta(\omega)$ (see Eq. (6.27)). Use Eq. (6.18) for the magnetic dipole and derive an expression for χ_m. Use Eq. (5.20) for $n(\omega)$, and sketch the result.

6.7. Develop a model for the molecules in Figure 6.2. Show that they rotate the polarization in opposite directions.

6.8. Consider an ensemble of dipole pairs, one of which is illustrated in Figure 6.6. In this case **s** is the vector between the dipoles labeled ℓ_1 and ℓ_2. Consider

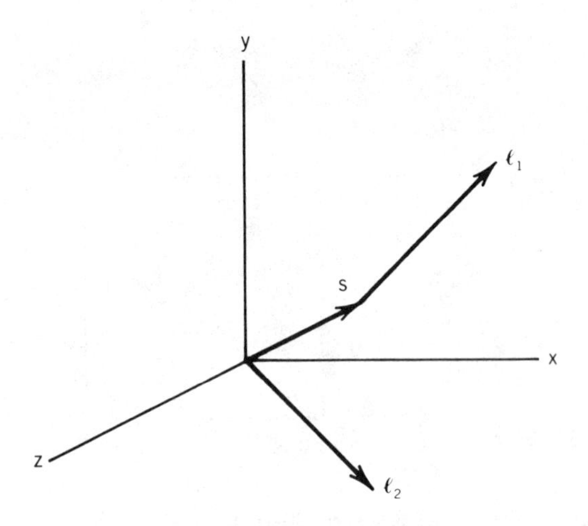

Figure 6.6. Illustration of $\boldsymbol{\ell}$ and **s** vectors for Problem 8

a crystalline solid in which the unit cell has three such pairs with $\hat{\mathbf{s}} = \hat{\mathbf{x}}, \hat{\mathbf{y}}$, and $\hat{\mathbf{z}}$. For the purpose of this exercise let the electric field be polarized in the $\hat{\mathbf{x}}$ direction and let the $\hat{\mathbf{k}}$ vector point in the $\hat{\mathbf{z}}$ direction. Set

$$\hat{\boldsymbol{\ell}}_1 = \frac{1}{\sqrt{2}}\,(\hat{\mathbf{x}} - \hat{\mathbf{y}}); \quad \hat{\boldsymbol{\ell}}_2 = \frac{1}{\sqrt{2}}\,(\hat{\mathbf{x}} + \hat{\mathbf{y}}).$$

Observe that the amplitude ℓ_α and the denominator $D_\alpha(\omega)$ are not explicit functions of α. Therefore the subscripts are dropped when appropriate.

(a) Repeat the development leading to Eq. (3.15) for this case. Assume that all resonant frequencies, decays and coupling strengths are the same, and show

$$Q_\alpha = \frac{1}{\sqrt{2}}\,\frac{\ell E}{mD(\omega)} \tag{6.28}$$

for the cases $\hat{\mathbf{s}} = \hat{\mathbf{z}}$ and $\hat{\mathbf{y}}$, and $Q_\alpha = 0$ for $\hat{\mathbf{s}} = \hat{\mathbf{x}}$. (Note that the vectors $\hat{\boldsymbol{\ell}}_1$ and $\hat{\boldsymbol{\ell}}_2$ have directions fixed with respect to $\hat{\mathbf{s}}$, and they change as $\hat{\mathbf{s}}$ changes.)

(b) Show that the polarization amplitude that results from this set of modes reads

$$P = \frac{2N\ell E}{mD(\omega)}\,(\hat{\mathbf{x}} + \hat{\mathbf{y}}\,\frac{iks}{4}) \tag{6.29}$$

Hint: You must take into explicit account the phase shift between the dipoles 1 and 2. Assume $ks \ll 1$.

7
Crystal Optics

In this chapter we deal with the case of electromagnetic radiation propagating through a periodic structure. In a crystal solid, all of the atoms or molecules are located at specific sites, and have a fixed orientation. For electromagnetic radiation with wavelengths greater than ten nanometers, there are many molecules in a cubic wavelength. Hence the details of the periodic structure average out as far as an electromagnetic wave is concerned, and we need not worry about effects such as Bragg reflections off crystal planes. The complication in crystals is that induced polarizations are not always parallel to the incident field, so that the relations between field quantities are described by tensors rather than scalars. Furthermore, the solutions to the wave equation may no longer be degenerate once the propagation direction is fixed.

7.1 PHYSICS OF CRYSTAL OPTICS

The property of crystals that differentiates them from isotropic media is that they are made up of periodic structures that are not locally isotropic. They are made up of unit cells whose position can be described in terms of a triplet vector (**a**,**b**,**c**), as shown in Figure 7.1a. The crystal is then described by translations of the unit cell along any vector of the triplet, which generates a three-dimensional structure.

The basic model for crystal optics is an electron bound in the unit cell by a set of springs. In Section 2.2 we showed that the potential associated with this set of springs is always symmetrical, and because of this it can be diagonalized. As indicated in Figure 7.1d, the coordinate system used to describe the displacement of the electron is just an ordinary orthonormal system. Therefore the normal coordinates used to describe the electronic normal modes q_α, $\alpha = 1,2,3$ correspond to an orthogonal system in the laboratory frame of reference (fixed with respect to the crystal). We use the subscripts i,j,k to label the axes in this laboratory frame where the allowed values of i, j and k can each be 1, 2 or 3. In this case, the vectors $\boldsymbol{\ell}_i$ that relate the direction of the induced dipoles to the normal coordinates q_i (see Sections 3.2 and A.3.1 for notation) point in the i'th direction, and hence the vector triplet $\hat{\boldsymbol{\ell}}_1,\hat{\boldsymbol{\ell}}_2,\hat{\boldsymbol{\ell}}_3$ is a set of orthogonal unit vectors. Since the magnitude of $\boldsymbol{\ell}_i$ is just the charge on the electron, we denote it by

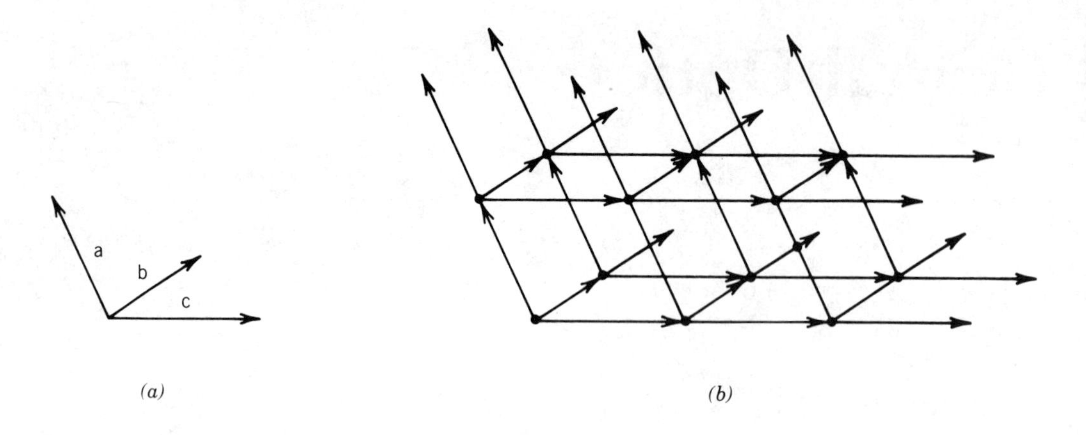

Figure 7.1. a) Vectors defining unit cell. b) Periodic structure obtained by translating the unit cell along **a**, **b**, and **c**. c) Springs binding electrons in unit cell. d) Coordinate system in which forces are diagonal. Note that there are no a priori relationships between c) and d) except for those imposed by the symmetry of the crystal.

ℓ_i = -e and its direction by $\hat{\boldsymbol{\ell}}_i = \hat{\mathbf{e}}_i$.

A field incident on a given unit cell causes a displacement of the electrons. Since we have not yet decided what form the propagating electromagnetic waves take in a crystalline solid, we do not assign a wave vector to the incident field at this point in the discussion. Thus we write

$$\mathbf{E} = \frac{1}{2}\,\boldsymbol{\mathcal{E}}(\mathbf{r},\omega)\,e^{-i\omega t} + \text{cc}.$$

For convenience we drop the arguments of the field in the following. The equation for the normal coordinate reads

$$\ddot{q}_i + \Gamma_i \dot{q}_i + \omega_i^2 q_i = -\frac{e}{m} E_i, \tag{7.1}$$

where m is the electronic mass. Following the analysis of Section 3.2 we write the solution for q_i as

$$q_i = \frac{1}{2} Q_i \, e^{-i\omega t} + cc$$

then

$$Q_i = -\frac{eE_i}{mD_i(\omega)}, \tag{7.2}$$

where $D_i(\omega)$ is defined in Eq. (3.14). Therefore the microscopic dipole induced points in the $\hat{\mathbf{e}}_i$ direction and is given by

$$\mathbf{p}_i = \frac{1}{2} \hat{\mathbf{e}}_i \frac{e^2 E_i}{mD_i(\omega)} e^{-i\omega t} + cc. \tag{7.3}$$

The microscopic polarizability $\boldsymbol{\alpha}$ as defined in Eqs. (3.34) and (3.35) is then diagonal and is given by (note that there is no Doppler broadening in solids, i.e., set $\mathbf{v} = 0$ in Eq. (3.35))

$$\alpha_{ii} = \frac{e^2}{mD_i(\omega)} \tag{7.4a}$$

$$\alpha_{ij} = 0, \qquad i \neq j. \tag{7.4b}$$

We have just shown that the polarizability tensor $\boldsymbol{\alpha}$ is diagonal in the force system that is diagonal. In Section 2.2 we showed that the force can always be diagonalized. Hence this result is completely general. In an arbitrary coordinate system, a diagonalizable matrix is Hermitian, so that

$$\alpha_{i'j'} = \alpha_{j'i'}^*, \tag{7.5}$$

where the general coordinate system is denoted with primes. In real unit cells, there is more than one electron, and it may happen that different electrons may have forces that are diagonal in different coordinate systems. Labeling the different electrons in the unit cell with the superscript u, and letting i,j,k again denote an arbitrary coordinate system, we write the macroscopic susceptibility and dielectric constant as

$$\chi_{ij} = \frac{N}{\varepsilon_0} \Sigma_u \alpha_{ij}^u \tag{7.6}$$

$$\varepsilon_{ij} = \varepsilon_0 \delta_{ij} + N \Sigma_u \alpha_{ij}^u \tag{7.7}$$

where δ_{ij} is the Kronecker delta

$$\delta_{ij} = 0, \qquad i \neq j \tag{7.8}$$

$$\delta_{ii} = 1, \qquad i = j, \tag{7.9}$$

which is just the unit matrix $\mathbf{I}$ written out in component form.

In Eqs. (7.6) and (7.7) the contribution of each electron to the dielectric tensor is itself Hermitian. When we add the effects of several such electrons in some common coordinate system, we describe the result with a sum of Hermitian matrices. It is left as an exercise to show that the sum of Hermitian matrices is itself Hermitian. Finally, we note that a Hermitian matrix can always be diagonalized in some frame of reference (which may not coincide with any of the normal coordinates used to describe the individual electrons). The key points which emerge from the preceding discussion are the following:

A. The electric susceptibility χ and the dielectric tensor ε are Hermitian matrices.
B. There is a coordinate system, called the principal axis system, in which $\boldsymbol{\chi}$ and $\boldsymbol{\varepsilon}$ are diagonal. Following the notation in Section 2.2 we drop the extra subscript in the diagonal frame and write $\varepsilon_{ii} = \varepsilon_i$, $\chi_{ii} = \chi_i$.

7.2 CRYSTAL CLASSES AND PRINCIPAL AXES

We always work in a principal axis system from now on. However, in triclinic and monoclinic crystals (the two lowest symmetry classes), the various electron forces are diagonalized in different coordinate systems, and therefore the principal axis system is a function of frequency. Hence all calculations which involve more than one frequency, such as those in nonlinear optics, must be done carefully in such crystals.

In all other crystal classes, the diagonal system of the forces is either determined by the crystal axes or is degenerate, in which case the forces are diagonal in any coordinate system. In such cases there is no real need to consider the detailed effects of several electrons and the coordinate axis system indicated in Figure 7.1d is the principal axis system. In other words, the polarizability is already diagonal in this system. Thus the diagonal elements of the susceptibility and dielectric constants are

$$\chi_i = \frac{N}{\varepsilon_0{}^2} \frac{e^2}{mD_i(\omega)} \tag{7.10}$$

$$\varepsilon_i = \varepsilon_0 + N \frac{e^2}{mD_i(\omega)}. \tag{7.11}$$

The thirty-two allowed crystal classes are listed in Table 7.1 along with the symmetry system to which they belong. The number of independent elements are listed for the linear susceptibility and, in addition, for the first two orders of nonlinear susceptibilities.

The significance of the number of independent elements in the linear susceptibility is discussed next.

One Independent Element

In this case all the spring force constants are equal. Therefore the elements χ_i and ε_i are also all equal to one another, i.e.,

$$\varepsilon_1 = \varepsilon_2 = \varepsilon_3, \tag{7.12}$$

and all orthogonal axes systems are principal axes. When two principal axes have the same value of ε_i, they are said to be degenerate, so that in this case all axis systems are degenerate. The material is said to be optically isotropic, insofar as the linear optical properties are indistinguishable from those of an isotropic medium. Note that this isotropy does not apply to the nonlinear susceptibilities.

Two Independent Elements

In this case two of the force constants are equal and one is different. We establish the following labeling convention, which is used almost universally in this case:

$$\varepsilon_1 = \varepsilon_2 \neq \varepsilon_3. \tag{7.13}$$

In this case the 1 and 2 axes are degenerate. Thus the 3 axis is a fixed principal axis, but the directions of the two remaining principal axes are arbitrary in a plane orthogonal to the 3 direction. This type of crystal is called uniaxial. It has one optic axis (this terminology is defined in Section 7.4).

Three Independent Elements

In this case all three spring force constants are different. But their directions are fixed in the crystal by crystal symmetry. In this, as well as in the next two cases,

$$\varepsilon_1 \neq \varepsilon_2 \neq \varepsilon_3 \neq \varepsilon_1. \tag{7.14}$$

Such a crystal is said to be biaxial, since it has two optic axes. In biaxials, the principal axis system is nondegenerate. Hence there is only one principal axis system, which is fixed in crystal. The directions of the principal axes are independent of frequency. This is the only case in which all of the crystal axes can be determined by linear optical measurements.

Four Independent Elements

In these biaxial crystals, one direction of force, and therefore one principal axis is fixed in the crystal. The directions of the others can change with frequency.

Table 7.1. Summary of Nomenclature and Optical Properties of Crystals

System	International	Schönflies	χ_{ij}	χ_{ijk}	χ_{ijkl}
Triclinic	1	C_1	6*	18	81
	$\bar{1}$	$S_2(C_i)$	6	0	81
Monoclinic	m	C_{1h}	4	10	41
	2	C_2	4*	8	41
	2/m	C_{2h}	4	0	41
Orthorhombic	2mm	C_{2v}	3	5	21
	222	$D_2(V)$	3*	3	21
	2/mm 2/mm 2/mm	$D_{2h}(V_H)$	3	0	21
Tetragonal	4	C_4	2*	4	21
	$\bar{4}$	S_4	2	4	21
	4/m	C_{4h}	2	0	21
	4mm	C_{4v}	2	3	11
	42m	$D_{2h}(V_d)$	2	2	11
	422	D_4	2*	1	11
	4/m 2/m 2/m	D_{4h}	2	0	11
Rhombohedral	3	C_3	2*	6	27
	$\bar{3}$	$S_6(C_{3i})$	2	0	27
	3m	C_{3v}	2	4	14
	32	D_3	2*	2	14
	32/m	D_{3d}	2	0	14
Hexagonal	$\bar{6}$ (3/m)	C_{3h}	2	2	19
	6	C_6	2*	4	19
	6/m	C_{6h}	2	0	19
	62m	D_{3h}	2	1	10
	6mm	C_{6v}	2	3	10
	622	D_6	2*	1	10
	6/m 2/m 2/m	D_{6h}	2	0	10
Cubic	23	T	1*	1	7
	2/m$\bar{3}$	T_h	1	0	7
	$\bar{4}$3m	T_d	1	1	4
	432	O	1*	0	4
	4/m $\bar{3}$ 2/m	O_h	1	0	4

Source: Bhagavantam, S. *Crystal Symmetry and Physical Properties*, Academic Press, New York, 1966.

Note: The classification system is based on the unit cell. The International and Schönflies notations refer to symmetry groups. The last three columns indicate the number of independent linear and nonlinear susceptibility components. A star in the χ_{ij} column indicates optical activity.

Six Independent Elements

These are biaxial crystals with a principal axis system which bears no necessary relationship to the crystal axes. All components of the principal axes may change with frequency.

7.3 OPTICAL PROPAGATION IN CRYSTALS

Here we determine the eigenvectors of an optical crystal. The term eigenvector refers to an electromagnetic field which can be written as a plane wave with a well-defined **k** vector. We take a nontraditional approach to this problem. We emphasize the role of the electric vector **E** and the response of the medium in the form of **P**. The conventional approach starts with the **k** vector, which is only of secondary importance in the problem. The advantage of our approach is that it focuses attention from the beginning on the directions of the electric field vectors, which govern the physics in all applications. This is of special importance in nonlinear optics where several fields, differing both in polarization direction and in the propagation direction, are present at the same time. For simplicity in notation we specialize our discussion to the case of transparent dielectrics. This means that the incident light frequency is far away from resonance ($\omega - \omega_i \ll \Gamma_i$) and $\boldsymbol{\chi}$ and $\boldsymbol{\varepsilon}$ are real.

The nature of the forces discussed in Section 7.2 clearly suggest that linear polarization is the candidate for an eigenvector. Note that we do not prejudice the answer merely by guessing this form, since the wave equation gives nonsense if the choice is inappropriate. Let us begin by considering Maxwell's equation from Section 1.1 to see what we can learn about these eigenvectors. First, we set **J** = 0, **M** = 0 for this case, which means that $\mathbf{B} = \mu_0 \mathbf{H}$. If we have a plane-wave field, then Eq. (1.4) implies that **E** is perpendicular to **B**, and Eq. (1.6) implies that **D** is perpendicular to H ($H = B/\mu_0$). From Eq. (1.8), this can be true only if **P** is perpendicular to **H**. Finally, Eq. (1.3) implies that **B** is perpendicular to **k**. (The proof of these assertions is left as an exercise.) In summary:

> A linearly polarized **plane-wave eigenvector** has the property that **E**, **D**, and **k** are coplanar. In addition, **P** is coplanar, but this is derivable from the assertion about **D** and **E**, so it is not an independent requirement.

To motivate this hypothesis, let us suppose that it were not true, and consider the possibility that **D**, **E**, and **k** are not coplanar. Equation (1.17) says that the **D** vector is driven by the projection of **P** on a plane perpendicular to **k**. This is depicted for our hypothetical case in Figure 7.2. Since either the direction of **D** or its sense of polarization must be changed, it is not a plane-wave eigenvector.

Let us now write down the wave equation we need to satisfy in order to check the eigenvectors. Equation (1.11), with **M** = **J** = 0, together with Eq. (5.4) and (A.58), gives

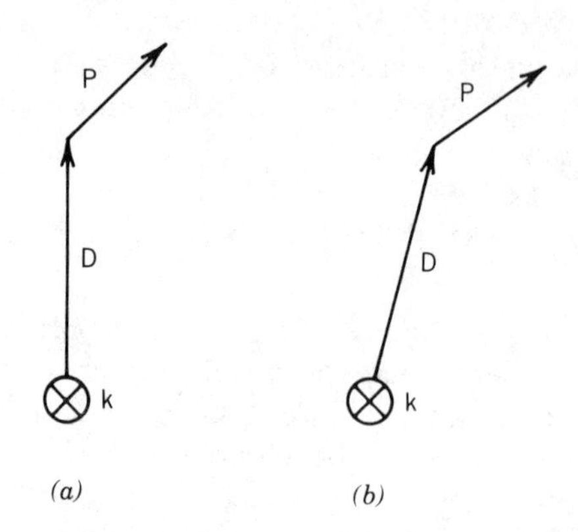

Figure 7.2. a) Hypothesis that plane-wave eigenvector can have **P**, **D**, and **k** (pointing out of plane of paper) in different planes. (b) Hypothesis rejected since **D** changes direction under propagation.

$$k^2\, \mathbf{O}(\hat{k})\cdot \boldsymbol{E} - \frac{\omega^2}{c^2 \varepsilon_0}\, \boldsymbol{\varepsilon} \cdot \boldsymbol{E} = 0, \tag{7.15}$$

where $\mathbf{O}(\hat{k}) = \mathbf{I} - \hat{\mathbf{k}}\hat{\mathbf{k}}$ is the projection operator defined in Eq. (1.48). We now prove the assertion we just made regarding plane-wave eigenvectors. Consider arbitrary vectors **u**, **v** and **w**. First note that $\mathbf{O}(\hat{u})\cdot\mathbf{u} = 0$, i.e., the projection of a vector on a plane perpendicular to itself is always zero. Let **v** be any vector perpendicular to **u**, and note that $\mathbf{O}(u)\cdot(\mathbf{O}(v)\cdot\mathbf{w}) = 0$ if and only if **u**, **v** and **w** are coplanar. Now dot $\mathbf{O}(\hat{\boldsymbol{D}})$ into Eq. (7.15). Since the second term is just $\boldsymbol{D}$, it gives zero by definition. The first term reads

$$\mathbf{O}(\hat{\boldsymbol{D}})\cdot(\mathbf{O}(\hat{k})\cdot \boldsymbol{E}) = 0,$$

which, since **D** and **k** are always orthogonal, can be true if and only if **E**, **D** and **k** are coplanar.

One case in which one gets an eigenvector is the following:

> An **o-configuration** plane-wave eigenvector occurs whenever **E** is parallel to a principal axis. This configuration has the following properties: 1) the velocity of the field is always determined by the index associated with the direction of the **E** vector; 2) any **k** vector normal to **E** is allowed; 3) **S** is always parallel to **k**.

An o-configuration eigenvector is illustrated in Figure 7.3. This configuration follows directly from the physics of the problem posed in the previous section. In this case the electric vector **E** and the polarization **P** are parallel to each other by definition. To see this formally, let us choose $\boldsymbol{E} = (E_1, 0, 0)$, where we can later replace 1 by 2 and 3 for complete generality. The general relation $\mathbf{D} = \boldsymbol{\varepsilon}\cdot\mathbf{E}$ reads

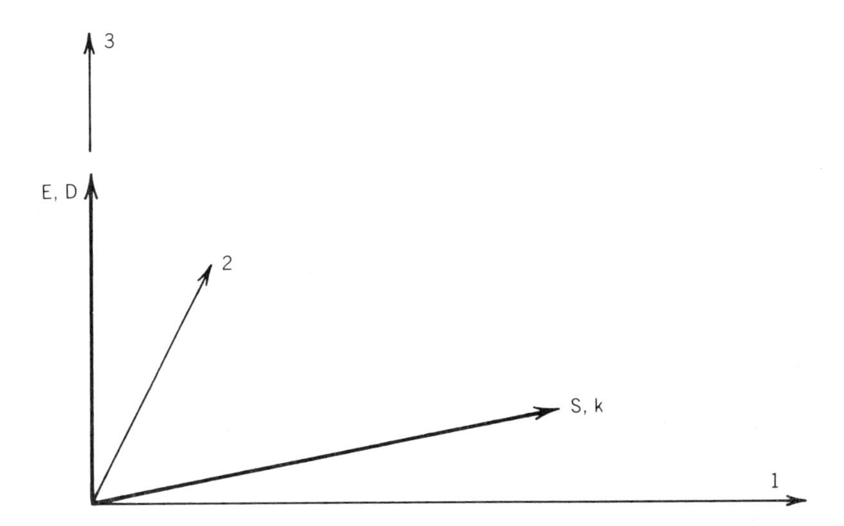

Figure 7.3. Illustration of an o-configuration eigenvector. This case is conventionally referred to as an e-ray in uniaxials and an o-ray in biaxials. In uniaxials, o-rays have **E** and **D** in the 1,2 plane.

$$\boldsymbol{D} = (\varepsilon_1 E_1, \varepsilon_2 E_2, \varepsilon_3 E_3). \tag{7.16}$$

In this case, $\boldsymbol{D} = (\varepsilon_1 E_1, 0, 0)$, and **D** and **E** (and hence **P**) are all parallel to each other. Because of this, we see that the choice of the **k** vector is arbitrary as long as it is perpendicular to **D** as demanded by Eq. (1.25). This arbitrary choice is clearly allowed, since any **k** so chosen is coplanar with **D**, **P**, and **E**. Since **E** and **D** are parallel, we have

$$\boldsymbol{E} \cdot \mathbf{k} = 0,$$

i.e., $\nabla \cdot \mathbf{E} = 0$ in this case. The direction of **S** is given by the direction of $\mathbf{E} \times \mathbf{H}$, and the direction of **k** is given by the direction of $\mathbf{D} \times \mathbf{B}$ (we confine our attention to cases in which **B** and **H** are parallel). Because **E** and **D** are parallel, it follows that the **S** vector is parallel to **k**.

Now let us prove that this choice satisfies Eq. (7.15). First $\mathbf{O}(\hat{\mathbf{k}})$ projects a vector onto a plane perpendicular to $\hat{\mathbf{k}}$. However, E is perpendicular to **k** already, so this projection operator can be dropped. Hence the terms in the $\hat{\mathbf{e}}_2$ and $\hat{\mathbf{e}}_3$ directions are identically zero and we are left with

$$\left(k^2 - \omega^2 \frac{\varepsilon_1}{\varepsilon_0 c^2}\right) E_1 = 0,$$

which can be true for nonzero E_1 only if the dispersion relation

$$\frac{k}{\omega} = \frac{n_1}{c} \tag{7.17}$$

is true. Otherwise E_1 is zero. Here we have defined

$$n_i = \left(\frac{\varepsilon_i}{\varepsilon_0}\right)^{1/2}. \tag{7.18}$$

Remember that the choice of the 1 direction was arbitrary and that 2 or 3 would work equally well. We have generalized the equations accordingly. We have the final property of these eigenvectors asserted above, namely that the velocity of the eigenvector is determined by the index associated with the direction of **E**.

It is important to always bear in mind that it is the direction of **E** and never the direction of **k** that determines the velocity of the field. This must be the case, since it is the motions of individual electrons that determine the velocities, and since the scale of the electrons is tiny compared to a wavelength, they cannot be influenced by the direction of propagation. The choice of **k** can, however, influence whether one has an eigenvector, which leads to the next case we consider:

> An **e-configuration** plane-wave eigenvector occurs whenever both **E** and **k** lie in a plane containing two principal axes with different n_i. This configuration has the following properties: 1) the velocity of the field is always determined by the indices associated with the principal plane (i.e., the plane containing the two principal axes) in which the electric field lies; 2) the velocity is a function of the field direction; 3) **S** is not in general parallel to **k**.

This configuration is illustrated in Figure 7.4. We choose $\mathbf{E} = (E_1, 0, E_3)$ to

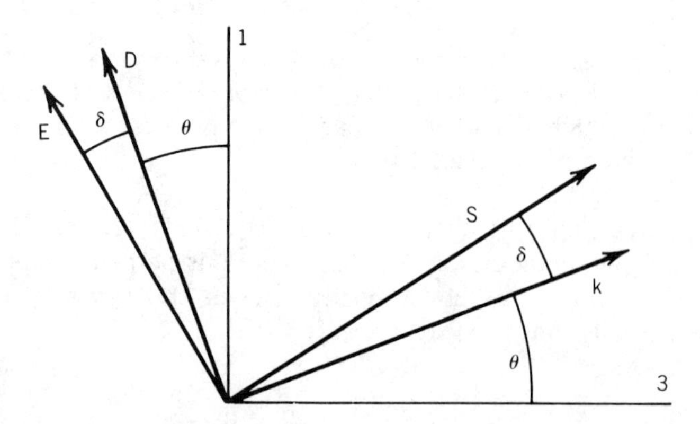

Figure 7.4. Illustration of an e-configuration eigenvector. In all contexts, this case is referred to as an e-ray.

illustrate the point. Since there is no field in the 2 direction, there is no polarization in the 2 direction, and likewise, **D** has the form (see Eq. (7.16))

$$\boldsymbol{D} = \left(\varepsilon_1 E_1, 0, \varepsilon_3 E_3\right). \tag{7.19}$$

Therefore **E** and **D** are coplanar. However, they are not necessarily parallel. Hence we must also take **k** to be coplanar, i.e.,

$$\mathbf{k} = (k_1, 0, k_3). \tag{7.20}$$

In this case, $\boldsymbol{E} \cdot \mathbf{k}$ is not, in general, zero and **S** is not parallel to k. We work out these relationships later.

For now we establish that these choices do, in fact, constitute a solution to Eq. (7.15). We do this somewhat indirectly, by constructing each of the vectors in turn. We use the coordinate system illustrated in Figure 7.4 from which

$$\mathbf{k} = k(\sin\theta, 0, \cos\theta) \tag{7.21}$$

and

$$\mathbf{O}(\hat{\mathbf{k}}) = \begin{pmatrix} \cos^2\theta & 0 & -\sin\theta\cos\theta \\ 0 & 0 & 0 \\ -\sin\theta\cos\theta & 0 & \sin^2\theta \end{pmatrix}. \tag{7.22}$$

Since $\boldsymbol{D} \cdot \mathbf{k} = 0$, **D** has the form

$$\boldsymbol{D} = D(\cos\theta, 0, -\sin\theta) \tag{7.23}$$

or, using Eq. (7.19),

$$\boldsymbol{E} = D\left(\frac{\cos\theta}{\varepsilon_1}, 0, -\frac{\sin\theta}{\varepsilon_3}\right). \tag{7.24}$$

Now let us evaluate Eq. (7.15) using these relations. In multiplying Eqs. (7.22) into (7.24), we treat the latter as a column vector. The first component of the equation reads

$$k^2 D\left(\frac{\cos^3\theta}{\varepsilon_1} + \frac{\sin^2\theta\cos\theta}{\varepsilon_3}\right) - \frac{\omega^2}{c^2\varepsilon_0} D\cos\theta = 0. \tag{7.25}$$

The second component is identically zero. The third component is

$$k^2 D\left(\frac{-\sin\theta\cos^2\theta}{\varepsilon_1} - \frac{\sin^3\theta}{\varepsilon_3}\right) + \frac{\omega^2}{c^2\varepsilon_0} D\sin\theta = 0. \tag{7.26}$$

If one divides Eq. (7.26) by $\sin\theta$ and Eq. (7.25) by $\cos\theta$, the two equations become identical. This identity must occur if these are the fields of an eigenvector. Let us denote by $n(\theta)$ the angle dependent index of refraction for this eigenvector, then if D is to be nonzero, we have from Eq. (7.25) that

$$\frac{ck}{\omega} = n(\theta). \tag{7.27}$$

We now use the definitions in Eq. (7.18) to write $n(\theta)$ as

$$n(\theta) = \frac{n_1 n_3}{(n_3^2\cos^2\theta + n_1^2\sin^2\theta)^{1/2}}. \tag{7.28}$$

This expression can now be generalized by an interchange of 1, 2, and 3, which gives

$$n_j(\theta_i) = \frac{n_i n_j}{(n_i^2\cos^2\theta_i + n_j^2\sin^2\theta_i)^{1/2}}. \tag{7.29}$$

The symbol $n_j(\theta_i)$ refers to the angle dependent index of an eigenvector in the i,j plane ($i \neq j$), where the angle θ_i is defined with respect to the i'th axis (in Figure 7.4 θ_i is θ_3). One important property that this index has is that

$$\min(n_i,n_j) \leqslant n_j(\theta_i) \leqslant \max(n_j,n_i), \tag{7.30}$$

i.e., $n_j(\theta_i)$ has a value that is always between the larger and smaller index. Within these bounds it varies sinusoidally as indicated in Figure 7.5.

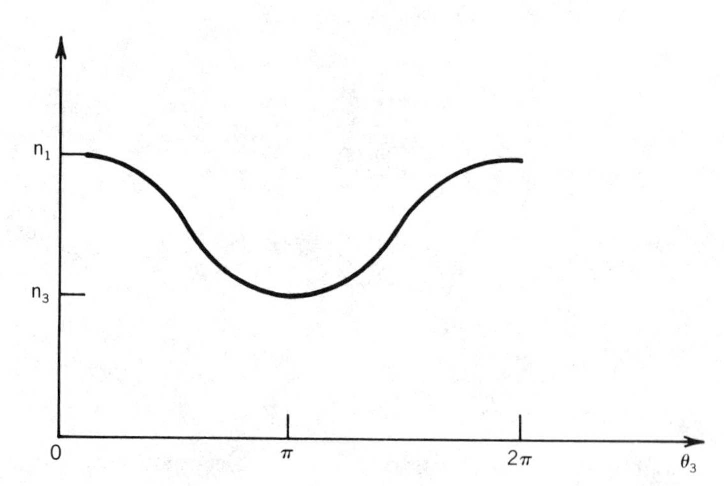

Figure 7.5. Index $n_1(\theta_3)$ as a function of θ_3. The subscripts refer specifically to the configuration in Figure 7.4.

Let us now consider the angle δ between E and D in Figure 7.4. It is also the angle between **S** and **k**, which follows from the fact that **D** and **k** are orthogonal and **S** and E are orthogonal, and they are all coplanar (i.e., orthogonal to H). The algebra of the derivation is left as an exercise, but one proceeds as follows: compute $\mathbf{D} \cdot \mathbf{E}$ to get $\cos\delta$ and $\mathbf{D} \times \mathbf{E}$ to get $\sin\delta$; then take the ratio of these and simplify to get

$$\tan\delta = \frac{n^2(\theta)}{2}\left[\frac{1}{n_3^2} - \frac{1}{n_1^2}\right]\sin 2\theta. \tag{7.31}$$

This angle is called the walkoff angle in nonlinear optics, for reasons that are explained in an exercise. In linear optics, the walkoff effect is incorporated in the overall phenomenon of birefringence.

All eigenvectors in uniaxial crystals are either e- or o- configuration eigenvectors and these are sufficient for virtually all applications in biaxial crystals. We therefore stop here and see what we can learn from them. One can find the other eigenvectors in biaxials by following the procedures outlined above. A general method is to construct D and E and note that the **k** vector is parallel to the projection of **E** that is orthogonal to D, i.e., in the direction $\mathbf{O}(\hat{\mathrm{D}})\cdot\mathbf{E}$. The easiest method, however, is to look up the answer in any optics text that deals with this case. Conventional methods diagonalize the determinant in Eq. (7.15), and follow the procedures used in Chapter 6. This leads to the index ellipsoid, which is a construction that we deliberately avoid. Problems in crystal optics are difficult enough without it. It is alleged that one can actually solve complex problems in crystal optics using the ellipsoid, but our experience indicates that the odds against success are too great to justify its use.

7.4 TERMINOLOGY

In this section, we discuss the terminology applied to the eigenvectors of the crystal. Let us start with the simple case of an optically isotropic crystal defined by Eq. (7.12). In this case, all orthogonal axes systems are principal axes, and all **E** fields are parallel to a principal axis and constitute o-configuration eigenvectors. Since $\mathbf{E}\cdot\mathbf{k} = 0$, this case is indistinguishable optically from the case of an amorphous medium.

The first case of interest is that of a uniaxial crystal. Let us take **k** as some arbitrary direction in the crystal. Since the 1 and 2 axes are arbitrary, we can, without loss of generality, let k lie in the 1,3 plane, as illustrated in Figure 7.4. There are then two eigenvectors that are obvious on inspection.

The **o-ray**; **E** is polarized in the 2 direction, so that $\mathbf{E}\cdot\mathbf{k} = 0$. This is the o-configuration eigenvector. The index is n_2 ($= n_1$) and is usually denoted n_o. The reason this is called an ordinary ray is that it obeys Snell's law. This must be the case, since the index is a constant, independent of θ, and the wave obeys all of the laws of isotropic media.

The **e-ray**; **E** is polarized in the 1-3 principle plane, which is the e-configuration eigenvector illustrated in figure 7.4. In this case, not only is the index dependent on angle, but the ray directions are different from the wave normals. The term "extraordinary ray" refers to the fact the light rays do not obey Snell's law.

These two eigenvectors are orthogonal to each other, i.e., $E_o \cdot E_e = 0$, $D_o \cdot D_e = 0$, which is a general property of eigenvectors. Note that if a general field $\mathbf{E}_{in}$, $\mathbf{k}_{in}$ is incident on the surface, the o component enters with a **k** vector given by Snell's law with $n = n_o$ and the e-ray enters with a k vector (but not **S** vector) given by Snell's law with $n = n(\theta)$. (Note that no special notation for $n(\theta)$ is needed in this case, since the choice in Eq. (7.29) of $i = 1$, $j = 2$ gives an o-ray, $i = 3$, $j = 1$ is the same as the case $i = 3$, $j = 2$, and the other choices are eliminated by the fact that Figure 7.4 defines the conventional choice for the angle.) This gives two refracted waves, which is the origin of the term birefringence applied to this and the biaxial case. Note that the nomenclature o and e applied to the rays is not quite synonymous with the nomenclature given to the eigenvectors. All e-configurations are e-rays, but in the case $\theta = 0°$ and $90°$ the e-ray is an o-configuration eigenvector. In nonlinear optics it is traditional to use the labels o and e in the traditional usage in uniaxials and to switch to a nomenclature given by the eigenvector configuration in biaxials (see Figure 7.4).

An optic axis is defined as the direction of **k** such that both eigenvectors move at the same velocity, or equivalently, that both have the same index. (The use of **k** rather than S to define the optic axis means we are using the wave optic axis versus the ray optic axis.) This means that $n_o = n(\theta)$, or in the earlier notation, $n_1 = n(\theta)$. From Eq. (7.30), and noting that n_2 ($n_2 = n_1$) is one of the boundaries of the range of $n(\theta)$, we see that there is one and only one angle for which $n_1 = n(\theta)$ is true. It is $\theta = 0°$ ($\theta = 180°$ is equivalent), for which Eq. (7.28) gives $n(\theta) = n_1 = n_2$. A wave propagating along the optic axis has

$$\mathbf{k} = (0,0,k_3). \tag{7.32}$$

The term uniaxial refers to the single optic axis. Note that when **k** points along an optic axis, the two eigenvectors are degenerate. Uniaxial crystals are said to be positive if $n_o < n_e$ and negative if $n_o > n_e$. A typical index curve for a negative uniaxial is shown in Figure 7.6a. The case illustrated is KDP (potassium dihydrogen phosphate), which is a widely used crystal in nonlinear and electro-optics.

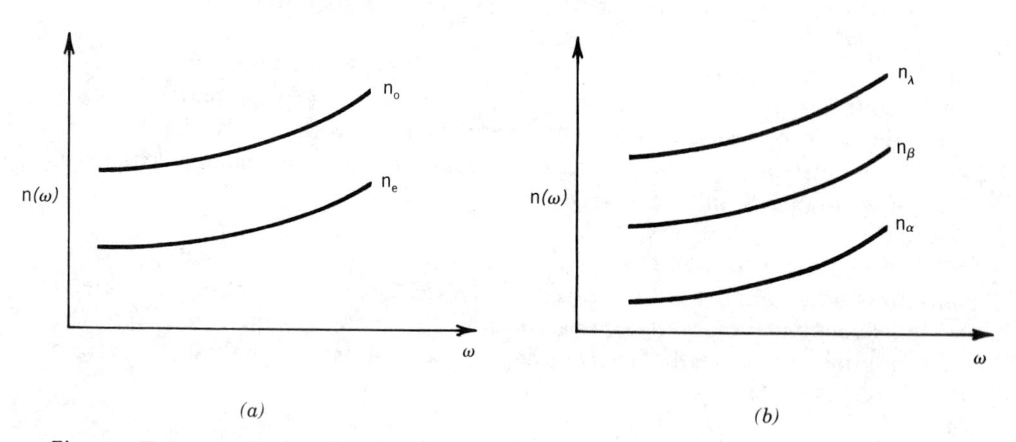

Figure 7.6. Indices of refraction as a function of frequency illustrating conventional nomenclature; a) negative uniaxial; b) biaxial.

Biaxial crystals are conventionally labeled by $n_\gamma > n_\beta > n_\alpha$ as illustrated in Figure 7.6b. The labels bear no necessary relationship to the 1, 2, 3 system or to the crystal axes. The same eigenvectors discussed for a uniaxial crystal work equally well in a biaxial, except that one is confined to planes defined by the principal axes. Let us use this to find the two optic axes which give the terminology biaxial. Note that if i,j corresponds to either the α and β axes or the β and γ axes in Eq. (7.30), then there is no way that the remaining index, i.e., n_γ or n_α, can be equal to $n(\theta)$. The only choice that is possible is for $i = \alpha$, $j = \gamma$ (or vice versa). Let us take $\theta = \theta_\gamma$ for definitiveness, and set $i = \alpha$ in Eq. (7.29). It is then straightforward to solve for the angle of the optic axes, and it is left as an exercise to show that

$$n_\alpha(\theta_\gamma) = n_\beta \tag{7.33}$$

is solved by

$$\sin\theta_\gamma = \pm \frac{n_\gamma}{n_\beta} \frac{n_\beta^2 - n_\alpha^2}{n_\gamma^2 - n_\alpha^2}. \tag{7.34}$$

The optic axes in uniaxial and biaxial crystals are illustrated in Figure 7.7. The α and γ axes are bisectors of the optic axes. The axes are illustrated as pointing in both directions. This is, in fact, the case since one can choose either the $\pm k$ direction and the optical consequences are unchanged.

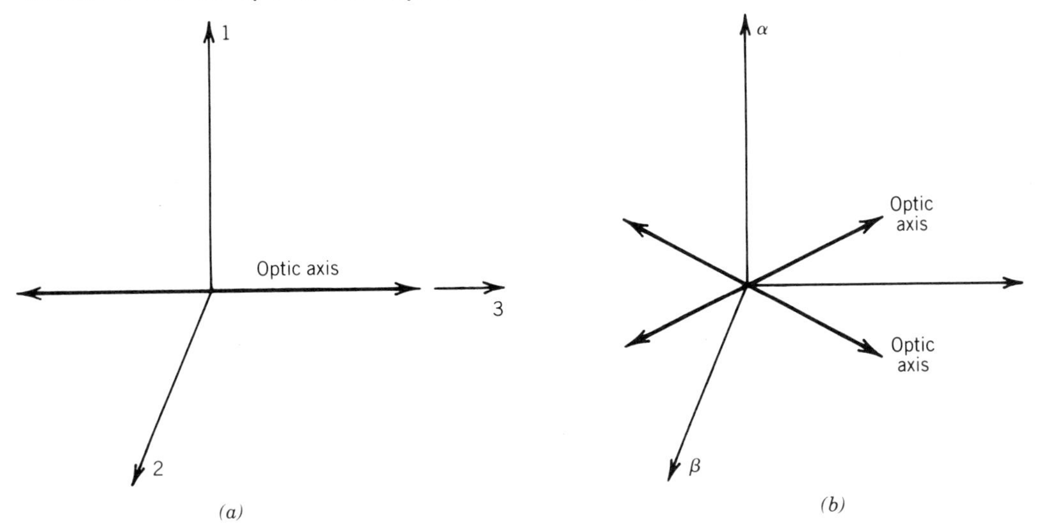

Figure 7.7. Illustration of optic (or optical) axes for a) uniaxial and b) biaxial crystals.

7.5 DEVICE APPLICATIONS

There are many optical devices such as polarizing prisms that are based on birefringence. These are discussed in optics texts, and are not discussed here. We discuss the problem of a wave plate, since it is a calculation we need later. We specialize to the cases $\theta = 90°$ (uniaxial) or $\theta = 0°$ (biaxial), so that we consider only o-configuration eigenvectors. Note that, in this case, the e-ray of a uniaxial crystal is polarized along a principal axis. Hence it is an o-configuration eigenvector and obeys Snell's law. However, the e-ray nomenclature is traditional and we use it anyway. By restricting the discussion to this case, we avoid annoying complications due to the walkoff effect.

The basic configuration for this calculation is illustrated in Figure 7.8. The surface is cut so that it contains two principal axes as illustrated in Figure 7.8b. We consider the case of normal incidence, and allow the incident **E**-field to have an arbitrary direction and sense of polarization, as illustrated in Figure 7.8a.

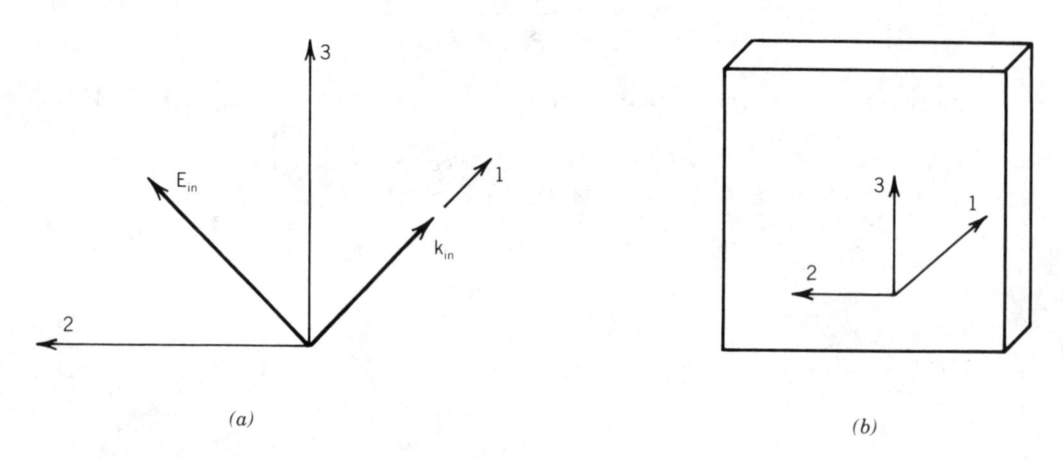

Figure 7.8. Setup for making a wave plate: a) input field configuration; b) principal axes inside crystal.

A general vector is written as a linear combination of eigenvectors

$$\mathbf{E} = \frac{1}{2}\left(\hat{\mathbf{e}}_2 E_2\, e^{i(k(2)z-\omega t)} + \hat{\mathbf{e}}_3 E_3\, e^{i(k(3)z-\omega t)} + cc\right), \tag{7.35}$$

where $z = x_1$ denotes the propagation direction and $\mathbf{k}(i) = (n_i\omega/c,0,0)$. At $z = 0$, the fields E_2 and E_3 match whatever the input field is at the boundary, i.e.,

$$E_i = \hat{\mathbf{e}}_i \cdot \boldsymbol{E}_{in}, \qquad i = 2, 3. \tag{7.36}$$

Writing $\Delta k = k(2)-k(3)$, Eq. (7.35) reduces to

$$\mathbf{E} = \frac{1}{2}\left(\hat{\mathbf{e}}_2 E_2 e^{i\Delta kz} + \hat{\mathbf{e}}_3 E_3\right) e^{i[k(3)z-\omega t]} + cc. \tag{7.37}$$

The expression exp(iΔkz) describes a phase shift of the 2 component relative to the 3 component. The phase shift causes a change in the direction of polarization as the wave propagates. The full wave condition corresponds to the case $\Delta z = 2m\pi$, where m is an integer, in which case $\exp(i\Delta kz) = 1$ and the wave recovers its original polarization. The half-wave condition is for $\Delta kz = 2\pi(m+1/2)$, so that $\exp(i\Delta kz) = -1$. In that case, the vector amplitude of the field is

$$E = -\hat{\mathbf{e}}_2 E_2 + \hat{\mathbf{e}}_3 E_3, \tag{7.38}$$

i.e., the 2 component of the field is reversed.

Let us consider the case where the input field is linearly polarized. Eqation (7.36) tells us that both E_2 and E_3 are real. Then from Eq. (7.38) the components are still real, so the field is still linearly polarized, but it is reflected about the 3 axis as shown in Figure 7.9. If $E_2 = E_3$, i.e., the polarization direction is 45° to

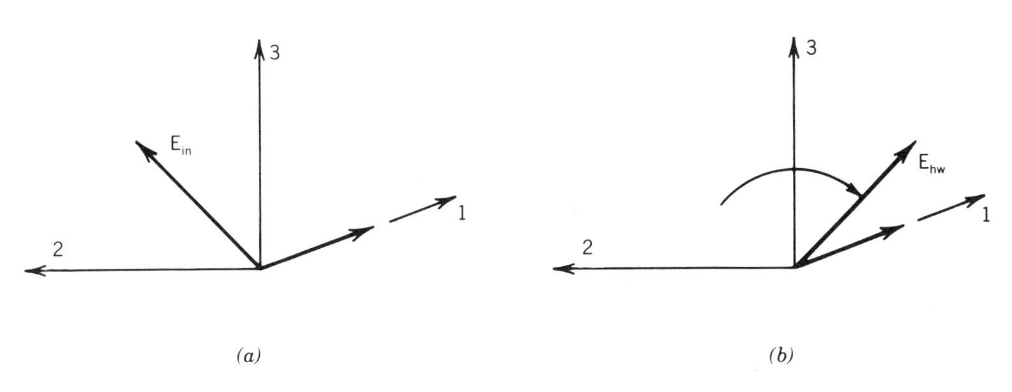

Figure 7.9. Result of propagation for a length that gives a π phase shift (half-wave condition) between eigenvectors. **E** field is reflected in the 3-1 plane.

the principal axis, then the output is polarized orthogonally to the input. If the initial field is circularly polarized, then E_2 is real (we choose E_2 positive), but $E_3 = iE_2$ is purely imaginary. The initial unit vector $\hat{\mathbf{e}}_2 + i\hat{\mathbf{e}}_3$ is changed to $-(\hat{\mathbf{e}}_2 - i\hat{\mathbf{e}}_3)$, which means that right circularly polarized is changed into left (or vice versa).

The quarter wave condition occurs when $\Delta z = 2m\pi + \pi/4$. (A three-quarter wave plate with $\Delta z = 2m\pi + 3(\pi/4)$ works very nearly the same.) Then $\exp(i\Delta z) = i$, or

$$\boldsymbol{E} = i\,\hat{\mathbf{e}}_2 E_2 + \hat{\mathbf{e}}_3 E_3.$$

Let us suppose that $\boldsymbol{E}_{in}$ is linearly polarized at 45° to the principal axes. Then $E_2 = \mp E_3$ or

$$\mathbf{E} = iE_2 (\hat{\mathbf{e}}_2 \mp i\hat{\mathbf{e}}_3),$$

i.e., the light is circularly polarized. When the crystal is cut to a length L such that L corresponds to one of these conditions at some wavelength λ, then it is called a quarter-wave plate or half-wave plate. Quarter-wave plates are used to convert linear polarization to circular and back again. Half-wave plates are primarily used to change the direction of polarization. Because of dispersion, and because Δ is a function of k and hence λ, a particular plate is restricted in the range of wavelengths over which it can be used.

7.6 OPTICAL ACTIVITY AND DICHROISM IN CRYSTALS

One of the most annoying features of birefringent crystals is that they are often optically active. This occurs even when the individual molecules are not spiral in their configuration (e.g., Figure 6.4), as in quartz where silicon dioxide has a linear structure. Several molecules can be arranged in a spiral formation and there is, of course, no statistical averaging over orientation in the crystal. Those crystal classes that have optical activity are indicated in Table 7.1. The magnetic effect is very tiny, so the eigenvectors are not changed much. In general, they are elliptically polarized, and the ratio of major to minor axes is of the order $1{:}10^{-3}$, i.e., the ratio of dipole to magnetic dipole fields. A good pair of the polarizers can achieve extinction ratios less than one in 10^4, so that it is more difficult to achieve extinction between crossed polarizers with an optically active crystal than it is with inactive crystals. When one looks down the optic axis, the polarization field does not define the specific directions of the eigenvectors (i.e., they are degenerate). In that case, an optically active crystal behaves like an optically active isotropic medium.

We limited our discussions to the case of transparent crystals, but the method of treating the propagation problem is unchanged if the polarizabilities are complex and the medium absorbs. One needs to use a **k** vector that is complex (see, e.g., Chapter 5), and one gets an absorption coefficient γ that depends on the direction of polarization of the field. This phenomenon is called dichroism. Polarized sunglasses are an example of dichroism that comes from an orientation of absorbing molecules (e.g., iodine molecules) in a plastic whose polymers are oriented in a particular direction during the manufacturing process.

ADDITIONAL READING

Index Ellipsoid

Born, M., and Wolfe, E. *Principles of Optics*, Pergamon Press, Oxford, 1970. Chapter 14.

Yariv, A., *Quantum Electronics*, Wiley, New York, pp. 82-90.

Fowles, G.R. *Introduction to Modern Optics*, Holt, Rinehart and Winston, New York, 1968. pp. 169-182.

Optical activity in crystals

Fowles, 1968, pp. 184-188.

Dichroism

Born and Wolfe, 1970, pp. 708-718.

Crystals

Nye, J.F. *Physical Properties of Crystals*, Clarendon Press, Oxford, 1957. Chapter 4.

PROBLEMS

7.1. (a) Obtain Eqs. (7.4a-b) from the argument given in the text. Show that a sum of Hermitian matrices is Hermitian.

(b) A diagonalization procedure involves taking a matrix **M** and multiplying it by a rotation matrix **R** and its inverse $\mathbf{R}^{-1}$ where $\mathbf{R}\cdot\mathbf{R}^{-1} = \mathbf{I}$. In the new coordinate system $\mathbf{M}' = \mathbf{R}^{-1}\cdot\mathbf{M}\cdot\mathbf{R}$ is diagonal. Show that if there is only one independent element in the χ_{ij} column in Table 7.1, then all orthogonal coordinate systems are principal axes.

(c) Show that if there are two independent elements in the χ_{ij} column in Table 7.1, then, using the conventions of Eq. (7.13), any pair of orthogonal axes in the 1,2 plane are principal axes.

7.2. Prove that if **M** = 0, **J** = 0, and if there is a unique **k** vector associated with a plane-wave **E** field, then **E**, **D**, **P**, **k**, and **S** are all orthogonal to **B**.

7.3. (a) Derive Eqs. (7.31) and (7.34).

(b) Verify that if n_i and n_j are not too different, then the walkoff angle (denoted here with the conventions of Eq. (7.29)) can be approximated as

$$\delta_j(\theta_i) \simeq \frac{n_i - n_j}{n}\sin(2\theta_i), \tag{7.39}$$

where either index can be used for n.

(c) The concept of walkoff as distinct from birefringence is important in nonlinear optics, and is illustrated here using linear optics. Suppose a biaxial crystal is cut with surfaces normal to an optic axis. Let a laser beam with a 1 mm diameter be incident normal to the surface. Show that the wave normals of the o- and e-rays are normal to the surface and are hence parallel to each other. The propagation properties of the two rays are illustrated in Figure 7.10. For n_α = 1.35, n_β = 1.4 and n_γ = 1.45, compute the propagation length at which the two rays have separated by one diameter. In cases in which the interaction of the two rays is important, the walkoff effect sets a limit on crystal length.

7.4. Consider the case in which the E vector lies in a plane in which the spring system has trigonal symmetry (i.e., the springs are symmetric under rotations in this plane). Assume the other springs are orthogonal to this plane (i.e., this direction is a principal axis which we take to be the 3 axis). Show that $\varepsilon_1 = \varepsilon_2$ (i.e., the optical interaction is isotropic in the trigonal plane).

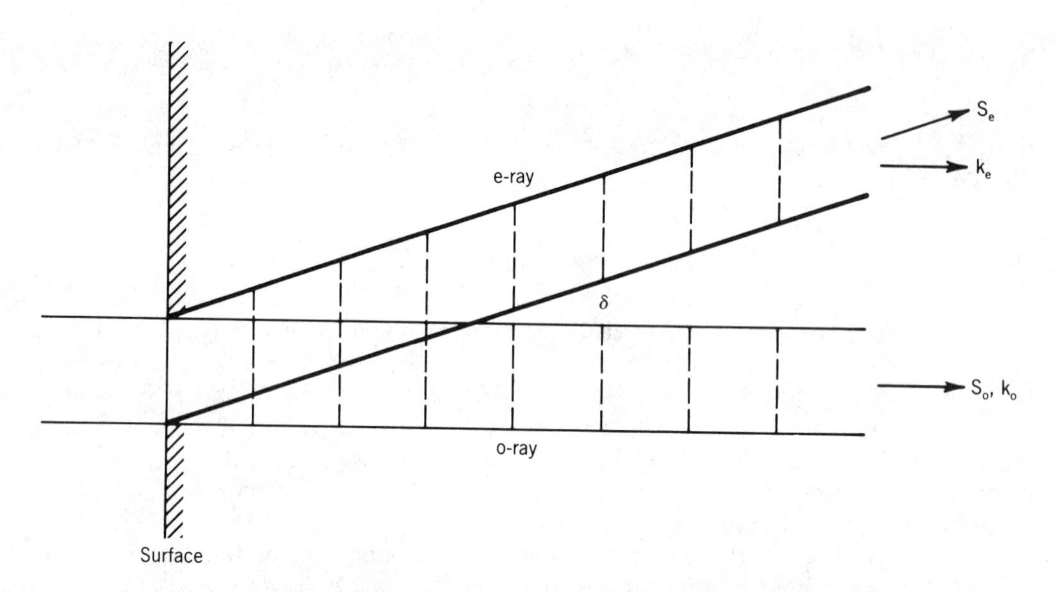

Figure 7.10. Illustration of walkoff. The o- and e-rays are indicated with a line denoting the location of the radius of a Gaussian laser beam. Vertical lines are wavefronts. The **k** vectors of the two rays are parallel but the **S** vectors are not parallel.

7.5. Consider a crystal of class 222, which has three orthogonal axes such that the crystal is symmetrical under 180° rotations about each axis. Show (this should be done pictorially rather than using formal group theory) that if there are two electrons per unit cell, each sketched as in Figure 7.1c, their forces must both be diagonal in the same coordinate system. Hint: It is easiest to prove that an arbitrary case is diagonalized in the symmetry axis system.

7.6. Consider a uniaxial crystal cut as shown in Figure 7.8b except that the 1 and 3 axes are interchanged, i.e., the optic axis is normal to the crystal surface. When the crystal is viewed through cross polarizers, one observes the pattern seen in Figure 7.11a, called a Maltese cross. In answering the following questions bear in mind that one is observing a cone of wave vectors about the optic axis as illustrated in Figure 7.11b.

(a) Show that the circular patterns of the Maltese cross of Figure 7.11a are caused by full waves of rotation, and that they are circular.

(b) Sketch what happens to the eigenvectors when the k vector is rotated in the azimuthal direction about the optic axis as drawn in Figure 7.11b. Use these sketches to determine why there is a cross pattern. Show why the vertical and horizontal bands in the cross get narrow near the optic axis.

7.7. Suppose you examine a biaxial crystal whose surface is cut normal to a principal axis. Show how to determine the orientation of the two principal axes that lie in the surface. Show why this is a poor technique for verifying that the principal axis is, in fact, normal to the surface.

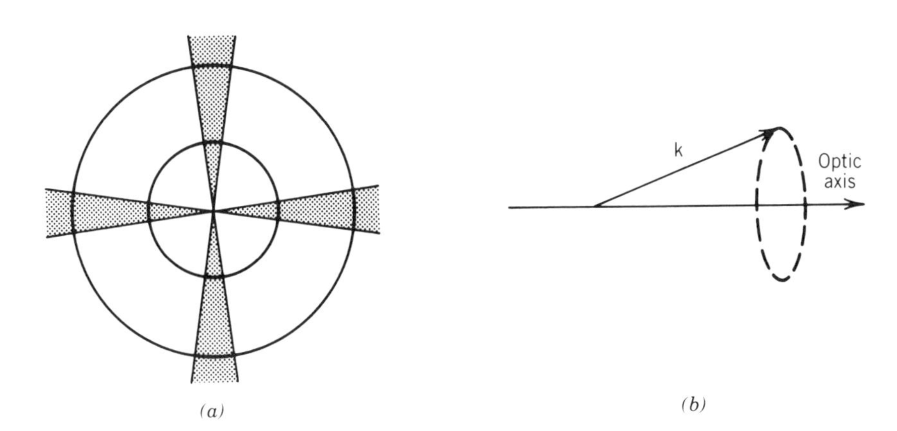

Figure 7.11. Illustration for Problem 7.6. (a) Maltese cross pattern; optic axis perpendicular to plane of paper. (b) Set of k vector directions that lead to the pattern in (a).

7.8. Supposing a uniaxial crystal has $n_o = 1.537$, $n_e = 1.496$ at a wavelength $\lambda = 1.06\ \mu m$. What is the shortest length needed to make a half-wave plate.

7.9. A high-quality pair of crossed polarizers can produce a null that transmits about 10^{-5} of an incident wave. Consider a case in which you wish to produce a null using the polarizers in parallel with a half-wave plate between. Using the criterion that the transmission should be less than 2×10^{-5}, express the range of wavelengths for which the plate is suitable as a function of the dispersions $\partial n_o/\partial \lambda$, $\partial n_e/\partial \lambda$.

7.10. Replace the three vectors which define the unit cell with three springs. Consider all combinations of equal and unequal spring constants and all conditions of orthogonality. Associate each with the number of independent components of χ_{ij} in Table 7.1.

8 Electro-Optics

In the preceding sections we dealt exclusively with cases in which the medium responds to the action of a single electric field. In this chapter we consider, for the first time, phenomena in which the polarization induced in the medium is created by a product of two or more fields. Such cases include all of nonlinear optics, as well as phenomena such as the Kerr, Pockels, and Cotton Mouton effects. In principle, they can all be deduced from a judicious choice of a free energy or an interaction potential. We skip this step and proceed directly to an expansion of the polarization in powers of the electric fields, DC and AC:

$$P_i = \varepsilon_0\, \chi_{ij}\, E_j + \varepsilon_0 \chi_{ijk}\, E_j\, E_k(DC) + \varepsilon_0\, \chi_{ijk}^{NL}\, E_j\, E_k + \varepsilon_0\, \chi_{ijkl}\, E_j\, E_k(DC)\, E_l(DC) + \varepsilon_0\, \chi_{ijkl}^{NL}\, E_j\, E_k\, E_l + \dots \tag{8.1}$$

where the χ_{ijkl}, etc. are the appropriate susceptibilities. The parameters superscripted NL refer to nonlinear terms used in the second volume of this book that deals with nonlinear optics. There is, in principle, nothing about the physics of the terms labeled NL that would differentiate them from the electro-optic terms. Both represent limiting cases of the same phenomena. The differences are mainly practical. In nonlinear optics, the power needed to exploit the nonlinearity is optical, while in electro-optics the power is electrical. Since in both cases the fields are electromagnetic, the distinction of where electro-optics stops and nonlinear optics begins is arbitrary.

In this chapter, we deal with the Pockels and Kerr effects, both of which involve DC electric fields. They are, respectively, the second and fourth terms in Eq. (8.1). They both fall under the general category of electro-optics, where one uses a DC electric field (or an AC field that is slowly oscillating compared to the optical frequency) either to induce an optical anisotropy in an otherwise isotropic medium or to alter the degree or nature of the anisotropy in a birefringent crystal. The Pockels effect involves a change in the refractive index that is linear in the DC electric field. This effect can occur only in those crystal classes with nonzero values in the χ_{ijk} column in Table 7.1. The Kerr effect causes changes in refractive index that are quadratic in the DC fields. This particular effect takes place to some degree in all materials (e.g., all crystal classes have nonzero values in the χ_{ijkl} column in Table 7.1).

8.1 THE POCKELS EFFECT

The Pockels effect can be described on a microscopic scale by the cubic term in the potential given in Eq. (2.10) (specifically the terms in Eqs. (2.12a) and (2.12b)). Therefore it can be written in terms of a nonlinear polarizability tensor $\langle\alpha_{ijk}\rangle$, which is linked to the macroscopic susceptibility tensor by

$$\chi_{ijk} = \frac{1}{\varepsilon_0} N \langle\alpha_{ijk}\rangle \tag{8.2}$$

where the α_{ijk} is the contribution of the individual molecule or unit cell. From Eq. (8.1) the macroscopic polarization (the quantity in Eq. (8.1)) is given by

$$P_i = \varepsilon_0 \sum_{jk} \chi_{ijk} E_j E_k(DC). \tag{8.3}$$

We now show how the parameter χ_{ijk} can be obtained from the cubic terms in the potential.

We start by examining the symmetry properties of the tensor χ_{ijk}. Let us prove that $\chi_{ijk} = 0$ for any medium with a center of inversion symmetry. Note that inversion symmetry can occur in two different ways. Either an individual molecule or a unit cell of a crystal can have this symmetry. In both cases, the cubic term in Eq. (2.10) is uniquely zero. Or, in an isotropic medium such as a gas, a liquid, or an amorphous solid, the individual molecules may have a cubic term, but they are randomly oriented so that the sum averages out to zero.

We now perform an exercise that determines whether a coefficient is zero in the case of inversion symmetry. Note that an inversion operation on a vector **v** takes it to **-v.** For a material with inversion symmetry, any expression that relates vectors must be equally valid if all vectors are inverted. If we now perform an inversion on each vector in Eq. (8.3), we obtain

$$-P_i = \varepsilon_0 \sum_{jk} \chi_{ijk} (-E_j) \big(-E_k(DC)\big),$$

that is,

$$P_i = -\varepsilon_0 \sum_{jk} \chi_{ijk} E_j E_k(DC). \tag{8.4}$$

Equations (8.3) and (8.4) can both be true only if

$$\chi_{ijk} = 0.$$

Therefore materials with a center of inversion symmetry do not exhibit a Pockels effect. For materials without inversion symmetry **v** need not go to **-v** on inversion, and the argument does not apply.

We restrict the ensuing discussion to crystals which lack a center of inversion. The pertinent crystal classes can be identified from Table 7.1, since they have at least one nonzero component for the χ_{ijk} tensor.

An important point to note is that the application of an electric field and the subsequent change in refractive index can effectively change the optical symmetry properties of a medium. For example, cubic crystals are optically isotropic, and therefore the linear forces on an electron are degenerate. Thus there need not be a relation between the crystal axes and the normal coordinates. However, when a strong DC electric field is applied, this is no longer the case. The crystal axes now define the 1, 2, and 3 axes associated with principal axes. In all cases other than triclinic and monoclinic, the linear force constants are diagonal in a coordinate system that can be sensibly related to the crystal axes and we can therefore use the i,j,k notation of normal coordinates as in Chapter 7.

The Pockels effect can be understood in terms of our basic model of electrons on springs, i.e., the electronic normal coordinates. The application of a strong electric field displaces the electron as shown in Figure 8.1. Because the springs become nonlinear when displaced a sufficiently large distance, small oscillations about the displaced position are not the same as they are about the undisplaced equilibrium position.

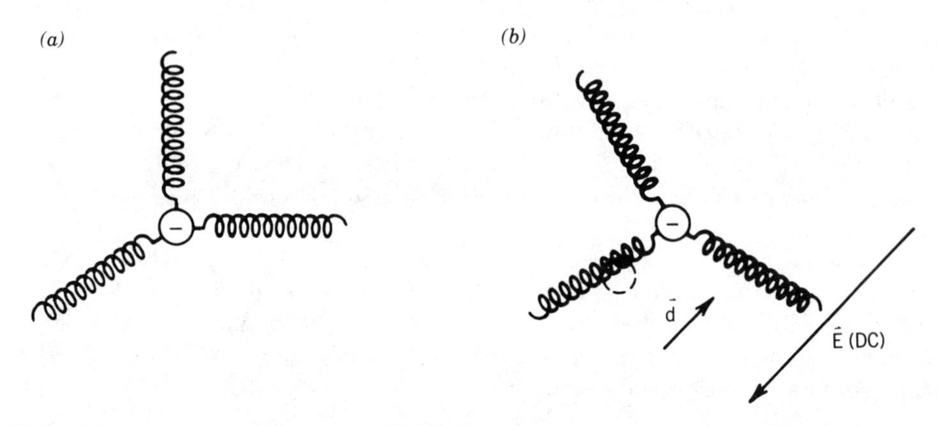

Figure 8.1. a) Undistorted symmetrical springs binding an electron. b) Electron displaced by strong field **E**(DC); symmetry of springs altered.

The total displacement of the normal coordinate is the sum of two terms,

$$q_i' = d_i + q_i \,, \tag{8.5}$$

where q_i' is the total displacement, d_i is the displacement due to the strong electric field (DC or AC, but with a frequency that is small compared to optical frequencies) and q_i is the small oscillation due to the optical field. In order to write down the equation of motion for the coordinate q_i (which gives the macroscopic polarization via $\mathbf{P} = N\Sigma_\beta \ell_\beta q_\beta$), we need to evaluate the nonlinear force from the cubic term in the potential. The force is given by

$$F_i = -\frac{\partial V}{\partial q_i'}.$$

The force arising from the cubic term, denoted $F_i^{(2)}$ for reasons explained later, reads

$$F_i^{(2)} = -\frac{1}{3}\sum_{jkl} k_{jkl}\left[\frac{\partial q_j'}{\partial q_i'} q_k' q_l' + q_j' \frac{\partial q_k'}{\partial q_i'} q_l' + q_j' q_k' \frac{\partial q_l'}{\partial q_i'}\right]. \tag{8.6}$$

Now

$$\frac{\partial q_j'}{\partial q_i'} = \delta_{ij}$$

where δ_{ij} is the Kronecker delta defined in Eqs. (7.8) and (7.9). Therefore

$$F_i^{(2)} = -\frac{1}{3}\sum_{jk} (k_{ijk} + k_{kij} + k_{jki})\, q_j' q_k'.$$

However, k_{ijk} has permutation symmetry, i.e.,

$$k_{ijk} = k_{kij} = k_{jki},$$

and therefore the nonlinear force term reads

$$F_i^{(2)} = -\sum_{jk} k_{ijk}\, q_j' q_k'. \tag{8.7}$$

The equation of motion for the normal coordinate contains forces linear in the DC electric field, the optical field and, in addition, the nonlinear term given by Eq. (8.7). Therefore

$$\ddot{q}_i' + \Gamma_i \dot{q}_i' + \omega_i^2 q_i' = \frac{e}{m} E_i(\mathrm{DC}) + \frac{e}{m} E_i - \sum_{jk} \frac{k_{ijk}}{m} q_j' q_k'. \tag{8.8}$$

Assuming that the nonlinear term is small and can be neglected in the first approximation, there are two driving terms, a DC term and another at the optical frequency. For the DC case ($q_i' = d_i$)

$$\omega_i^2 d_i = \frac{e}{m} E_i(\mathrm{DC}),$$

which gives

$$d_i = \frac{e}{m\omega_i^2} E_i(\mathrm{DC}). \tag{8.9}$$

Therefore Eq. (8.5) now becomes

$$q_i' = \frac{e}{m\omega_i^2} E_i(DC) + q_i,$$

which, when substituted into Eq. (8.8), gives

$$\ddot{q}_i + \Gamma_i \dot{q}_i + \omega_i^2 q_i = \frac{e}{m} E_i - 2 \sum_{jk} k_{ijk} q_j \frac{e}{m^2\omega_k^2} E_k(DC). \tag{8.10}$$

In Eq. (8.10) we retain only the nonlinear terms that oscillate at the optical frequency ω, which are proportional to $d_i q_j$. The other terms are quadratic in the q_i's, and they are discussed at length in the second volume dealing with nonlinear optics.

The equation of motion, Eq. (8.10), can now be solved in a perturbative fashion. First, we note that the original cubic term, which is converted to the quadratic force term in Eq. (8.8), is now linear in q_j. Hence it serves to change the spring constant. Since this term is proportional to q_j (and not q_i), it not only modifies the magnitude of the spring constant but also changes the coordinate system in which it is diagonalized. However, the nonlinear term is small and we can solve Eq. (8.10) by a perturbation series so that we do not actually need to diagonalize the equation in order to obtain a useful solution. The lowest order term is obtained by dropping the nonlinear term:

$$q_i = \frac{1}{2} (Q_i^{(1)} + Q_i^{(2)} + \cdots) e^{-i\omega t} + cc$$

where $Q_i^{(1)} \gg Q_i^{(2)} \cdots$. The standard solution for $Q_i^{(1)}$, given in Eq. (3.15), takes the form

$$Q_i^{(1)} = \frac{e}{mD_i(\omega)} \mathcal{E}_i.$$

Evaluating $Q_i^{(2)}$ with the nonlinear driving force term,,

$$Q_i^{(2)} = - \frac{2}{D_i(\omega)} \sum_{jk} k_{ijk} \frac{e}{m\omega_k^2} E_k(DC) \frac{e\mathcal{E}_j}{m^2 D_j(\omega)}. \tag{8.11}$$

This displacement leads to a macroscopic polarization

$$P_i^{(2)} = \mathcal{N} e Q_i^{(2)} = \frac{-2\mathcal{N}e^3}{m^3} \sum_{jk} \frac{k_{ijk} E_k(DC)}{\omega_k^2 D_i(\omega) D_j(\omega)} \mathcal{E}_j. \tag{8.12}$$

Comparing with Eq. (8.3) gives

$$\chi_{ijk} = \frac{-2\mathcal{N}e^3 k_{ijk}}{\varepsilon_0 m^3 D_i(\omega) D_j(\omega) \omega_k^2}. \tag{8.13}$$

The conventional terminology is to express this result in terms of the dielectric displacement so that

$$D_i = \sum_j \varepsilon_{ij} E_j - \varepsilon_0^{-1} \sum_{jk} \varepsilon_{ii} \varepsilon_{jj} r_{ijk} E_j E_k(DC) \tag{8.14}$$

where

$$r_{ijk} = \frac{2Ne^3 k_{ijk} \varepsilon_0}{m^3 \omega_k^2 D_i(\omega) D_j(\omega) \varepsilon_{ii} \varepsilon_{jj}}. \tag{8.15}$$

Note that r_{ijk} is symmetric under the interchange of the first two indices, i.e., $r_{ijk} = r_{jik}$, but is not symmetric under interchanges involving the third index. The fact that the interchange with the k index is not allowed is not surprising since i and j refer to optical fields, whereas k refers to the DC field.

It is standard notation to contract the first two indices of r_{ijk} into one index. The symmetry of the coefficients under interchange of i and j means that there are six, rather than nine possible combinations of these two indices. The contractions are shown in Table 8.1 and the contracted notation is called the Voigt notation. The contracted tensor is written as $r_{\mu j}$ with $\mu = 1,2,\cdots,6$ and j = 1,2,3, and has the form of a 6 × 3 matrix, as shown in Table 8.1. The nonzero tensor components for the crystal classes which exhibit a Pockels effect are listed in Table 8.2. One important feature of Table 8.2 is that the coefficients are defined with respect to crystal axes rather than principal axes. Thus the labels 1 and 2 in uniaxial crystal classes refer to a particular pair of principal axes, which cannot be determined by linear optical measurements alone. It is customary for suppliers of crystals to provide information as to what these axes are, so that one knows how to use them properly.

Table 8.1. Index Contraction

Index Pair and Contracted Index: ij	μ	Tensor $r_{\mu,j}$		
		r_{11}	r_{12}	r_{13}
11	1	r_{21}	r_{22}	r_{23}
22	2	r_{31}	r_{32}	r_{33}
33	3	r_{41}	r_{42}	r_{43}
32, 23	4	r_{51}	r_{52}	r_{53}
31, 13	5	r_{61}	r_{62}	r_{63}
12, 21	6			

Table 8.2. Nonzero Tensor Elements of the $r_{i\mu}$ Tensor

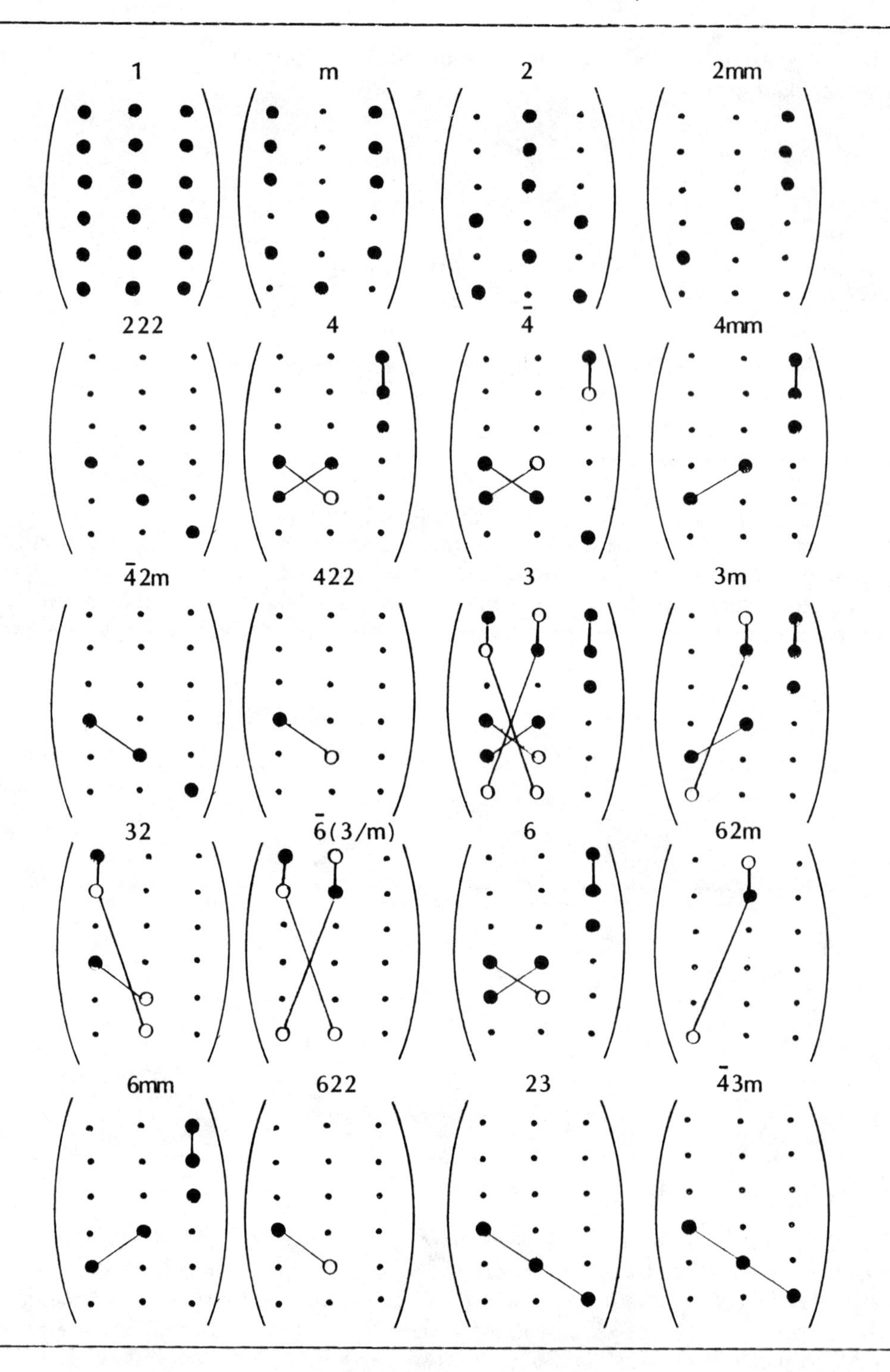

Source: Kaminow, I.P. *An Introduction to Electro-optic Devices*, Academic Press, New York, 1974. pp. 43-48.

Note: The conventions for the elements read: small dots are zero elements; open and dark circles are nonzero elements; whenever lines connect dark circles to dark circles or open circles to open circles the elements are equal; open circles are equal and opposite to dark circles if connected by lines.

The proper value to use for $r_{\mu j}$ depends on how the crystal is mounted. The change in the position of the electron illustrated in Figure 8.1b changes the binding forces among the atoms, i.e., a stress (second rank tensor) is produced. This effect is called piezoelectricity, i.e., the production of a stress linear in the applied field. (There is always an electrostrictive effect proportional to E^2.) If the crystal is allowed to deform in response to this stress, the optical properties change. This change in the optical properties has precisely the same symmetry properties as the $r_{\mu j}$ that came directly from the nonlinear response of the electrons. Hence it is simply a correction to coefficients that are already phenomenological in nature.

8.2 THE KERR EFFECT

In the Kerr effect, there is a change in refractive index which is proportional to the incident field squared. It has two contributions, namely, the "electronic" and "nuclear" terms. The first arises from nonlinear potential and force terms acting on the electrons. The second occurs in anisotropic molecules. The electronic term is proportional to χ_{ijkl}, as indicated in Eq. (8.1), and occurs in all materials. Media with intrinsic isotropy have four independent components of χ_{ijkl}, as do the simplest cubics. In triclinic or monoclinic crystals, the number of independent elements is so large that, for all practical purposes, they are not measurable.

8.2.1 Electronic Kerr Effect

The electronic Kerr effect arises from the quartic term in the potential given by Eq. (2.10). It is calculated in exactly the same way as the cubic term in the previous section. The principle difference in the algebra comes in computing the contribution of the nonlinear force term

$$F_i^{(3)} = -\sum_{jkl} k_{ijkl}\, q_j' q_k' q_l'. \tag{8.16}$$

The important term is the one that oscillates at the incident light frequency, which is

$$q_j' q_k' q_l' \simeq q_j\, d_k\, d_l,$$

And the force linear in q reads

$$F_i^{(3)} = -3 \sum_{jkl} k_{ijkl}\, q_j \frac{e E_k(DC)}{m\omega_k^2} \frac{e E_l(DC)}{m\omega_l^2}. \tag{8.17}$$

This term is quadratic in the DC field. (As with the Pockels effect, the coefficient k_{ijkl} depends on whether the crystal is or is not permitted to respond to the electrostrictive force exerted on it by the DC field.)

The contribution to the dielectric tensor is usually expressed in terms of a tensor S_{ijkl}. Therefore,

$$D_i = \sum_j \left(\varepsilon_{ij} - \sum_{jkl} \frac{\varepsilon_{ii}\varepsilon_{jj}}{\varepsilon_0} S_{ijkl}\, E_k(DC)\, E_l(DC) \right) \mathcal{E}_j, \tag{8.18}$$

where we have simplified the overall expression by leaving out the Pockels term. Here

$$S_{ijkl} = N \frac{3e^4 k_{ijkl}}{\varepsilon_0 m^4 \omega_k{}^2 \omega_l{}^2 D_i(\omega) D_j(\omega)} \frac{\varepsilon_0}{\varepsilon_{ii}\varepsilon_{jj}}. \tag{8.19}$$

This tensor has symmetry properties by which the first and second indices (i and j) can be permuted, and the third and fourth indices (k and l) can be permuted. Thus

$$S_{ijkl} = S_{jikl} = S_{ijlk} \tag{8.20}$$

and the standard Voigt contractions involve the first and last pairs of indices, i.e., $\mu \leftrightarrow ij$ and $\nu \leftrightarrow kl$, according to the scheme in Table 8.1. For cubics, the nonzero elements are

$$\begin{aligned} S_{11} &= S_{22} = S_{33} \\ S_{12} &= S_{23} = S_{13} \\ S_{21} &= S_{31} = S_{32} \\ S_{66} &= S_{55} = S_{44}. \end{aligned} \tag{8.21}$$

8.2.2 Kerr Effect in Liquids

The quadratic electro-optic effect was discovered by Kerr in liquid carbon disulfide. This molecule and benzene continue to be widely used materials in applications involving cubic nonlinearities. Before the advent of modern crystal growing techniques, the "Kerr shutter" was the only electro-optic device available.

The key to understanding the Kerr effect in liquids is to note that the molecules are highly anisotropic and, because of double bonds, also highly polarizable. A sketch of the carbon disulfide and benzene molecules is given in Figure 8.2. The interaction of such molecules with the electric field was discussed in Section 3.4. We note here that there is a difference between the DC and optical polarizabilities, i.e., the optical values do not include contributions from the dipole active vibrational and rotational modes. The optical and DC polarizabilities are identified here as $\boldsymbol{\alpha}(\omega)$ and $\boldsymbol{\alpha}$, respectively, wherever the distinction needs to be made.

We start by defining a unit vector $\hat{\mathbf{s}}_a$ which points along the symmetry axis of the a'th molecule. The polarizability in the laboratory frame of reference has the structure

$$\boldsymbol{\alpha} = \alpha_\perp\ \mathbf{O}(\hat{\mathbf{s}}_a) + \alpha_\parallel\, \hat{\mathbf{s}}_a\, \hat{\mathbf{s}}_a \tag{8.22}$$

where the interaction potential has the general form of Eq. (3.37), and the angle θ_a is measured from E(DC) to the direction of the $\hat{\mathbf{s}}_a$ axis. Assuming that $\mathbf{E}$(DC) lies along the $\hat{\mathbf{x}}$ axis,

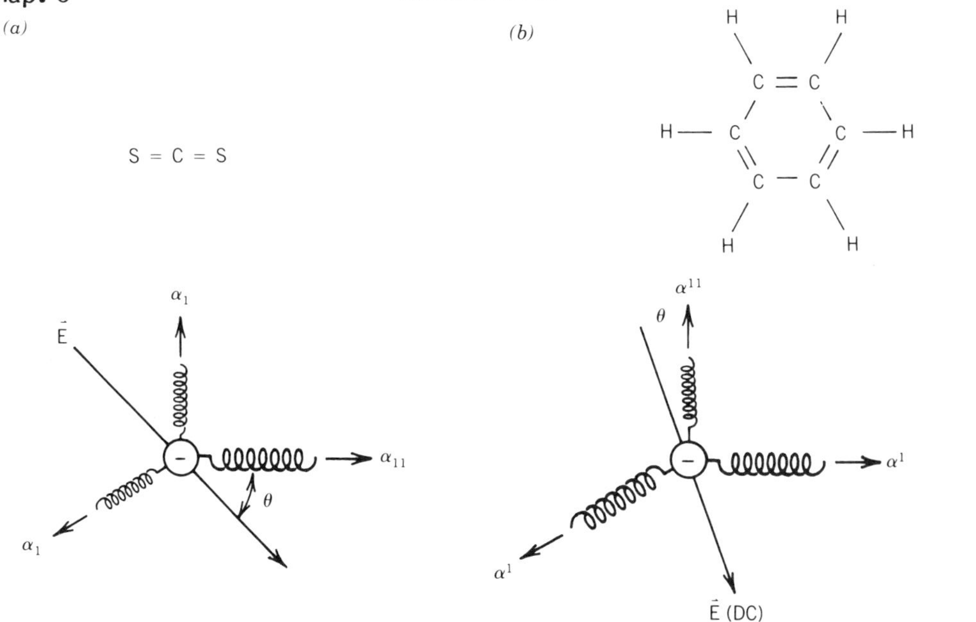

Figure 8.2. Anisotropic polarizable molecules. a) Carbon disulfide, linear molecule. b) Benzene, planar molecule. Symmetry axis (denoted $\hat{s}_a$ in text) is labeled with its polarizability $\alpha_\parallel$.

$$V_{int} = -\frac{1}{2}\left(\alpha_\perp(1-\cos^2\theta_a) + \alpha_\parallel\cos^2\theta_a\right)E^2(DC). \tag{8.23}$$

This can be rewritten as

$$V_{int} = -\frac{1}{2}\left(\alpha_\perp + (\alpha_\parallel - \alpha_\perp)\cos^2\theta_a\right)E^2(DC). \tag{8.24}$$

The use of the phenomenological potential in Eq. (8.24) (note that we do not use the special notation of Section 3.3 for the potential since none is found in the literature) indicates that we are dealing with Raman active modes. The role of the dipole active modes is contained implicitly in the polarizabilities, which is the condition under which potentials of the form of Eq. (3.37) are used. The phenomenon we are discussing is conventionally referred to as a DC Kerr effect, which implies a contrast with an AC Kerr effect. This contrast is somewhat misleading since, in current usage of the terms, the two phenomena are, with one minor change, exactly the same. One replaces $E^2(DC)$ by E^2, where E is an optical field. The AC Kerr effect arises from the DC component of E^2. This effect is described by Eq. (8.24) provided one substitutes the appropriate field amplitude terms and evaluates the polarizabilities at the frequency of the optical field. Both the AC and DC Kerr effects arise because the potential in Eq. (8.24) causes a statistical alignment of the molecules relative to the strong field. This is illustrated for carbon disulfide in Figure 8.3.

There are two contributions in Eq. (8.24). The first term does not involve the angle θ_a. It does, however, contribute to the electro-optic effect since the total interaction is the sum over all atoms, which gives rise to a term proportional to

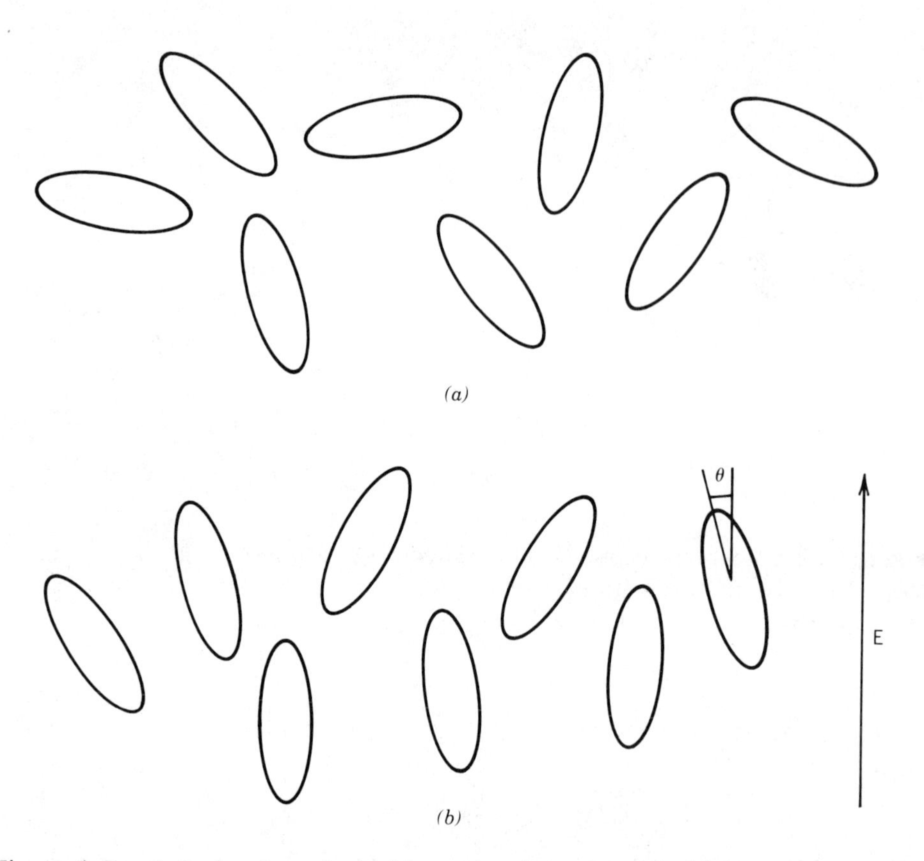

Figure 8.3. Anisotropic polarizable molecules. a) No field, b) with applied field. The molecules align statistically along the field direction (greatly exaggerated in the figure), causing the liquid to acquire the optical properties of a uniaxial crystal whose 3 axis is the direction of the DC field.

$N\alpha_{\perp} E^2(DC)$. The corresponding force increases the density of the medium and is called electrostriction. It is the contribution (electronic) to the Kerr effect calculated in the previous section for crystalline media. In liquids, its contribution is small and can be neglected in comparison to the angular effect.

The second contribution to Eq. (8.24) arises from molecular alignment by the incident electric fields. It is difficult to keep track of the motions of individual molecules as they are knocked about in the liquid. Instead, we recognize that V_{int} is part of the molecular Hamiltonian and it therefore must be part of the general Boltzmann distribution in the liquid. We therefore write

$$P(\theta_a) = \frac{e^{(\alpha_\parallel - \alpha_\perp)\cos^2\theta_a E^2(DC)/2KT}}{Z} \tag{8.25}$$

where $P(\theta_a)d\cos\theta_a d\phi$ is the fraction of molecules in the angle element $d\cos\theta_a d\phi$ at angles θ_a and ϕ, and Z is a normalization coefficient

$$Z = \int_0^{2\pi} d\phi \int_{-1}^{1} d\cos\theta_a \, e^{(\alpha_\parallel - \alpha_\perp)\cos^2\theta_a E^2(DC)/2KT} . \tag{8.26}$$

In Eqs. (8.25) and (8.26) we leave out the constant term $\exp(\alpha_\perp E^2(DC)/2KT)$, which cancels.

The dipole induced in a molecule is given by

$$\mathbf{p}_a = \left(\alpha_\perp(\omega)\, \mathbf{O}(\hat{s}_a) + \alpha_\parallel(\omega)\hat{s}_a\hat{s}_a\right)\cdot(\hat{x}E_x + \hat{y}E_y), \tag{8.27}$$

where we have written the optical field so it has components both parallel (E_x) and perpendicular (E_y) to the DC field. The polarizabilities in Eq. (8.27) are expressed explicitly as a function of the frequency of the optical field to distinguish them from the DC polarizabilities. The macroscopic polarization in the x and y directions, which we label P_x and P_y, is

$$P_x = \sum_a \hat{x}\cdot\mathbf{p}_a \tag{8.28a}$$

$$P_y = \sum_a \hat{y}\cdot\mathbf{p}_a. \tag{8.28b}$$

This can be rewritten as

$$P_x = N\,\langle\hat{\mathbf{x}}\cdot\mathbf{p}_a\rangle, \tag{8.29a}$$

$$P_y = N\,\langle\hat{\mathbf{y}}\cdot\mathbf{p}_a\rangle, \tag{8.29b}$$

where we now average over all possible orientations for a single molecule (just as in Chapter 5). To carry out the calculation, we note that $\hat{s}_a$ in component form reads

$$\hat{s}_a = (\cos\theta,\ \sin\theta\cos\phi,\ \sin\theta\sin\phi) \tag{8.30}$$

and therefore $\hat{s}_a\hat{s}_a$ is

$$\hat{s}_a\hat{s}_a = \begin{pmatrix} \cos^2\theta & \sin\theta\cos\theta\cos\phi & \sin\theta\cos\theta\sin\phi \\ \sin\theta\cos\theta\cos\phi & \sin^2\theta\cos^2\phi & \sin^2\theta\sin\phi\cos\phi \\ \sin\theta\cos\theta\sin\phi & \sin^2\theta\sin\phi\cos\phi & \sin^2\theta\sin^2\phi \end{pmatrix} . \tag{8.31}$$

Note that $P(\theta_a)$ is independent of ϕ, so that averaging $\hat{s}_a\hat{s}_a$ over ϕ means integrating the individual terms in Eq. (8.31) from 0 to 2π, it is useful to define a projection tensor averaged over ϕ, i.e.,

$$\int_0^{2\pi} d\phi \hat{s}_a\hat{s}_a = \begin{pmatrix} 2\pi\cos^2\theta & 0 & 0 \\ 0 & \pi\sin^2\theta & 0 \\ 0 & 0 & \pi\sin^2\theta \end{pmatrix} . \tag{8.32}$$

Furthermore, the normalization term Z reads

$$Z = 2\pi \int_{-1}^{1} d\cos\theta_a \, e^{(\alpha_\parallel - \alpha_\perp)\cos^2\theta_a E^2(DC)/2KT} .$$

As a result, the polarization amplitude is

$$P = \hat{\mathbf{x}} N \left(\alpha_\perp(\omega) + 2\pi [\alpha_\parallel(\omega) - \alpha_\perp(\omega)] \int_{-1}^{1} P(\theta_a)\cos^2\theta_a \, d\cos\theta_a \right) E_x$$

$$+ \hat{\mathbf{y}} N \left(\alpha_\perp(\omega) + \pi [\alpha_\parallel(\omega) - \alpha_\perp(\omega)] \int_{-1}^{1} P(\theta_a)\sin^2\theta_a \, d\cos\theta_a \right) E_y . \tag{8.33}$$

The quantity $(\alpha_\parallel - \alpha_\perp)E^2(DC)/KT$ is typically small compared to unity, which is why the Kerr effect is quadratic in the field and not a more complicated function of the DC field.

Equation (8.33) is difficult to evaluate in general (see, e.g., Section 5.5.1). Because the exponent is small, it is useful to expand the probability function in a Taylor's series. After much algebra

$$P(\theta_a) \simeq \frac{1}{4\pi} - \left(\frac{1}{12\pi} - \frac{\cos^2(\theta_a)}{4\pi}\right) \frac{\alpha_\parallel - \alpha_\perp}{2KT} E^2(DC) . \tag{8.34}$$

Substituting Eq. (8.34) into (8.33),

$$P = \hat{\mathbf{x}} N \left(\frac{2\alpha_\perp(\omega)}{3} + \frac{\alpha_\parallel(\omega)}{3} + \frac{2}{45}(\alpha_\parallel - \alpha_\perp)(\alpha_\parallel(\omega) - \alpha_\perp(\omega)) \frac{E^2(DC)}{2KT} \right) E_x$$

$$+ \hat{\mathbf{y}} N \left(\frac{2\alpha_\perp(\omega)}{3} + \frac{\alpha_\parallel(\omega)}{3} - \frac{1}{45}(\alpha_\parallel - \alpha_\perp)(\alpha_\parallel(\omega) - \alpha_\perp(\omega)) \frac{E^2(DC)}{2KT} \right) E_y . \tag{8.35}$$

The first part of each term gives the linear susceptibility, i.e.,

$$\chi_{ii} = \frac{1}{\varepsilon_0} N \left(\frac{2}{3}\alpha_\perp(\omega) + \frac{1}{3}\alpha_\parallel(\omega) \right) \tag{8.36a}$$

$$\chi_{ij} = 0 \tag{8.36b}$$

in terms of molecular polarizabilities. The second term is the Kerr effect contribution with the S_{ijkl} coefficients

$$S_{xxxx} = -\frac{\varepsilon_0}{\varepsilon^2}\frac{2}{45}\frac{N}{2KT}[\alpha_\parallel - \alpha_\perp][\alpha_\parallel(\omega) - \alpha_\perp(\omega)] \tag{8.37a}$$

$$S_{xxyy} = \frac{\varepsilon_0}{\varepsilon^2}\frac{1}{45}\frac{N}{2KT}[\alpha_\parallel - \alpha_\perp][\alpha_\parallel(\omega) - \alpha_\perp(\omega)]. \tag{8.37b}$$

We are limited to these coefficients by virtue of our choice of the direction of the DC field. Since the choice of the coordinates x and y is arbitrary, we have

$$S_{xxxx} = S_{yyyy} = S_{zzzz} \tag{8.38a}$$

$$S_{xxyy} = S_{xxzz} = S_{yyzz} = S_{yyxx} = S_{zzxx} = S_{yyxx}. \tag{8.38b}$$

Note that for our case, $S_{xxxx} = -2S_{xxyy}$, which implies that these coefficients are not independent of each other, in contradiction of our earlier assertion that they are independent. The reason for this contradiction is that we neglected the electrostrictive contribution in Eq. (8.24). It is small, but it does have different symmetry from the reorientation, insofar as it is isotropic. It adds an identical term to S_{xxxx} and S_{xxyy} such that $S_{xxxx} \neq -2S_{xxyy}$. The third coefficient, S_{xyxy}, is significant for Raman effects discussed in the second volume.

The Kerr effect is tabulated in the literature in terms of the Kerr constant denoted K. It is defined by

$$n_x - n_y = K\lambda E^2(DC). \tag{8.39}$$

From Eqs. (8.18) and (8.35) to (8.37)

$$n_x = n - \frac{n^3}{2} S_{xxxx}E^2(DC) \tag{8.40a}$$

$$n_y = n - \frac{n^3}{2} S_{xxyy}E^2(DC) \tag{8.40b}$$

Therefore

$$K\lambda = -\frac{n^3}{2}(S_{xxxx} - S_{xxyy}), \tag{8.41}$$

that is,

$$K = \frac{1}{30}\frac{(\alpha_\parallel - \alpha_\perp)[\alpha_\parallel(\omega) - \alpha_\perp(\omega)]}{n\lambda 2KT\varepsilon_0}. \tag{8.42}$$

Since $K\lambda E^2(DC)$ is dimensionless, Eq. (8.39) is independent of units. Typical values of K are given in Table 8.3.

The Kerr effect can be used to rotate the plane of polarization of a plane wave. In Section 7.5 (Eq. (7.37)) we showed that a phase lag $\Delta kL = \pi$ can be used to rotate the plane of polarization by 90° where $\Delta k = [k(3)-k(2)]$. In this case $k(3) = 2\pi n_x/\lambda$ and $k(2) = 2\pi n_y/\lambda$ and therefore

Table 8.3. Typical Values of K (m/V^2)

Benzene	0.7×10^{-14}
Carbon Disulfide	3.5×10^{-14}
Nitrobenzene	4.4×10^{-12}

$$\Delta kL = \frac{2\pi L}{\lambda}(n_x - n_y),$$

i.e.,

$$\Delta kL = 2\pi KLE^2(DC). \tag{8.43}$$

For a liquid such as nitrobenzene, a field of the order of 3×10^6 V/m is required to produce a π phase shift for a length $L \simeq 1$ cm.

In the next sections we discuss in some detail the applications of the Pockels effect in laser physics. In modern optical applications, Pockel cells have replaced devices such as the Kerr shutter (modulator). The Kerr shutter played an important role in early twentieth century optics and technology and was used, for example, in speed of light measurements. A modern version of this shutter uses the AC Kerr effect with optical pulses 10^{-12} second in duration, replacing the DC field. This is called the Duguay shutter and has been used to freeze a light pulse in space.

8.3 POCKELS EFFECT IN CRYSTALS WITH TETRAHEDRAL STRUCTURE

Although there are twenty crystal classes which exhibit a Pockels effect (see Table 8.2), relatively few are used in practice. The most commonly used crystals are potassium dihydrogen phosphate (KDP) and its analogs (ADP, RDA, etc.), in which the only nonzero coefficients are $r_{63} = r_{123}$, and $r_{41} = r_{52} = r_{63}$. Because they involve cross terms in the indices, it is difficult to see intuitively how such terms occur, and it is also difficult to understand their significance. We discuss their physics in this section.

The skew coefficients r_{63}, r_{52} and r_{41} are characteristic of crystals with a tetrahedral structure. A schematic of this case is shown in Figure 8.4, which is representative of KDP. The four white circles describe identical atoms arranged as a tetrahedron. By themselves they describe a cubic crystal like gallium arsenide. The addition of two more atoms (black circles) represents the extra atoms that define the optic axis (3 direction, using the notation of Chapter 7). From symmetry, the two axes perpendicular to this axis are optically equivalent.

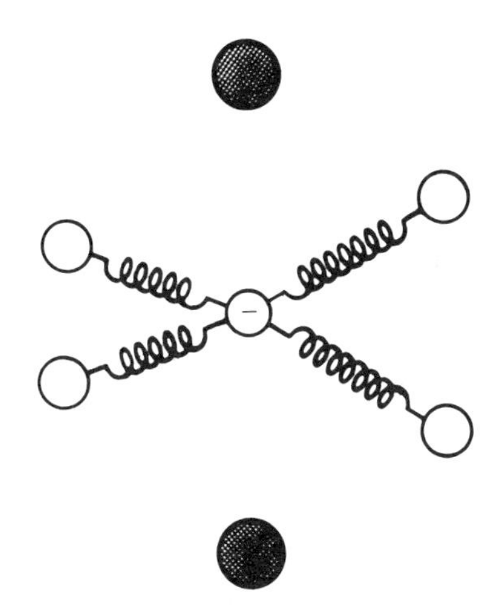

Figure 8.4. Tetrahedral configuration in a uniaxial. Solid and open circles represent two different kinds of atoms. Open circles are arranged as a tetrahedron with the electron at the center. Axis connecting dark circles is the optic axis.

Now, let us consider the influence of a DC field applied along the optic (3) axis. This corresponds to the subscript 3 in r_{63}. It is illustrated in Figure 8.5. The atoms A are so designated because they lie above the plane of symmetry, which is orthogonal to the optic axis; atoms B lie below this plane. The electron, in the absence of an applied field, is located symmetrically with respect to the tetrahedron. When the DC field is in the downwards direction (Figure 8.5a), the electron is pushed up between the A atoms. Assume for simplicity that it is now coplanar with these atoms. It cannot move freely along the axis connecting the A atoms, but it can move much more freely along a direction perpendicular to this line, i.e., at right angles to the springs connecting it to the A atoms. With respect to the B atoms, the electron now forms a triangle with these atoms and can more easily be displaced in the plane containing the B atoms and electrons than if they were all colinear. The net result is that in this position the electron can be displaced more easily perpendicular to an axis joining the A atoms than parallel to it. These two directions are fixed in the crystal at 45° with respect to the crystal axes 1 and 2. If the electric field is reversed (Figure 8.5b), the electron is pushed down between the B atoms and the easy and hard axes are reversed. The application of the DC field induces an optical anisotropy in the material.

We now consider the effects of this anisotropy in the response of the electron to a combined DC and optical field. From examination of crystal class $\bar{4}2m$ in Table 8.2 and taking $\mathbf{E}(DC) = (0,0,E_3(DC))$, all of the pertinent r's are zero except r_{63}. Then

$$D_i = \sum_j [\varepsilon_{ij} - \varepsilon_0 n_o^4 r_{ij3} E_3(DC)] E_j, \tag{8.44}$$

where $\varepsilon_0 n_0^2 = \varepsilon_{11} = \varepsilon_{22} \neq \varepsilon_{33}$, $\varepsilon_{ij} = 0$ when $i \neq j$, and n_0 is the ordinary refractive index. The total effective dielectric tensor is

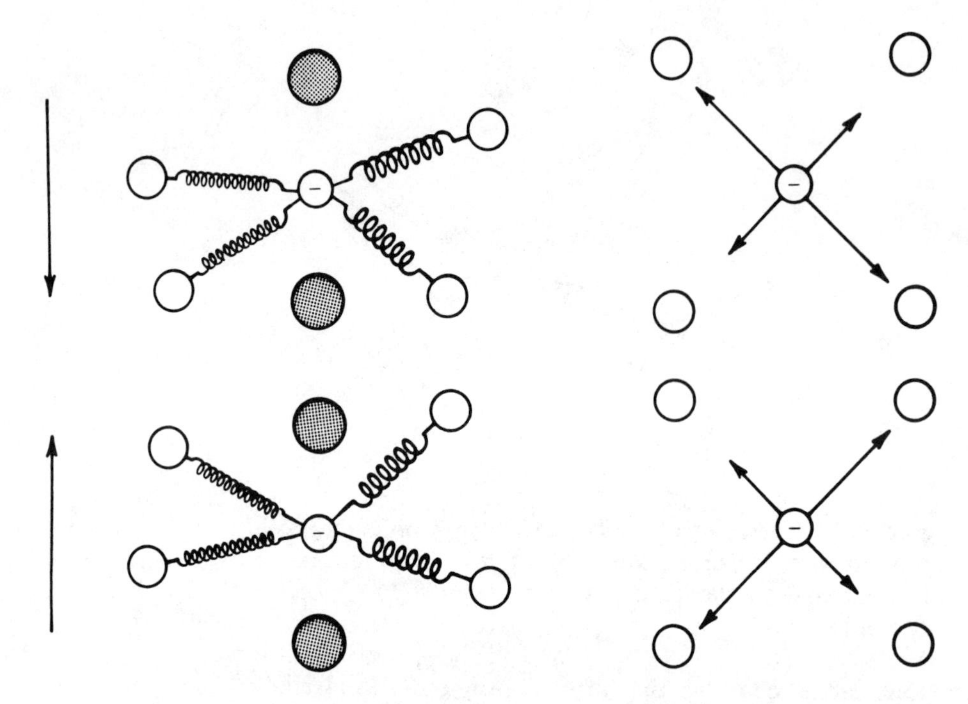

Figure 8.5. Illustration of the effect of the skew coefficients r_{63}, r_{52} or r_{41} under the action of a DC field E pointing along the optic axis (i.e., the axis connecting the dark circles). To the right is the top view of the tetrahedron (optic axis orthogonal to the plane of the paper). The arrows to the right point in the 1' and 2' directions of Figure 8.6.

$$\boldsymbol{\varepsilon} = \begin{pmatrix} \varepsilon_1 & -\varepsilon_0 n_o^4 r_{63} E_3(DC) & 0 \\ -\varepsilon_0 n_o^4 r_{63} E_3(DC) & \varepsilon_1 & 0 \\ 0 & 0 & \varepsilon_3 \end{pmatrix} . \tag{8.45}$$

To obtain the new principal axes, it is necessary to rediagonalize this matrix. This is accomplished by rotating the coordinate system about the 3 axis. Formally, one needs a rotation matrix $\mathbf{R}(|\mathbf{R}| = 1)$ such that

$$\boldsymbol{\varepsilon}' = \mathbf{R}^{-1} \cdot \boldsymbol{\varepsilon} \cdot \mathbf{R}, \tag{8.46}$$

where $\boldsymbol{\varepsilon}'$ is now diagonal, and $\mathbf{R}^{-1}$ is the inverse matrix of $\mathbf{R}$. For rotation about the 3 axis, these matrices have the general form

$$\mathbf{R}(\mathbf{R}^{-1}) = \begin{pmatrix} \cos\beta & (-)+\sin\beta & 0 \\ (+)-\sin\beta & \cos\beta & 0 \\ 0 & 0 & 1 \end{pmatrix} , \tag{8.47}$$

where β is the angle of rotation about the 3 axis. It is relatively straightforward to show that

$$\cos\beta = \sin\beta = \frac{1}{\sqrt{2}} ,$$

i.e., $\beta = 45°$, and that the diagonal form $\boldsymbol{\varepsilon}'$ is

$$\boldsymbol{\varepsilon}' = \begin{pmatrix} \varepsilon_1 - \varepsilon_0 n_o^4\, r_{63} E(DC) & 0 & 0 \\ 0 & \varepsilon_1 + \varepsilon_0 n_o^4\, r_{63}\, E(DC) & 0 \\ 0 & 0 & \varepsilon_3 \end{pmatrix} . \quad (8.48)$$

The net effect of applying the DC field is to change the crystal symmetry class. This change is small and results in substantial optical effects only when the fields propagate over distances of many wavelengths. The crystal has now become, in effect, biaxial with principal axes fixed in the crystal. These axes are illustrated in Figure 8.6, along with the standard notation for a biaxial crystal appropriate to KDP.

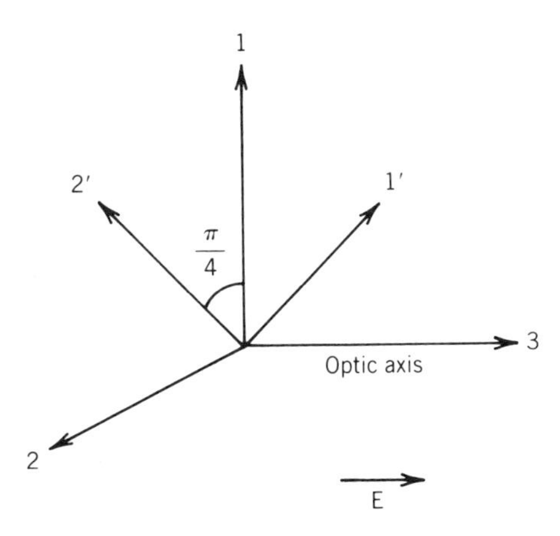

Figure 8.6. Axis systems used to solve electro-optical problems involving skew coefficients. The 1, 2 and 1', 2' axes systems are at 45° to each other. The physical meaning of the 1', 2' system is illustrated in Figure 8.5.

8.4 ELECTRO-OPTIC DEVICES

This creation of a biaxial crystal by the application of a DC field has numerous uses in optics and lasers. For the geometry discussed in the last section, the birefringence induced by the electric field can be utilized by propagating light along the optic axis. The electrodes required to create the DC field parallel to the optic axis can be deposited in the two configurations shown in Figure 8.7. The first requires transparent electrodes on the input and output faces. Since such electrodes always absorb energy, they cannot be used for high power laser applications. More appropriate are ring electrodes with cylindrically shaped crystals, which are in fact used in electro-optic Q-switches (discussed later in this section).

The voltages required to establish the DC fields must frequently be switched on or off quickly. Switching of high voltages requires expensive equipment. It is therefore desirable to use smaller voltages and to apply the fields transverse to a

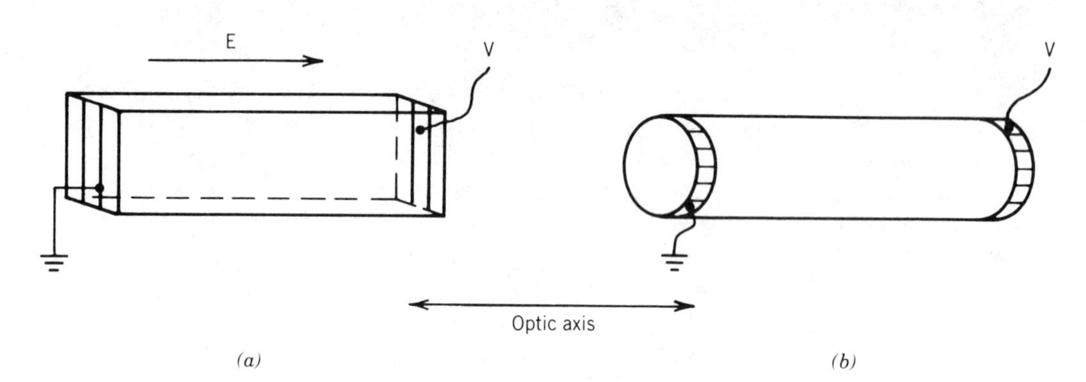

Figure 8.7. Crystal and electrodes for an electro-optic device in which the long axis is the optic (3) axis. a) Electrodes on input and output surface; b) ring electrodes on cylindrically cut crystal.

long, thin crystal. It is left as an exercise to show that orienting the field perpendicular to the optic axis (i.e., using r_{41}) produces a response that is quadratic in the applied field and is therefore unsuitable. What one needs to do is to orient the optic axis perpendicular to the long axis of the crystal, as illustrated

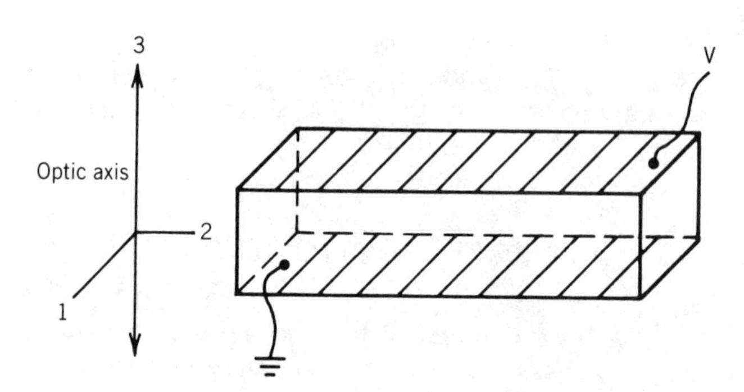

Figure 8.8. Crystal and electrodes for an electro-optic device in which the long axis is perpendicular to the optic (3) axis.

in Figure 8.8. The light propagates along the long axis, i.e., 2' axis of the crystal. Therefore one can use the extraordinary ray for one field polarization and the field modified ordinary ray for the second field polarization. Usually, one relies on the difference between the refractive indices for these two polarizations and the problem is that the natural birefringence of KDP is very temperature sensitive. Therefore, to avoid spurious effects, the crystal temperature must be fixed. This is difficult to achieve when lasers are involved, since the crystal absorbs some light and is heated as a consequence.

It is possible to minimize this effect by using two crystals of different orientation. To cancel out the permanent birefringence effect, one orients the two crystals orthogonally to each other, as shown in Figure 8.9. In this configuration, the field-induced effects are additive, and the permanent birefringence effects cancel if the crystals are identical and have the same length.

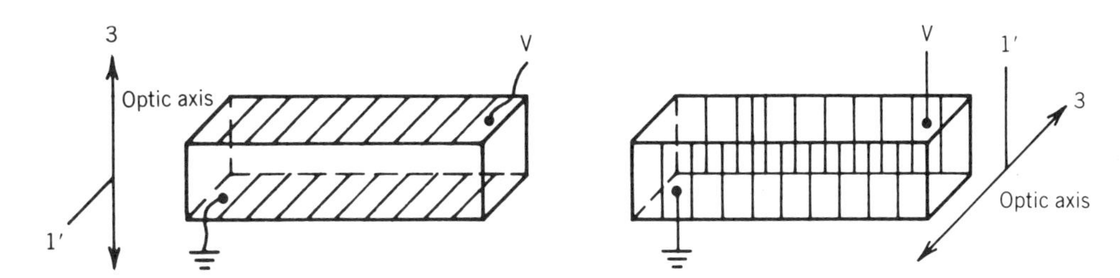

Figure 8.9. Generalization of case in Figure 8.8 needed to eliminate temperature sensitivity of device.

In cases where the optic axis is at an angle other than 90° to the propagation direction, it is customary to use four crystals. They are arranged in pairs to cancel out the displacement of the e- and o-rays. The pairs are then orientated relative to each other as indicated in Figure 8.9. The remaining problem is now temperature gradients rather than temperature stability, and it is important that care be taken to ensure temperature homogeneity.

Devices based on multiple crystal setups are expensive and cumbersome. In general, acousto-optic devices (discussed in Section 11.3) are superior for low speed applications by virtue of their simplicity. However, if large bandwidths (high speeds) are required, electro-optic devices must be used.

Although, as discussed in the preceeding section, there are various useful orientations of an electro-optic crystal, we concentrate on the simplest one in discussing the various device applications. Consider the geometry shown in Figure 8.6 with the electrodes arranged as indicated in Figure 8.7. For optical fields polarized along the new principal axes 1' and 2', an applied DC field changes the index of refraction. For example, for a wave polarized along the 1' axis, from Eq. (8.48)

$$\varepsilon_{11} = \varepsilon_1 - \varepsilon_0 n_o^4 r_{63} E(DC).$$

Since $\varepsilon_1 = n_o^2 \varepsilon_0$ and $\varepsilon_{11} = (n_o - \delta n)^2 \varepsilon_0$, the change δn in index along the 1' axis reads (assuming $n_o \gg \delta n$)

$$\delta n = \frac{1}{2} n_o^3 r_{63} E(DC). \tag{8.49a}$$

For a wave polarized along the 2' axis,

$$\delta n = -\frac{1}{2} n_o^3 r_{63} E(DC). \tag{8.49b}$$

If the incident wave has an arbitrary polarization, and therefore has components along both the 1' and 2' axes, the difference in refractive index experienced by the two resulting orthogonally polarized waves is $2\delta n$. (For the geometry of Figure 8.8, the field dependent δn is given by Eq. (8.49a), and the total difference in refractive indices between the two orthogonally polarized eigenmodes is $n_e - n_o - \delta n$.)

Most devices use AC fields and not DC fields, i.e., $E_3(DC) \rightarrow E_3(t)$, but the time dependence is slow compared to an optical period of oscillation. For fast applications, the material must be free from strains which may arise in mounting the crystal. This case is sometimes denoted by S in tables that give numerical values for the r coefficients.

The simplest application of an electro-optic device is to use the E field to control the rotation of the plane of polarization. Consider an input wave, plane polarized along the 2 axis, as indicated in Figure 8.6. In the crystal, eigenmodes polarized along the 1' and 2' are excited with equal amplitudes, i.e., $\varepsilon_{1'} = \varepsilon_{2'}$. Again defining Δk as the difference in the optical wave vectors, $\Delta k = k(2') - k(1')$, and from Eqs. (8.49a) and (8.49b)

$$\Delta k = -\frac{2\pi}{\lambda_{vac}} n_o^3 r_{63} E_3(DC). \tag{8.50}$$

For propagation distance L, the total phase difference $\delta\phi$ between the two-waves is

$$\delta\phi = \Delta k L = \frac{2\pi L}{\lambda_{vac}} n_o^3 r_{63} E_3(DC). \tag{8.51}$$

If the two-waves are now recombined at the end of the crystal, the resulting wave is elliptically polarized in the general case. Thus the device acts as a wave plate.

Now we consider the geometry of Figure 8.10 where the crystal is placed between two polarizers. Here, LP1 orients the polarization in the 1 direction, and LP2 is, for example, oriented to pass radiation polarized along the 2 axis. Writing

$$E_{1'} = \frac{1}{2} \frac{in}{\sqrt{2}} e^{i[k(1')z - \omega t]} + cc \tag{8.52a}$$

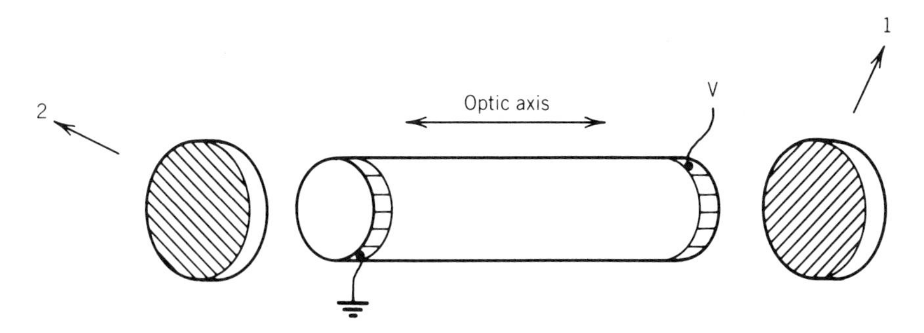

Figure 8.10. Electro-optical device between crossed polarizers.

$$E_{2'} = \frac{1}{2}\frac{\mathcal{E}_{in}}{\sqrt{2}}\, e^{i[k(2')z-\omega t]} + cc \tag{8.52b}$$

for the fields inside the crystal, and then recombining the fields after a propagation distance L,

$$E_1 = \frac{1}{2}\frac{\mathcal{E}_{in}}{2}\, e^{-i\omega t}\left(e^{ik(1')L} + e^{ik(2')L}\right) + cc$$

$$E_2 = \frac{1}{2}\frac{\mathcal{E}_{in}}{2}\, e^{-i\omega t}\left(-e^{ik(1')L} + e^{ik(2')L}\right) + cc. \tag{8.53}$$

In this geometry, only E_2 is passed by the polarizer and

$$E_2 = \frac{1}{2}\frac{\mathcal{E}_{in}}{2}\, e^{ik(1')L-i\omega t}\,(e^{i\Delta kL} - 1) + cc. \tag{8.54}$$

If $E_3(t)$ is an AC field and ΔkL is small, then Eq. (8.54) describes an amplitude modulated field in which the carrier (E_1) has been suppressed, i.e., for the field amplitude

$$\mathcal{E}_2 = -\frac{i\pi\mathcal{E}_{in}}{\lambda_{vac}}\, e^{ik(1')L}\, n_o^3\, r_{63}\, E_3(t)L. \tag{8.55}$$

The factor iexp(ik(1')L) is just an overall phase shift. If $E_3(t)$ is a sine wave of frequency Ω, it is left as an exercise to show that the transmitted field amplitudes occur at the sideband frequencies $\omega \pm \Omega$.

For large values of ΔkL, the output is modulated in both amplitude and phase. If one is interested in just the output power (no phase information), the device is simply an intensity modulator. From Eq. (8.54), the transmitted intensity is just

$$S_{out} = S_{in} \sin^2(\Delta kL/2). \tag{8.56}$$

A special case of this type of application is the electro-optic Q-switch, which is a way of controlling the effective reflectivity of a mirror through polarization changes. The pertinent geometry is shown in Figure 8.11. Normally, the incident field polarization is parallel to the polarizer, and the loss in the elements in front

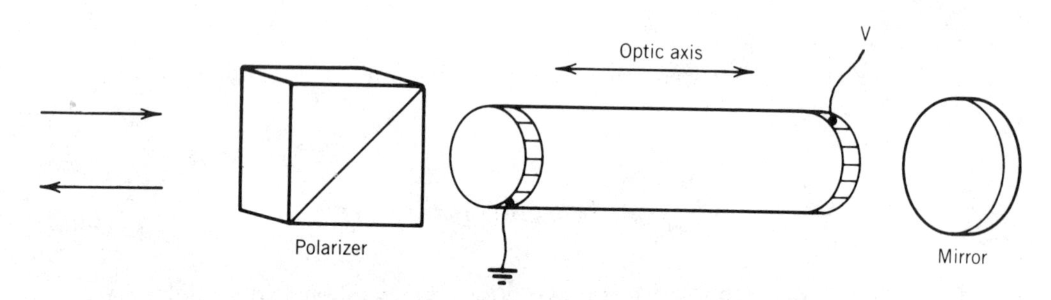

Figure 8.11. Electro-optic device configured as a Q-switch.

of the mirror is small if no voltage is applied to the crystal. (The losses are minimized by anti-reflection coatings.) Suppose now that a voltage across the crystal is maintained so that the crystal acts as a quarter-wave plate. After one round trip, the plane of polarization is rotated through 90° and the return beam is blocked by the polarizer. The reflectivity of the Q-switch assembly is therefore zero when the quarter-wave voltage is applied. If one now short circuits the input to the crystal as quickly as possible, the Q-switch reflectivity is turned on rapidly and these elements can act as one of the reflecting mirrors of a high power laser cavity. Note that in this application it is not necessary to turn the high voltage on quickly, and therefore crystals requiring high voltages do not affect the speed of the device, i.e., we are not limited by the power supply.

In practice, the crystal is immersed in a liquid cell. There are two reasons for doing this. First, it is more economical to put an anti-reflection coating on the cell windows rather than the crystal, since they are more easily replaced when damaged. Second, KDP and its analogues are hygroscopic and the liquid protects them from attack by water. Obviously, the liquid is closely index-matched to the crystal for low loss.

One can produce a frequency or phase modulation by orienting the two polarizers in Figure 8.10 parallel to each other and to the 1' (or 2') axis. In this case, the field inside the crystal is simply given by Eq. (8.52a) with $\mathcal{E}_{in}/\sqrt{2}$ replaced by $\mathcal{E}_{in}$ with $k(1') = (n_o - \delta n)\omega/c$. Therefore the amplitude of the output wave passed by the second polarizer is

$$E_L = E_{in}\, e^{-i\,\delta n L\,\omega/c}. \tag{8.57}$$

The distinction between phase and frequency modulation is somewhat arbitrary. For phase modulation, the field is in the form of a step function and one talks about the phase shift which results when the field is turned on. That is, if the field is zero at t = 0 and is suddenly changed to its DC value E(DC), the total phase shift $\delta\phi$ is

$$\delta\phi = \frac{\pi n_o^3}{\lambda_{vac}} r_{63} L\, E(DC). \tag{8.58}$$

If, on the other hand, the field is sinusoidal, then the phase changes in time as does its derivative, the frequency. This second case is usually called frequency modulation.

Electro-optical devices continue to find new applications. For example, the phase shift can be used to generate short optical pulses, as illustrated in Figure 8.12. We briefly discuss this case by first reasserting some of the basic results of

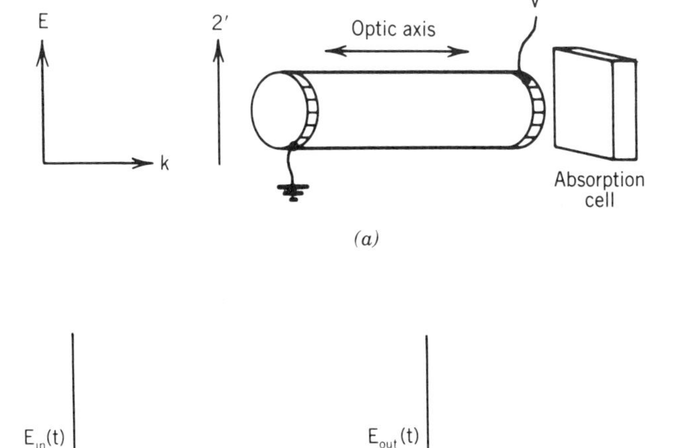

Figure 8.12. a) Electro-optical device configured to produce short optical pulses using free-induction decay in the absorption cell. b) Input amplitude and c) output amplitude as a function of time.

Chapter 4 with respect to absorption. There it was shown that the reradiated fields interfered destructively with the incident fields to produce absorption. Suppose that the phase of the incident field is suddenly switched by 180° in a time much shorter than Γ_α^{-1} of the normal modes. For a short period of time, the induced dipoles continue to radiate with the phase imparted to them by the original (pre-switching) incident optical field. Now the reradiated field interferes constructively with the incident (switched) field, and the dipoles emit radiation rather than absorbing it. In the time interval Γ_α^{-1} the dipoles readjust their phase such that they once again absorb, but during that time interval a pulse is emitted whose width is about Γ_α^{-1}. Since the radiated field from the dipoles prior to switching is $-E_{in}$ (it cancels the preswitched incident field), and because the switched incident field is now itself $-E_{in}$, the total field during this time Γ_α^{-1} can be as large as $-2E_{in}$. Therefore the peak power emitted in the pulse is four times the incident power. A similar result occurs when using a modulator that shuts the incident field off, in which case the radiation emitted during the time Γ_α^{-1} is called free induction decay. In that case, the peak radiated field is just $-E_{in}$ and the peak power obtained is just the incident wave power. One application for which this technique is well suited is to obtain synchronous pulses with different carrier frequencies.

8.5 LIMITATIONS OF THE ELECTRO-OPTIC EFFECT

It is always tempting in any application of electro-optics, or indeed any of the cases we deal with later in magneto-optics or nonlinear optics, to pick a material based primarily on how large a coefficient it has. This is an important consideration, and in some cases decisive insofar as it gives the biggest effect for the smallest amount of applied field. On the other hand, there are other considerations of great importance. For example, one must be able to obtain large crystals of decent quality. In laser applications, damage from large optical fields can be a decisive limitation.

A material with a large coefficient does not, in general, produce a maximum electro-optic effect that is larger than one which is obtainable with a material with a smaller coefficient. There is a basic limitation that says that one cannot change an index by more than about 0.001 no matter what the coefficient is, and all electro-optic effects do about equally well in this regard. The reasons for this upper bound are basic, and the key to it is illustrated in Figure 8.13. The electron

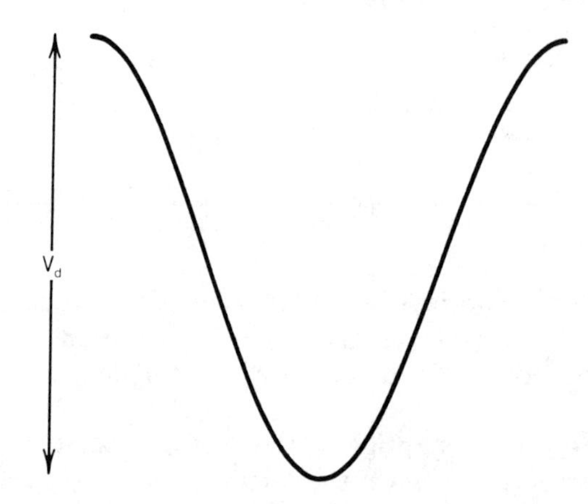

Figure 8.13. Potential well for electrons.

sits in an effective potential well whose depth V_d is set by quantum mechanics. Near the bottom, the well is quadratic and the forces are those of a linear spring. The nonlinear terms become important as the electron is pushed by the external fields up to the edge of the well. At that point, the electron is close to escaping from the well entirely, and if it does so it starts an electrical arc that causes damage to the medium. The depth of the well is thus both a measure of how large a field one needs to make a nonlinearity (small V_d's imply larger r_{ijk}'s and S_{ijkl}'s) and is a measure of the damage threshold (small V_d's imply a low damage threshold and more severe restrictions on E(DC)). A safety factor against damage is equivalent to staying a certain distance below the rim of the well, and all wells are about equally nonlinear in that case.

One reason that acousto-optics is such an effective technique in spite of the fact that it involves quartic terms (like S_{ijkl}) is that it uses the internal electric fields of the medium to change the index. The internal field is on the order of 10^{10} V/m, which is significantly higher than one can apply externally. A relatively small motion of an atom can thus produce a large effect. These types of perturbations do not damage the medium, and hence one can get significant changes in indices. A similar principle is involved when one changes the indices by changing the temperature of the crystal, which can produce larger changes in an index than can be obtained from external electric fields. The modification of indices by temperature plays an important role in nonlinear optics.

ADDITIONAL READING

General discussion

Yariv, A., *Quantum Electronics*, 2nd Edition, Wiley, New York, 1975. pp. 327-356.

Details of devices, coefficients

Kaminow, I.P. *An Introduction to Electro-optic Devices*, Academic Press, New York, 1974.

Coefficients

CRC Handbook of Lasers with Selected Data on Optical Technology, R.J. Presley, ed., Chemical Rubber Co., Cleveland, 1971, Chapter 15.

PROBLEMS

8.1. Consider a crystal of class 2, which has a single axis of symmetry such that the crystal is symmetrical under a rotation about that axis. Show that the zero elements in Table 7.2 are, in fact, zero.

8.2. Write down the dielectric tensor up to terms including $r_{\mu k}$ (a) for crystal classes 622, 3, 4 using a field $\mathbf{E}$(DC) = $(0,E_2,0)$, and (b) for crystal classes 4mm, 2mm, $\bar{4}$3m using a field $\mathbf{E}$(DC) = $(0,0,E_3)$.

8.3. For crystal class $\bar{4}$2m, show that the use of a DC field $\mathbf{E}$(DC) = $(0,E_2,0)$ and an optical field propagating along the optic axis gives rise to an electro-optic effect that is quadratic in the DC field.

8.4. Which uniaxial crystal classes allow a linear electro-optic effect when the DC field is orthogonal to the optic axis and the optical field is propagating along the optic axis?

8.5. Derive Eqs. (8.34) and (8.35).

8.6. Suppose that the AC field in Eq. (8.50) is of the form

$$E(t) = E_0 \cos\Omega t$$

Show that the output of the device has a spectrum consisting of two sidebands at $\nu \pm \Omega$.

8.7. Derive Eqs. (8.19) and (8.20). Note that the argument is nearly identical to the one that gets the r_{ijk} coefficients. Use references to that argument whenever possible.

8.8. For the case of trigonal crystals (3, 3m, 32) consider only DC and AC fields with $E_3 = 0$, and only the effect of the dielectric tensor in the 1,2 plane.

(a) Show that in the 1,2 plane, all trigonal crystals have a dielectric tensor

$$\boldsymbol{\varepsilon} = \begin{pmatrix} \varepsilon - \dfrac{\varepsilon^4}{\varepsilon_0} r_{11} E_1 + \dfrac{\varepsilon^4}{\varepsilon_0} r_{22} E_2 & \dfrac{\varepsilon^4}{\varepsilon_0} r_{22} E_1 + \dfrac{\varepsilon^4}{\varepsilon_0} r_{11} E_2 \\ \dfrac{\varepsilon^4}{\varepsilon_0} r_{22} E_1 + \dfrac{\varepsilon^4}{\varepsilon_0} r_{11} E_2 & \varepsilon + \dfrac{\varepsilon^4}{\varepsilon_0} r_{11} E_1 - \dfrac{\varepsilon^4}{\varepsilon_0} r_{22} E_2 \end{pmatrix}$$

(b) Using the rotation tensor in Eq. (8.46) show that this is diagonalized with an angle β such that

$$\tan 2\beta = \frac{r_{22}E_1 + r_{11}E_2}{r_{11}E_1 - r_{22}E_2}$$

Using the trigonometric identity for the tangent of the sums of angles, show that

$$2\beta = \tan^{-1}\frac{r_{22}}{r_{11}} + \tan^{-1}\frac{E_2}{E_1}$$

(c) Consider $E = \mathcal{E}(\cos\Omega t, \sin\Omega t, 0)$ and an electromagnetic field propagating in the 3 direction. Show how to produce an output that shifts the input frequency by a constant (this is called a frequency shifter). Hint: The procedure involves taking a circularly polarized incident field and producing an output with opposite sense of circular polarization. The electro-optical phase lag is $\pi/2$ (e.g., for class 3m, $r_{22}EL = \pi/2$). This problem is much more easily solved with the techniques of Chapter 10.

9
Magneto-Optics

In this chapter we discuss the changes in optical properties that occur when an external magnetic field **B** is applied to a dielectric. Just as with the electro-optic effect, there are effects that are linear and quadratic in **B**. The linear phenomenon is called the Zeeman effect on resonance, and the Faraday effect off resonance. The quadratic effect is called the Cotton Mouton effect. An important difference from the electro-optic effect is that no special medium symmetries are required for magneto-optic effects, and therefore they occur in all materials. Because the linear effects are significantly larger than the quadratic ones, the Zeeman and Faraday effects are the ones of interest for useful magneto-optic phenomena.

We can now augment our general expansion of the polarization in powers of the electromagnetic fields to also include magnetic fields. For magneto-optics

$$P_i = \varepsilon_0 \chi_{ijk} E_j B_k + \varepsilon_0 \chi_{ijkl} E_j B_k B_l. \tag{9.1}$$

Again, it is possible to relate the pertinent susceptibility coefficients to basic properties of the medium.

9.1 FARADAY AND ZEEMAN EFFECTS

We concentrate on the case of primary practical interest, namely isotropic media. The electronic normal coordinates are described by three equal and orthogonal springs. Since they are degenerate, we can, without loss of generality, orient them as shown in Figure 9.1. A DC magnetic field **B** is aligned along the 3 axis, and we also assume an optical field of still unspecified orientation. The electron coordinates and their velocities are written as (r_1,r_2,r_3) and $(\dot{r}_1,\dot{r}_2,\dot{r}_3)$, and the force on them is given by Eq. (1.9) as

$$\mathbf{F} = e\,\mathbf{E} + e(\hat{\mathbf{e}}_1\dot{r}_2 - \hat{\mathbf{e}}_2\dot{r}_1)B. \tag{9.2}$$

In component form

$$m(\ddot{r}_1 + \Gamma\dot{r}_1 + \omega_1{}^2 r_1) = e\dot{r}_2 B + eE_1 \tag{9.3a}$$

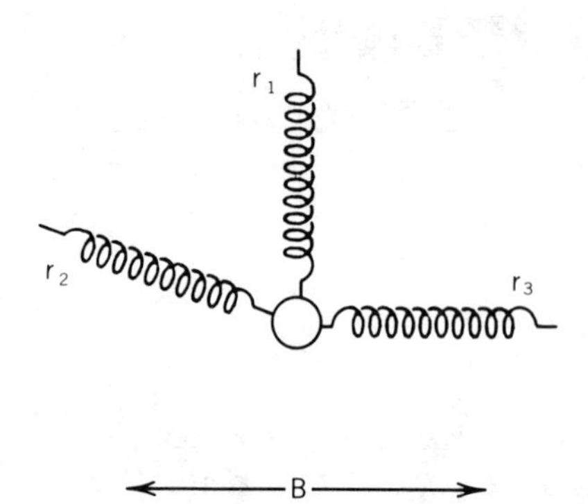

Figure 9.1. Spring system for magneto-optic effects.

$$m(\ddot{r}_2 + \Gamma\dot{r}_2 + \omega_1{}^2 r_2) = -e\dot{r}_1 B + eE_2 \tag{9.3b}$$

$$m(\ddot{r}_3 + \Gamma\dot{r}_3 + \omega_1{}^2 r_3) = eE_3. \tag{9.3c}$$

Since the equation for r_3 does not depend on the magnetic field, we set $E_3 = 0$ and concentrate on magnetic field related effects. This implies, for an isotropic medium, that $\mathbf{k} = k\hat{\mathbf{e}}_3$, i.e., the light propagates along the 3 axis.

The appropriate normal optical modes are not linearly but circularly polarized. We define normal coordinates for the medium as shown in Figure 9.2:

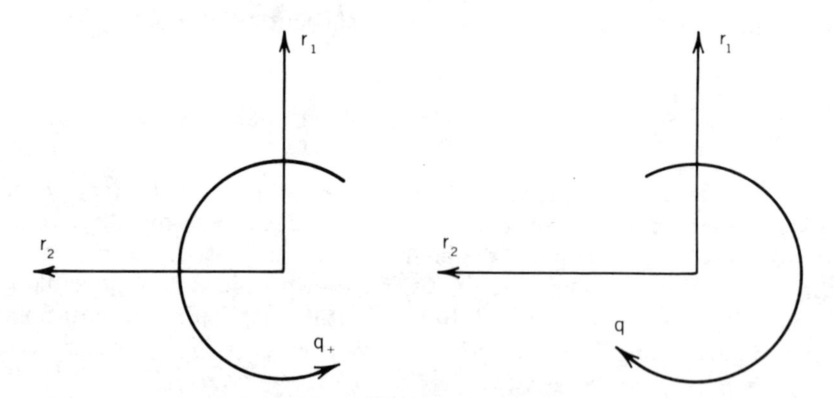

Figure 9.2. Coordinates $q_\pm$.

$$q_\pm = \frac{r_1 \pm ir_2}{\sqrt{2}}, \tag{9.4}$$

as well as consider the optical field to be circularly polarized,

$$E_{\pm} = \frac{E_1 \pm iE_2}{\sqrt{2}} \, . \tag{9.5}$$

Substituting Eqs. (9.4) and (9.5) into Eqs. (9.3a,b) gives

$$m(\ddot{q}_{\pm} + \Gamma\dot{q}_{\pm} + \omega_1{}^2 q_{\pm}) = \mp ieB\dot{q}_{\pm} + eE_{\pm}. \tag{9.6}$$

We can now solve for the polarization associated with the normal coordinates. Writing

$$q_{\pm} = \frac{1}{2}\,(Q_{\pm}\, e^{i(k_{\pm}z - \omega t)} + cc) \tag{9.7}$$

one obtains

$$Q_{\pm} = \frac{eE_{\pm}}{m[\omega_1{}^2 - \omega^2 \pm \omega eB/m - i\omega\Gamma]}. \tag{9.8}$$

The quantity eB/m has the units of frequency and is called the Larmor precession frequency. The absorption resonances associated with Eq. (9.8) are found by setting the real part of the denominator to zero. These resonances occur for

$$\omega \simeq \omega_1 \pm \frac{eB}{2m} \tag{9.9}$$

and the corresponding absorptions are sketched in Figure 9.3. This splitting of the resonance is called the Zeeman effect. When the incident light frequency is far

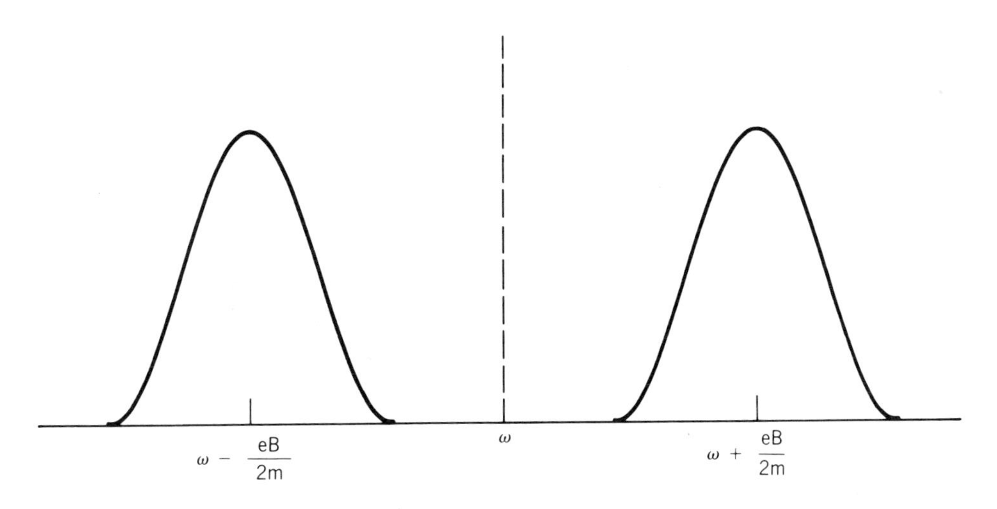

Figure 9.3. Absorption resonances in Zeeman effect.

away from the resonance, this is termed the Faraday effect. For light polarized along the 3 direction, i.e., along the magnetic field **B**, the absorption is not affected by the magnetic field and it peaks at $\omega = \omega_1$.

In most applications in optics, this phenomenon is used far off resonance, i.e., as the Faraday effect. In this case, we can now expand the denominator in a Taylor series. For $\omega_1^2-\omega^2 \gg \omega\Gamma$ and $\omega_1^2-\omega^2 \gg \omega eB/m$,

$$Q_\pm = \frac{eE_\pm}{m(\omega_1^2-\omega^2)}\left(1 \mp \frac{\omega eB}{m(\omega_1^2-\omega^2)}\right). \tag{9.10}$$

The corresponding polarization fields, $P_\pm = NeQ_\pm$, are now given by

$$P_\pm = \varepsilon_0\left(\chi \mp \frac{2n\gamma cB}{\omega}\right)E_\pm, \tag{9.11}$$

where γ is called the Verdet coefficient. Thus

$$\gamma = \frac{Ne^3\omega^2}{2n\varepsilon_0 m^2(\omega_1^2-\omega^2)^2 c}. \tag{9.12}$$

Define the dielectric constant as

$$D_\pm = \varepsilon_\pm E_\pm; \tag{9.13}$$

then

$$\varepsilon_\pm = \varepsilon_0\left(n^2 \mp \frac{2n\gamma Bc}{\omega}\right), \tag{9.14}$$

and the refractive indices for the two circular polarizations are

$$n_\pm = n \mp \frac{\gamma Bc}{\omega}. \tag{9.15}$$

9.2 PRACTICAL APPLICATIONS

This effect can be used to rotate the plane of polarization of an incident plane polarized light wave. It was shown in Chapter 6 that plane polarized light incident on an optically active medium propagates as two circularly polarized waves of unequal velocity, and then recombines to produce plane polarized light at the exit surface with a polarization different from the incident one. The same phenomenon occurs here. The angle of rotation of the plane of polarization is given by γBL (see Eqs. (6.20) through (6.27)) for a propagation distance L. (Note that all of the coefficients introduced in Eq. (9.11) cancel out to give a simple result for the angle of rotation.

Table 9.1 gives some representative values for the Verdet coefficients. Unlike the case of optical activity, the rotation achieved by passing through a Faraday cell in the $+x_3$ direction is not undone when passing back in the $-x_3$ direction. Instead, the cumulative rotation is $2\gamma BL$, where L is the length of the Faraday cell. We shall now verify this point. Summarized in Table 9.2 is the notation we use to describe the sense of circular polarization, the wave propagation direction and the field amplitude. We consider the case where light is

Table 9.1. Verdet Coefficients

Material	γ(radians/Tesla/m)
Water	3.81
CS_2	12.3
NaCl	10.4
Phosphorous	38.6

Source: Jenkins, F.A. and White, H.E. *Fundamentals of Optics*, McGraw-Hill, New York, 1976. p. 686.

reflected by a mirror on passing through a Faraday cell, and then reintroduced back through the exit face of the cell. On reflection off a mirror, the sense of polarization is reversed (right to left, or vice versa), so that the amplitude (E_+ or E_-) is used in either direction. Hence reversing the direction with a mirror preserves the eigenvector. (In optical activity the eigenvectors are reversed.) The plane of polarization rotates in one round trip because E_+ moves at a different speed from E_-, and this continues to be the case on both passes.

Table 9.2. Association of Field Amplitudes and Wave Vectors with Sense of Polarization

Amplitude	Wave Vector	Sense of Circular Polarization
E_+	$+k_+$	right hand
E_-	$+k_-$	left hand
E_-	$-k_-$	right hand
E_+	$-k_+$	left hand

9.3 FARADAY ISOLATORS

The classic application of magneto-optics is to the Faraday isolator illustrated in Figure 9.4. A Faraday cell is inserted into the center of a solenoid which generates a uniform axial **B** field. The linear polarizers have their axes set 45° apart and the Faraday cell parameters are tuned so that $\gamma BL = \pi/4$. The return beam is rotated again by 45°, so the total angle of rotation is 90°. Hence the return beam is blocked by the first polarizer. There can therefore be no beam

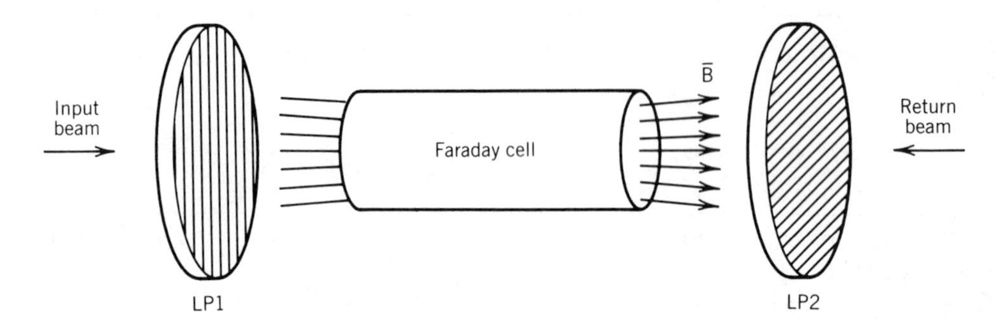

Figure 9.4. Faraday isolator.

returning along the incident beam direction, hence the device in Figure 9.1 isolates the system to the left from the system to the right.

There are easier ways of accomplishing such a blocking function using wave plates. However, neither birefringent crystals nor electro-optic cells are readily available with very large apertures. Since Faraday cells can be made from glass or liquids, there are no limitations on aperture size. Apertures of up to 20 × 20 cm are used in the isolation stages of large fusion lasers.

Because the eigenvectors do not depend on the direction of propagation, it is possible to express the propagation aspects in terms of an effective dielectric tensor. Equation (9.13) can easily be reexpressed in terms of Cartesian coordinates, which gives the effective dielectric tensor

$$\varepsilon = \varepsilon_0 \begin{pmatrix} n^2 & ig & 0 \\ -ig & n^2 & 0 \\ 0 & 0 & n^2 \end{pmatrix} , \tag{9.16}$$

where

$$g = -\frac{2\gamma n B c}{\omega}. \tag{9.17}$$

This tensor can now be used to find the general eigenvectors. For general wave vectors **k**, in addition to the effects discussed already, one finds birefringence effects (called Voigt birefringence) and indices that are quadratic in B (the Cotton Mouton effect). These are of little practical interest and are left as exercises.

ADDITIONAL READING

Verdet Coefficient

Fowles, G.R. *Introduction to Modern Optics*, Holt, Rinehart and Winston, New York, 1968. p. 89.
Jenkins, F.A. and White, H.E. *Fundamentals of Optics*, McGraw-Hill, New York, 1976. p. 597.
CRC Handbook of Lasers with Selected Data on Optical Technology, R.J. Presley ed., Chemical Rubber Co., Cleveland, 1971, Chapter 16.
CRC Handbook for Chemistry and Physics, 52nd ed., R.C. Weast, ed., Chemical Rubber Co., Cleveland, 1971-72, pp. E227-E228.

Magneto-Optical Discussion

Fowles, G.R., pp. 188-191.
Strong, J.M. *Radiation and Optics*, McGraw-Hill, New York, 1963. pp. 445-455.

PROBLEMS

9.1. (a) Consider the case of optical propagation in a direction perpendicular to the **B** field. Take the case **k** = (0,k,0) and write the dispersion relation for light propagation.

(b) Find the eigenvectors and velocities for this case. Show that this case leads to an induced birefringence quadratic in the **B** field (the Cotton Mouton effect).

9.2. Derive Eq. (9.6).

9.3. Show that media with large refractive indices should have large Verdet coefficents.

9.4. Use Table 9.2 to show that the Faraday isolator in Figure 9.4 works as described in the text. Use Table 6.1 to show that if an optically active medium replaces the Faraday cell in Figure 9.4, no isolation occurs.

9.5. Derive the Faraday effect from Eqs. (9.16) and (9.17).

10
Electrodynamics of Weak Polarizations

Many problems in electrodynamics involve cases in which the phenomenon of interest produces a small perturbation on the susceptibility. To be useful or measurable, such weak effects require propagation distances large compared to an optical wavelength. In each case, the polarization of the medium can be separated into two terms. One is associated with the response of the medium in the absence of the perturbation and the second results from the perturbation itself. This type of problem appears in Chapters 11 and 12, and is especially prevalent in nonlinear optics. In this chapter we develop a systematic approach for including the perturbation effects in the solutions. We develop what is called the "slowly varying amplitude and phase approximation," which we abbreviate as SVEA (the abbreviation stands for the alternative terminology "slowly varying (complex) envelope approximation").

10.1 SLOWLY VARYING AMPLITUDE AND PHASE APPROXIMATION

The SVEA is a powerful technique for treating a variety of problems. Unfortunately it is not a thoroughly understood approximation, and it can produce erroneous results even in regimes in which it appears to be applicable. Hence we give the subject a careful introduction. We assume that the macroscopic polarization can be written as two terms,

$$\mathbf{P} = \mathbf{P}_S + \mathbf{P}_W, \tag{10.1}$$

where $\mathbf{P}_S$ is the strong polarization, usually the medium response leading to the refractive indices and the associated optical eigenvectors, as considered in detail in Chapters 5 and 7. The weak polarization, $\mathbf{P}_W$, can arise from any one of a number of interactions. In Chapter 4 the entire interaction was treated as if it were weak. Optical activity (Chapter 6), electro-optics (Chapter 8) and magneto-optics (Chapter 9) can be regarded as arising from weak effects.

We now consider the wave equation as expressed by Eq. (1.17) with $\mathbf{M} = 0$, which now reads

$$-\nabla^2 \mathbf{D} + \frac{1}{c^2}\frac{\partial^2 \mathbf{D}}{\partial t^2} = \nabla \times (\nabla \times \mathbf{P}_s) + \nabla \times (\nabla \times \mathbf{P}_w). \tag{10.2}$$

The contribution from $\mathbf{P}_s$ cannot be approximated. Instead we assume that $\mathbf{D}$ is approximately a plane-wave eigenvector of the wave equation, so that an index of refraction is well defined. With this assumption, Eq. (10.2) becomes

$$-\nabla^2 \mathbf{D} + \frac{n^2}{c^2}\frac{\partial^2 \mathbf{D}}{\partial t^2} = \nabla \times (\nabla \times \mathbf{P}_w). \tag{10.3}$$

Note that Eq. (10.3) is really an abbreviation for a pair of equations describing two orthogonal eigenvectors with potentially different indices. The proper equations are used whenever such considerations are important.

Now let us take the amplitude of the eigenvector to be weakly dependent on the propagation direction, which we denote as the z axis. The relation between the 1, 2, and 3 principal axes and the propagation coordinates is illustrated in Figure 10.1 for a biaxial or uniaxial crystal. In this case, waves of the form $(D_x,0,0)$ and $(0,D_y,0)$ correspond to e-configuration and o-configuration eigenvectors respectively. In this axis system, the wave vector is $\mathbf{k} = (0,0,k)$.

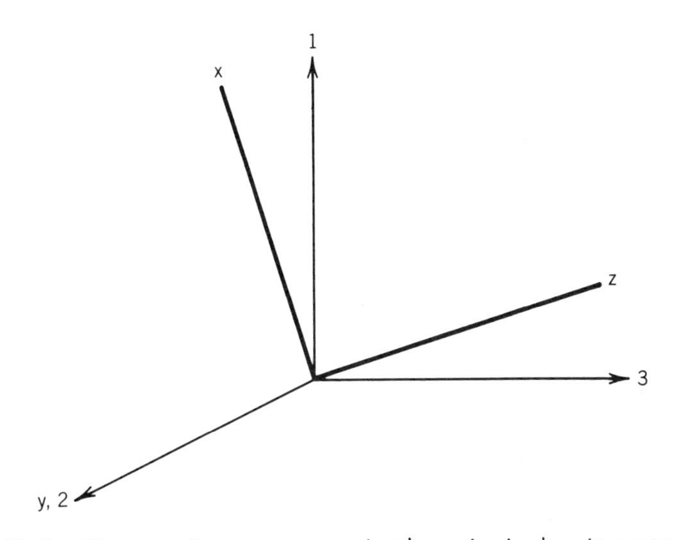

Figure 10.1. Propagation axes x,y,z in the principal axis system 1,2,3.

We now define the slowly varying amplitude as

$$\mathbf{D} = \frac{1}{2}\left(\boldsymbol{D}(z;\omega,\mathbf{k})\, e^{i(kz-\omega t)} + \mathrm{cc}\right) \tag{10.4}$$

$$\mathbf{E} = \frac{1}{2}\left(\boldsymbol{E}(z;\omega,\mathbf{k})\, e^{i(kz-\omega t)} + cc\right) \tag{10.5}$$

with analogous expressions for **B** and **H**. We have explicitly allowed the vectors $\boldsymbol{D}$, $\boldsymbol{E}$, etc. to vary with propagation distance.

From Eq. (10.3), we see that the weak polarization term acts as a source for the wave equation. The polarization source is usually driven by some interaction at the frequency ω with a characteristic wave vector $\mathbf{k}_p$. This wave vector need not correspond to the wave vector associated with solutions to the homogeneous wave equation at frequency ω, i.e., Eq. (10.3) with $\mathbf{P}_w = 0$. Thus we write the weak polarization as

$$\mathbf{P}_w = \frac{1}{2}\left(\boldsymbol{P}_w(z;\omega,\mathbf{k}_p)\, e^{i(k_p z-\omega t)} + cc\right). \tag{10.6}$$

Although $\mathbf{k}$ is not necessarily equal in magnitude to $\mathbf{k}_p$, Eq. (10.6) does define the direction of $\hat{\mathbf{k}}$, i.e.,

$$\hat{\mathbf{k}} = \hat{\mathbf{k}}_p. \tag{10.7}$$

We now define the slowly varying condition as

$$\left|\frac{\partial \boldsymbol{D}(z;\omega,\mathbf{k})}{\partial z}\right| \ll k\,\left|\boldsymbol{D}(z;\omega,\mathbf{k})\right|. \tag{10.8}$$

Equation (10.8) requires that the electromagnetic field amplitudes vary slowly over distances comparable to an optical wavelength. For Eq. (10.8) to be valid, it is not necessary that $\boldsymbol{P}_w$ vary slowly, but it is necessary that any susceptibility associated with the weak polarization must be much less than one (see Problem 10.17). However, for all cases of interest that we are aware of, $\boldsymbol{P}_w$ does vary slowly and we can write

$$\nabla \times (\nabla \times \mathbf{P}_w) = \frac{1}{2} k_p^2\, \mathbf{O}(\hat{\mathbf{k}}_p)\cdot\left(\boldsymbol{P}_w(z;\omega,\mathbf{k}_p)\, e^{i(k_p z-\omega t)} + cc\right) \tag{10.9}$$

where we have expressed the curl-curl operation in the form of Eq. (A.58), and we have ignored all spatial derivatives of $\boldsymbol{P}_w$. The operator $\mathbf{O}(\hat{\mathbf{k}})$ projects out the component of $\boldsymbol{P}_w(z;\omega,\mathbf{k}_p)$ in a plane orthogonal to $\hat{\mathbf{k}}_p$ (see Eq. (1.48) for definition). It is obviously this component which affects the eigenmodes.

We now evaluate the left-hand side of Eq. (10.3). Substituting Eq. (10.4) into (10.3) gives

$$-\nabla^2 D + \frac{n^2}{c^2}\frac{\partial^2 D}{\partial t^2} = \frac{1}{2}\left(\frac{-\partial^2 \boldsymbol{D}}{\partial z^2} - 2ik\frac{\partial \boldsymbol{D}}{\partial z} + \left(k^2 - \frac{n^2\omega^2}{c^2}\right)\boldsymbol{D}\right) e^{i(kz-\omega t)} + cc. \tag{10.10}$$

The terms involving k^2 and $n^2\omega^2/c^2$ are large and must be eliminated completely if any approximation is to be valid. One can either define

$$k = \frac{n\omega}{c} \tag{10.11}$$

or one can regard Eq. (10.11) as following from the assumption that the unperturbed eigenvectors are solutions of the unperturbed equations. If, as Eq. (10.8) asserts, the derivative of the amplitude is small, it follows that $|\partial^2 \boldsymbol{D}/\partial z^2|$ is small relative to $|2ik\partial \boldsymbol{D}/\partial z|$. In dropping the second derivative it is important to recognize that the derivatives do not appear in the equation in the form of absolute values. They are complex numbers, and it is not logically straightforward to assert that if one complex number has a much larger amplitude than another, it follows that one can neglect the real and imaginary parts of the smaller number relative to the real and imaginary parts of the larger. Hence, while it is intuitively plausible to drop the second derivative term in Eq. (10.10) and write the SVEA approximation as

$$-\nabla^2 D + \frac{n^2}{c^2}\frac{\partial^2 D}{\partial t^2} \simeq \frac{1}{2}\left(-2ik\frac{\partial \boldsymbol{D}}{\partial z}\, e^{i(kz-\omega t)}\right) + \text{cc},$$

it is not a logical consequence of the wave equation, and thus the approximation must be used with caution. We now complete the development by substituting this result and Eq. (10.9) into (10.3), which gives

$$-2ik\frac{\partial \boldsymbol{D}}{\partial z} = k_p^2\, \mathbf{O}(\hat{\mathbf{k}}) \cdot \boldsymbol{P}_w\, e^{i(k_p - k)z}. \tag{10.12}$$

The phase match condition discussed in Section 10.4 restrict the cases of interest to those in which $k_p \simeq k$. Hence we can replace k_p^2 by k^2 in the coefficient multiplying the right-hand side of Eq. (10.12). We also define

$$\Delta\mathbf{k} = \mathbf{k}_p - \mathbf{k} = (k_p - k)\hat{\mathbf{z}} = \Delta k\hat{\mathbf{z}}. \tag{10.13}$$

Rewriting Eq. (10.12),

$$\frac{\partial \boldsymbol{D}}{\partial z} = i\,\frac{n\omega}{2c}\, \mathbf{O}(\hat{\mathbf{k}}) \cdot \boldsymbol{P}_w\, e^{i\Delta kz}. \tag{10.14}$$

There are always two orthogonal eigenvectors for D with unit vectors $\hat{\mathbf{e}}_a$ and $\hat{\mathbf{e}}_b$. Therefore the right-hand side of Eq. (10.14) can always be decomposed along the directions $\hat{\mathbf{e}}_a$ and $\hat{\mathbf{e}}_b$, which results in the two eigenvectors D_a and D_b being generated. (The refractive indices n_a and n_b are associated with these two modes.) Thus

$$\frac{\partial D_a}{\partial z} = i\,\frac{n_a\omega}{2c}\, \hat{\mathbf{e}}_a \cdot \mathbf{O}(\hat{\mathbf{k}}) \cdot \boldsymbol{P}_w\, e^{i\Delta k_a z} \tag{10.15}$$

$$\frac{\partial D_b}{\partial z} = i\,\frac{n_b\omega}{2c}\, \hat{\mathbf{e}}_b \cdot \mathbf{O}(\hat{\mathbf{k}}) \cdot \boldsymbol{P}_w\, e^{i\Delta k_b z} \tag{10.16}$$

where $\Delta k_a = k_p - k_a$ and $\Delta k_b = k_p - k_b$ involve using the appropriate refractive index. Since $\boldsymbol{P}_w$ is frequently written as a function of electric fields, it is usually convenient to reexpress Eqs. (10.15) and (10.16) in terms of $\boldsymbol{E}$-fields. Thus from $\boldsymbol{E}_a = \boldsymbol{D}_a/n_a^2\varepsilon_0$ and $\boldsymbol{E}_b = \boldsymbol{D}_b/n_b^2\varepsilon_0$

$$\frac{\partial E_a}{\partial z} = i \frac{\omega}{2n_a \varepsilon_0 c} \hat{e}_a \cdot \mathbf{O}(\hat{k}) \cdot \boldsymbol{P}_w e^{i\Delta k_a z} \tag{10.17}$$

$$\frac{\partial E_b}{\partial z} = i \frac{\omega}{2n_b \varepsilon_0 c} \hat{e}_b \cdot \mathbf{O}(\hat{k}) \cdot \boldsymbol{P}_w e^{i\Delta k_b z}. \tag{10.18}$$

Equations (10.17) and (10.18) correspond to standard usage in the literature. In tricky circumstances, it is best to use Eqs. (10.15) and (10.16) despite the fact that it is necessary to express $\mathbf{P}_w$ in terms of $\mathbf{D}$. We return to this point later.

10.2 A SIMPLE EXAMPLE: ABSORPTION

One of the most straightforward applications of SVEA is in solid and liquid materials in which there is a weak dilutant that absorbs light. In that case it is the imaginary part of the susceptibility (refractive index, etc.) that is responsible for the absorption. We express this as

$$P_w = i\varepsilon_0 \, \text{im}(\chi) \, E, \tag{10.19}$$

and in this case $k_p = k$, since it is the $\mathbf{E}$-field which induces the polarization. Let us further consider an isotropic medium so that linearly polarized waves are eigenvectors and have indices denoted by n. The vector multiplications are straightforward and give

$$\frac{\partial E}{\partial z} = - \frac{\omega}{2nc} \, \text{im}(\chi) \, E.$$

Beer's law of absorption is given by Eq. (4.17) and the corresponding present result is

$$\frac{\partial E}{\partial z} = - \frac{\gamma}{2} E \tag{10.20}$$

where

$$\gamma = \frac{\omega}{nc} \, \text{im}(\chi). \tag{10.21}$$

It is left as an exercise to show that this result is identical to the one in Eq. (4.18).

This case of absorption is a good example of what we mean by the polarization being small or weak. Equation (10.8) requires that

$$\frac{\gamma}{2} \ll k.$$

We use Eqs. (10.21) and (10.11) to write this result as

$$\text{im}(\chi) \ll 2\,n^2,$$

or, since $2n^2 \sim 1$, this inequality is essentially the same as requiring that

$$\text{im}(\chi) \ll 1. \tag{10.22}$$

In the case of magneto-optics, electro-optics, etc., the susceptibilities, or the equivalent field-dependent terms, are less than 10^{-3}. Hence the inequality (10.22) is almost always satisfied. However, in special cases such as when primary absorption bands occur in solids or liquids (e.g., metals, semiconductors), one can encounter large absorptions and the approximations break down. In such cases, SVEA cannot be used.

10.3 ANOTHER EXAMPLE: THE FARADAY EFFECT

We now reexamine the Faraday effect discussed in Chapter 9 from the point of view of the SVEA. This case provides an example in which some of the subtler aspects of SVEA can be exploited. The susceptibility which gives rise to the weak polarization comes from the off-diagonal elements of the tensor given by Eq. (9.16). In terms of the 1,2,3 coordinate system, we have

$$\boldsymbol{P}_{\text{w}} = ig\varepsilon_0(\hat{\mathbf{e}}_1 E_2 - \hat{\mathbf{e}}_2 E_1). \tag{10.23}$$

In order to use Eqs. (10.15) and (10.16) for the SVEA, the preceding equation must be expressed in terms of the $\boldsymbol{D}$ vectors. It is left as an exercise to show that

$$\varepsilon_0\boldsymbol{P}_{\text{w}} = \left(\hat{\mathbf{e}}_1\,(-\,\frac{g^2}{\varepsilon^2 n^2}\,D_1 + \frac{ig}{\varepsilon n^2}\,D_2) - \hat{\mathbf{e}}_2\,(-\,\frac{g^2}{\varepsilon^2 n^2}\,D_2 + \frac{ig}{\varepsilon^2 n^2}\,D_1)\right). \tag{10.24}$$

We consider first the case discussed in Chapter 9 to show that the SVEA approach yields the correct result. The light is assumed to propagate along the direction of the B-field, i.e., the 3 direction and $x \equiv x_1$, $y \equiv x_2$, $z \equiv x_3$. Since the quadratic terms in Eq. (10.24) are negligible, either the E or D form of the SVEA can be used, and we choose the **E**-field form. In this case the eigenvectors are degenerate with $n_a = n_b = n$, and $\Delta k = 0$ (i.e., P_w is induced by **E**). Thus from Eqs. (10.17) and (10.18) we have

$$\frac{\partial E_x}{\partial z} = -\,\gamma B\, E_y \tag{10.25}$$

$$\frac{\partial E_y}{\partial z} = \gamma B\, E_x. \tag{10.26}$$

Taking the derivative of Eq. (10.25) with respect to z and substituting (10.26) gives

$$\frac{\partial^2 E_x}{\partial z^2} = -\,(\gamma B)^2\, E_x,$$

whose solution is a linear combination of $\cos\gamma Bz$ and $\sin\gamma Bz$. Assuming a boundary condition of $E = \hat{e}_x E_x(0)$ at $z = 0$, the solution to Eqs. (10.25) and (10.26) is simply

$$E_x = E_x(0) \cos\gamma Bz$$

$$E_y = E_x(0) \sin\gamma Bz.$$

As expected, this is identical to the result obtained in Chapter 9.

Now let us consider the case where propagation takes place along the 2 direction in Figure 9.1. Then $x_1 \equiv y$, $x_2 \equiv z$, and $x_3 \equiv x$. Here it is most convenient to use the D-form of the SVEA. Since $\mathbf{O}(\hat{k}) \cdot \hat{e}_2 = 0$, the 2 component of Eq. (10.24) is irrelevant. We simplify the problem by assuming that the medium boundaries lie in the (1,3) plane and that we are dealing with normal incidence. Hence $D_2 = 0$ at the boundary (since it is the normal component) and remains zero for all z. The y component of D is obtained from Eq. (10.24) and Eq. (10.15),

$$\frac{\partial D_y}{\partial z} = -\frac{i\omega g^2}{2n^3 c} D_y$$

or

$$D_y = D_y(0)\, e^{-i(\omega g^2/2n^3c)z}. \qquad (10.27)$$

The D_x is unaffected by $\mathbf{P}_w$, since it points in the direction of $\mathbf{B}$. The net result is an induced birefringence which is quadratic in the DC magnetic field. It is left as an exercise to show that this result is identical to that obtained in Problem 9.1. We have encountered difficulties in obtaining this result from the E-field form of the SVEA. Such fine details are rather rare, but it is useful to keep the D form available for special cases.

10.4 RADIATION FROM A PHASED ARRAY OF DIPOLES

In both of the preceding cases, the polarization was induced at the same frequency as the incident field. There are, however, many cases in nonlinear optics and acousto-optics in which the weak polarization generates an optical field at a frequency different from that of the incident field. A test problem for this case is to treat the amplitude of $\mathbf{P}_w$ as a constant over some region of space, for example the interval $0 \leq z \leq L$, and to assume that the wave radiated by P_w is zero at $z = 0$. The general case usually involves only one eigenvector and we use Eq. (10.17) (and drop the subscript a). We define the effective polarization as

$$P_{eff} = \hat{e} \cdot \mathbf{O}(\hat{k}) \cdot P_w. \qquad (10.28)$$

Since all of the terms on the right-hand side of Eq. (10.17) are constant, except for the term $\exp(i\Delta kz)$, we can write

$$\frac{\partial E}{\partial z} = i \frac{\omega}{2nc\varepsilon_0} P_{eff}\, e^{i\Delta kz}$$

so that

$$E(z) = i \frac{\omega}{2nc\varepsilon_0} P_{eff} \int_0^z e^{i\Delta kz'} dz', \tag{10.29}$$

where we have used the boundary condition $E(z=0) = 0$. From this integral

$$E(z) = i \frac{\omega}{2nc\varepsilon_0} P_{eff} \frac{e^{i\Delta kz}-1}{i\Delta k}$$

or

$$E(z) = i \frac{\omega}{4nc\varepsilon_0} P_{eff}\, z\, e^{i\Delta kz/2} \frac{\sin(\Delta kz/2)}{\Delta kz/2}. \tag{10.30}$$

From Eq. (1.30), the time average flux is

$$\overline{\mathbf{S}} = \frac{\omega^2}{32nc\varepsilon_0} |P_{eff}|^2 z^2 \operatorname{sinc}^2 \frac{\Delta kz}{2} \hat{\mathbf{z}}, \tag{10.31}$$

where we use the usual definition for the sinc(x) function, i.e.,

$$\operatorname{sinc}(x) = \frac{\sin x}{x}. \tag{10.32}$$

Except for the case $\Delta k = 0$, the flux varies sinusoidally as illustrated in Figure 10.2. The distance at which the power reaches a maximum is called the coherence length, i.e.,

$$\frac{\Delta k L_{coh}}{2} = \frac{\pi}{2},$$

or

$$L_{coh} = \frac{\pi}{\Delta k}. \tag{10.33}$$

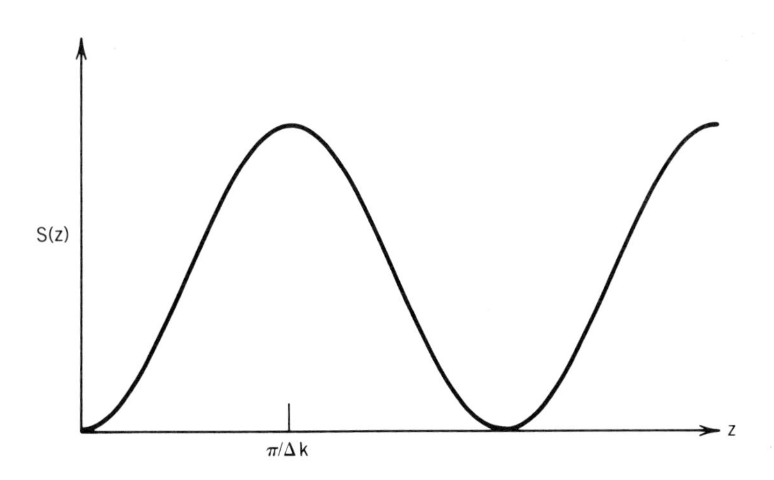

Figure 10.2. Flux as a function of distance for $\Delta k \neq 0$.

For arbitrary Δk, L_{coh} is usually of the order of a few wavelengths of light, and the flux is extremely small. On the other hand, if the interaction is phase matched, which can be loosely defined as

$$\Delta k = 0, \tag{10.34}$$

or more precisely as either

$$L_{coh} \gg L \tag{10.35}$$

or

$$\Delta kL \ll \pi, \tag{10.36}$$

then the radiation can become quite large. In that case, the flux grows as

$$\overline{S} = \frac{\omega^2}{32n^2c\varepsilon_0} |P_{eff}|^2 z^2 . \tag{10.37}$$

The case where Eqs. (10.35) and (10.36) fail to hold is called phase mismatch. The associated sinusoidal flux oscillation is due to interference effects: The light emitted at the position z is exactly out of phase with the light emitted at the position $z + L_{coh}$. (This can most easily be deduced from Eq. (10.29).) Hence, for $L = 2L_{coh}$, for every portion of the medium which emits radiation, there is another portion which emits exactly out of phase with the first section and assures complete destructive interference. We return to this point in subsequent chapters.

10.5 PHASED ARRAYS OF SUSCEPTIBILITY: BRAGG REFLECTION

In this section we show the equivalence of phase matching to Bragg reflection. Let us suppose that the weak polarization arises in the following way:

$$\boldsymbol{P} + \boldsymbol{P}_W = \varepsilon_0\left(\chi\, \mathbf{I} + \frac{1}{2}\, \boldsymbol{\chi}_W(\mathbf{K},\Omega)\, e^{i(\mathbf{K}\cdot\mathbf{r}-\Omega t)} + cc\right) \cdot \boldsymbol{E}_I . \tag{10.38}$$

Here χ is the space-time average susceptibility (taken to be isotropic) and the second term represents a phased array of susceptibility. This type of susceptibility is used to describe the acousto-optic interaction, as well as a variety of nonlinear optics phenomena. For the present case, we simply take Eq. (10.38) as given. The incident field is given in the usual way by

$$\mathbf{E}_I = \frac{1}{2}\left(\boldsymbol{E}_I\, e^{i(\mathbf{k}_I\cdot\mathbf{r}-\omega_I t)} + cc\right) \tag{10.39}$$

and therefore

$$P_W = \frac{\varepsilon_0}{4}\Big(\boldsymbol{E}_I \cdot \boldsymbol{\chi}_W(\mathbf{K},\Omega)\, e^{i(\mathbf{k}_I+\mathbf{K})\cdot\mathbf{r}-i(\omega_I+\Omega)t}$$

$$+\ \boldsymbol{E}_I \cdot \boldsymbol{\chi}_W^*(\mathbf{K},\Omega)\, e^{i(\mathbf{k}_I-\mathbf{K})\cdot\mathbf{r}-i(\omega_I-\Omega)t} + cc\Big). \qquad (10.40)$$

This polarization is now substituted into the SVEA.

The phased array of susceptibility gives rise to two weak polarization sources, each of which can radiate (but may not do so efficiently). Let us choose the second term and discuss it in some detail. Then

$$\boldsymbol{P}_W = \frac{1}{2}\varepsilon_0 \boldsymbol{E}_I \cdot \boldsymbol{\chi}_W^* \qquad (10.41)$$

$$\mathbf{k}_p = \mathbf{k}_I - \mathbf{K} \qquad (10.42)$$

and

$$\omega = \omega_I - \Omega. \qquad (10.43)$$

We explicitly assume that $\omega_I > \Omega$ so that $\omega > 0$. If the opposite were the case ($\Omega > \omega_I$), the complex conjugate from Eq. (10.40) would be taken to define the weak polarization amplitude in Eq. (10.41) (i.e., $1/2\, \varepsilon_0 \boldsymbol{E}_I^* \cdot \boldsymbol{\chi}_W$). From Eq. (10.7), the $\mathbf{k}_p$ in Eq. (10.42) defines the z-direction, i.e., the direction in which the generated field grows. Therefore

$$\Delta\mathbf{k} = \mathbf{k}_I - \mathbf{K} - k\hat{\mathbf{z}}. \qquad (10.44)$$

If we now assume an isotropic medium, all eigenvectors $\boldsymbol{E}$ are degenerate and so we can choose $\hat{\mathbf{e}}$ to be parallel to $\mathbf{O}(\hat{k}) \cdot \boldsymbol{P}_W$. Hence

$$\hat{\mathbf{e}}\, P_{eff} = \mathbf{O}(\hat{k}) \cdot \boldsymbol{P}_W. \qquad (10.45)$$

Equations (10.40) to (10.45) define a phased array of dipoles which radiates according to the principles outlined in the preceding section.

We now consider what the phase match condition $\Delta\mathbf{k} = 0$ means in this context. In Figure 10.3a we show the wave vector and the wavefronts of the phased array of susceptibility. The wave vectors corresponding to phase matching are illustrated in Figure 10.3b. In general, the angles of incidence θ_i, and reflection θ_r drawn in Figure 10.3 are different. However, in the limit appropriate to scattering from sound waves, $\Omega \ll \omega_I$ and $\omega \simeq \omega_I$. From Eq. (10.11) $k \simeq k_I$, which implies that the figure in Figure 10.3b describes an isosceles triangle, i.e.,

$$\theta_i = \theta_r = \theta.$$

If we write $\mathbf{K} = 2\pi/\Lambda$ and $k = 2\pi/\lambda$, then $\Delta\mathbf{k} = 0$, and from elementary geometry

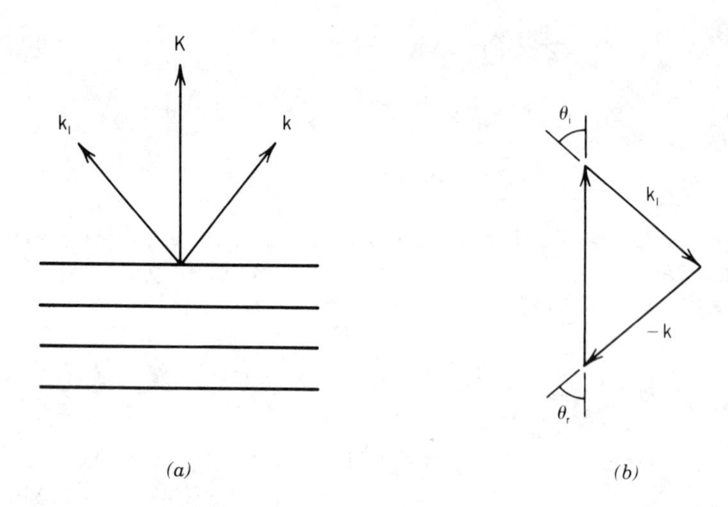

Figure 10.3. a) Wavefronts of the phased array of susceptability. b) Phase matching with incident angle θ_i and reflection angle θ_r.

$$\lambda = 2\Lambda \cos\theta, \tag{10.46}$$

where λ is the optical wavelength in the medium. The conventional Bragg condition is

$$m\lambda = 2\Lambda \cos\theta, \quad m = 1,2,3,\cdots \tag{10.47}$$

and the reason that only $m = 1$ is allowed is that the wave is assumed to be sinusoidal. (Equation (10.47) is valid for an arbitrary grating profile with periodicity Λ.) Therefore, the phase match and Bragg reflection conditions are equivalent. It can also easily be shown that the frequency shift $\omega - \omega_l = \Omega$ is just the Doppler shift when scattering occurs from a moving grating. This is left as an exercise.

ADDITIONAL READING

The slowly varying envelope approximation

Yariv, A., *Quantum Electronics*, Wiley, New York, 1967. pp. 356-360.

Sargent, M., Scully, M.O., and Lamb, W.E. Jr. *Laser Physics*, Addison-Wesley, Reading, MA, 1974. pp. 99-101.

Zernike, F., and Midwinter, J.E. *Applied Nonlinear Optics*, Wiley, New York, 1973. p. 42.

PROBLEMS

10.1. (a) Derive the SVEA approximation for the case where the fields vary slowly both in position and in time. Simplify your considerations to an isotropic medium in which D and $\mathbf{O}(\hat{k}) \cdot \boldsymbol{P}_w$ are parallel and $\Delta k = 0$. Show that

$$\frac{\partial E}{\partial z} + \frac{1}{c}\frac{\partial E}{\partial t} = i\,\frac{\omega}{2nc}\,\hat{\mathbf{e}}\cdot\mathbf{O}(\hat{\mathbf{k}})\cdot\boldsymbol{P}_w. \tag{10.48}$$

(b) Write down and solve the problem of absorption in this case using Eq. (10.19).

10.2. Derive Eqs. (10.25) and (10.26) from the definitions of g and γ in Chapter 9.

10.3. Compare the result of Eq. (10.27) with that of Problem 9.1. Show explicitly that they are the same.

10.4. Express the SVEA in Eqs. (10.17) and (10.18) in the general coordinate system $\mathbf{r} = (x_1,x_2,x_3)$ of Figure 10.1. Show that the equations read

$$\hat{\mathbf{k}}_a\cdot\nabla E_a = \frac{i\omega}{2n_a c\varepsilon_0}\,\hat{\mathbf{e}}_a\cdot\mathbf{O}(\hat{\mathbf{k}})\cdot\boldsymbol{P}_w\, e^{i\Delta\mathbf{k}_a\cdot\mathbf{r}} \tag{10.49}$$

$$\hat{\mathbf{k}}_b\cdot\nabla E_b = \frac{i\omega}{2n_b c\varepsilon_0}\,\hat{\mathbf{e}}_b\cdot\mathbf{O}(\hat{\mathbf{k}})\cdot\boldsymbol{P}_w\, e^{i\Delta\mathbf{k}_b\cdot\mathbf{r}}. \tag{10.50}$$

10.5. Use SVEA to express the propagation of a wave traveling along the optic axis of a KDP crystal in the case where the DC field also points along the optic axis. Verify Eq. (8.51).

10.6. Consider the case where the susceptibility is of the form

$$\chi = \chi + f(\mathbf{K}\cdot\mathbf{r}-\Omega t)$$

for $\Omega \ll \omega_l$ and general f(x). Use the phase match condition to obtain the higher integers in Eq. (10.47). Hint: Make a Fourier expansion of f(x).

10.7. Verify that Eq. (10.21) and Eq. (4.18) are the same.

10.8. (a) Show that Eq. (10.46) follows from $\Delta k = 0$ for the case $k \simeq k_l$.

(b) Derive the Bragg reflection condition in Eq. (10.46) from the conventional arguments using the diagram sketched in Figure 10.4. One requires that $b + c - a = n\lambda$ so that there is constructive interference.

(c) Show that Eq. (10.43) follows from considering the Doppler shift off the grating in Figure 10.4 when it moves at a velocity $v = \Omega/2K$.

10.9. (a) Set $\mathbf{K} \to -\mathbf{K}$ in Eq. (10.38). Derive the equations governing emission.

(b) Show that for $\Omega \ll \omega$, the emitted wave from part (a) goes in the same direction as the one discussed in the text, and that the frequency is upshifted.

(c) Interpret this shift in terms of Doppler shifts.

10.10. Use Eq. (10.29) to show explicitly that a phase mismatch corresponds to an interference of the fields radiated by the polarization wave.

10.11. Obtain Eqs. (5.54) and (5.55) using SVEA techniques.

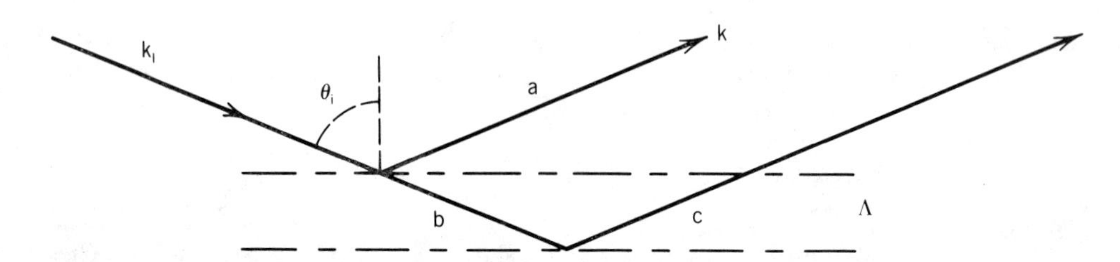

Figure 10.4. Diagram for Problem 10.8. The lines (- • - • -) are the wavefronts.

10.12. Derive Eq. (6.27) by treating the right-hand side of Eq. (6.8) as a weak polarization.

10.13. Solve Problem 8.8 using the SVEA. Simplify the discussion to the crystal class 3m so that you can verify explicitly that optimal performance is achieved for $r_{22}E_1 = \pi/2$. The weak susceptibility in this case is

$$\boldsymbol{\chi}_w = \begin{pmatrix} -r_{22}E\cos\Omega t & r_{22}E\sin\Omega t & 0 \\ r_{22}E\sin\Omega t & r_{22}E\cos\Omega t & 0 \\ 0 & 0 & 0 \end{pmatrix}$$

and the propagation is along the optic axis.

10.14. Consider a dielectric coefficient that has a weak susceptibility of the form

$$\boldsymbol{\chi}_w = \chi_0 \begin{pmatrix} \cos Kz & \sin Kz & 0 \\ \sin Kz & -\cos Kz & 0 \\ 0 & 0 & 0 \end{pmatrix}$$

(χ_0 is a constant) in a medium that is otherwise isotropic (this is the case for some liquid crystals). Use SVEA to compute the propagation of two orthogonal fields propagating in the z direction such that

$$\mathbf{E} = (E_a, E_b, 0)$$

$$\mathbf{P}_w = \varepsilon_0 \boldsymbol{\chi} \cdot \mathbf{E}.$$

10.15. (a) A field $E(z,t)$ can be written in terms of a Fourier amplitude $E(k,\omega)$ using Eq. (1.19). The Fourier amplitude of a pulse traveling in the z direction reads

$$E(k,\omega) = E(\omega)\delta(k_X)\delta(k_Y)\delta(k(\omega) - k_Z),$$

where $k(\omega) = \omega n(\omega)/c$, $k(-\omega) = -k(\omega)$, and $E(-\omega) = E^*(\omega)$. Show that

$$E(z,t) = \int d\omega\, E(\omega)\, e^{i(k(\omega)z-\omega t)} + \text{cc}. \qquad (10.51)$$

(b) Let $E(\omega)$ be a function that is zero except in a region near ω_0. Use this to expand $k(\omega)$ as a Taylor's series. Show that

$$E(z,t) = \left[\int d\omega\, E(\omega)\, e^{i(k'(\omega_0)z-t)(\omega-\omega_0)}\right] e^{i(k_0 z-\omega t)} + \text{cc}. \qquad (10.52)$$

Show that the expression in square brackets describes a field amplitude that moves at the velocity

$$v_g = \frac{1}{k'(\omega_0)} \equiv \left(\frac{dk(\omega)}{d\omega}\Big|_{\omega=\omega_0}\right)^{-1}, \qquad (10.53)$$

where v_g is the group velocity mentioned in Chapter 3. Note that if the amplitude in Eq. (10.48) moves at the velocity v_g, then the flux in Eq. (1.30) (just replace $E(k,\omega)$ by $E(z,t)$) moves at v_g.

10.16. (a) Derive the group velocity in an isotropic medium using SVEA techniques. Write the total polarization amplitude as

$$P(\omega) = \varepsilon_0\chi(\omega_0)E(\omega) + \varepsilon_0(\chi(\omega) - \chi(\omega_0)E(\omega)).$$

The first term is a large polarization that should be used in a rederivation of Eq. (10.48) in which the phasor is $\exp(i(k_0 z - \omega_0 t))$ and $1/c$ is replaced by $n(\omega_0)/c$. Treat the remaining term as a weak polarization. Show that

$$P_W(z,t) \simeq -i\varepsilon_0 \frac{\partial\chi}{\partial\omega}\Big|_{\omega=\omega_0} \frac{\partial E(z,t)}{\partial t}.$$

Write the SVEA equation, and show that this equation is solved by any function $f(t - z/v_g)$. By now you have learned that the SVEA is not always the easiest technique for solving problems.

(b) Expand $k(\omega)$ up to second order in $(\omega-\omega_0)$. The term in $(\omega-\omega_0)^2$ causes a short pulse to broaden as it propagates. (For those who are unfamiliar with this dispersive effect, it is easily derived by using Eq. (10.51) with $k(\omega)$ expanded to second order and $E(\omega)$ in the form of a Gaussian centered at $\omega = \omega_0$.) Attempt to derive this dispersive effect using the SVEA. You will probably be unsuccessful. This is a case in which weak polarizations have observable consequences that are not easily found using the SVEA.

10.17. The SVEA requires an approximation for the second derivative of the weak polarization. It is easiest to consider this approximation in the derivation of Eqs. (10.17) and (10.18) from Eq. (1.15). Assume medium isotropy and a P_W that is parallel to E for simplicity (i.e., $\nabla \cdot E = 0$). Write $P_W = \varepsilon_0 \chi_W E$. Let E describe an optical pulse in the form used in the previous two problems. Show that the approximation

$$\frac{\partial^2 P}{\partial t^2} \simeq -\omega^2 P$$

is valid if

$$\chi_W \ll 1.$$

Hint: This comes from requiring that the next set of terms in the expansion of the right- and left-hand sides of Eq. (1.15) must be of the same order. This requires that

$$\left|\frac{\partial^2 E}{\partial t^2}\right| \sim \omega\, \mu_0 \left|\frac{\partial P_W}{\partial t}\right|.$$

Then use a modification of Eq. (10.8) that reads

$$\left|\frac{\partial^2 E}{\partial t^2}\right| \ll \omega \left|\frac{\partial E}{\partial t}\right|$$

to obtain the result.

10.18. Suppose there are no strong polarization terms. Show that corrections due to local fields can be neglected. Note that the contributions of weak polarizations are inherently negligible relative to the strong polarizations. Hence in cases in which the SVEA is valid, the local field correction can only effect phenomenological coefficients.

10.19. Develop an SVEA for the amplitude of a normal coordinate driven by a force with a slowly-varying amplitudes $\mathcal{F}_\alpha(t)$. Use the approximation in Eq. (A.70) to obtain

$$\frac{\partial Q_\alpha}{\partial t} + (\frac{\gamma}{2} + i(\omega_\alpha - \omega)Q_\alpha = i\,\frac{\mathcal{F}_\alpha}{\omega m}, \tag{10.54}$$

in which case the polarization amplitude reads (homogeneous broadening only)

$$\frac{\partial \mathcal{P}_\alpha}{\partial t} + (\frac{\gamma}{2} + i(\omega_\alpha - \omega))\mathcal{P}_\alpha = i\langle\frac{\ell_\alpha \mathcal{F}_\alpha}{\omega m}\rangle. \tag{10.55}$$

11
Bulk Acousto-Optics

This chapter deals with the scattering of light by sound waves generated by sonic transducers. We concentrate on the case in which a sound wave sinusoidally modulates the refractive index inside a material so that the incident light encounters a bulk grating. This interaction scatters light much more efficiently than does a sound wave on the surface (called ripple scattering) and is the one most commonly encountered in applications.

The scattering of light by sound waves is essentially the same process as the diffraction of light by a grating. For example, a sound wave travelling along a surface generates a structure that differs from a grating ruled on the surface only insofar as it is moving. Since the velocity of light is about 10^5 times larger than that of sound, the sound wave is effectively stationary when traversed by the light. The small acoustic velocity manifests itself in a Doppler shift of the scattered light, which changes the optical frequency by less than 1 part in 10^5. The light is scattered by the surface wave into diffraction orders, both on reflection and on transmission, just as it is with a ruled grating, as indicated in Figure 11.1a. The scattering of a sound wave due to a bulk grating is similar to scattering on a surface except that one has to take explicit account of the transverse dimension of the grating. As indicated in Figures 11.1b and 11.1c, the sound wave appears as a distributed, or thick, grating to the light and some of the light is deflected into diffraction orders. How many orders appear is determined by the Raman-Nath parameter (defined later) which separates the Raman-Nath regime (multiple diffraction orders as illustrated in Figure 11.1b) from the Bragg regime (a single deflected beam as illustrated in Figure 11.1c). Most devices operate in the Bragg limit, which is discussed at length in Chapter 10, since it provides the most efficient light deflection.

11.1 WEAK POLARIZATION

In the spirit of the preceding chapters on electro-optics and magneto-optics, the polarization can be written as a product of acoustic and optical fields. In general

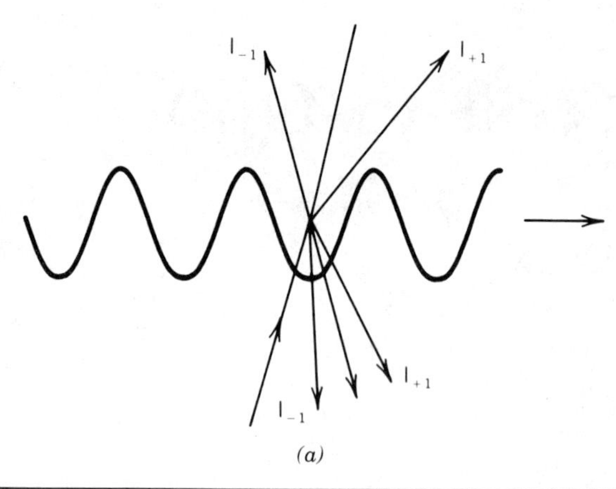

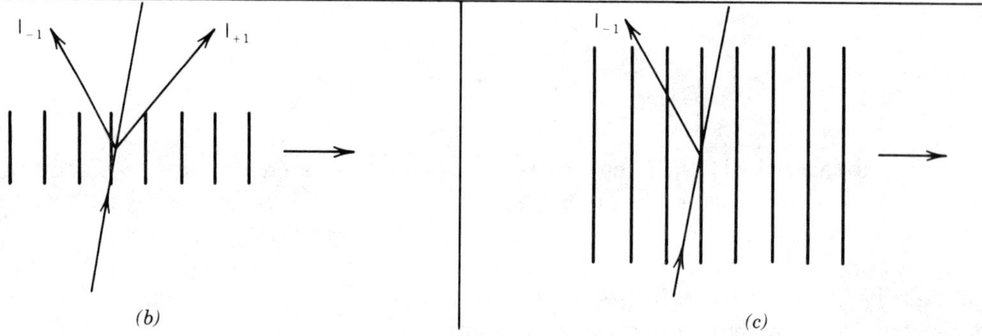

Figure 11.1. Deflection of optical beams by sound waves: a) deflection at surface; b) and c) deflection in bulk. b) Raman-Nath regime, transverse dimensions of sound wave small, interaction length short, more than one scattered wave. c) Bragg regime, transverse dimensions of sound wave large, interaction length long, one scattered wave.

$$P_i = -\frac{1}{\varepsilon_0} \sum_{jkl} \varepsilon_{ii}\, \varepsilon_{jj}\, p_{ijkl}\, S_{kl}\, E_j \,, \tag{11.1}$$

where P_i is the i'th component of the weak polarization induced via the elasto-optic interaction. The electromagnetic field **E** couples to the acoustic strain tensor **S** via the elasto-optic tensor coefficients p_{ijkl}. The indices i,j refer to optical fields and are clearly interchangeable. As we show later, the same is true for the k,l indices and therefore, in the Voigt notation, the tensor can be written as $P_{\mu,\nu}$ ($\mu = 1,\cdots,6$, $\nu = 1,\cdots,6$). This tensor has elements which depend on the symmetry properties of the bulk medium. For an isotropic solid

$$\begin{gathered} p_{11} = p_{22} = p_{33} \neq 0 \\ p_{12} = p_{21} = p_{31} = p_{13} = p_{23} = p_{32} \neq 0 \\ p_{44} = p_{55} = p_{66} = (p_{11}-p_{12})/2 \end{gathered} \tag{11.2}$$

and for a liquid

$$p_{11} = p_{22} = p_{33} = p_{12} = p_{21} = p_{31} = p_{13} = p_{23} = p_{32} \neq 0$$

$$p_{44} = p_{55} = p_{66} = 0. \tag{11.3}$$

In principle, one can now take the polarization in Eq. (11.1), which is weak under the definitions of Chapter 10, and use the SVEA to calculate the radiated fields. Before carrying out this exercise, we examine the nature of strain fields in a medium.

11.1.1 Acoustic Strain Fields

The problem of collective normal modes, i.e., sound waves, is discussed in Chapter 2. There the motion of a linear chain of atoms is analyzed. The development in Chapter 2 is confined to longitudinal compression of a one-dimensional lattice. When generalized to three dimensions, such motions are called longitudinal waves, and they are illustrated in Figure 11.2a. In a three-dimensional lattice there are two transverse oscillations, which are referred to

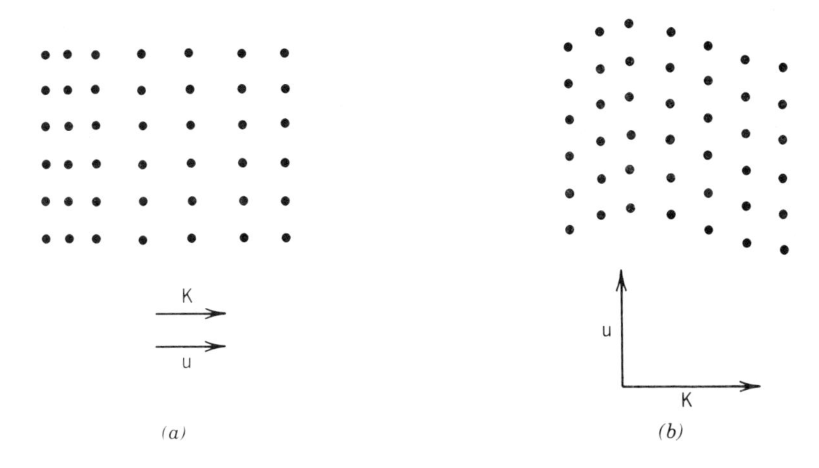

Figure 11.2. a) Longitudinal wave, $\mathbf{K} \times \mathbf{u} = 0$. b) Shear wave, $\mathbf{K} \cdot \mathbf{u} = 0$.

as shear waves, one of which is illustrated in Figure 11.2b. Many solids have normal modes that are not purely longitudinal or purely shear, in which case the one most like the ideal cases is referred to as quasi-longitudinal and quasi-shear. While we write down formulae that apply in general, we emphasize throughout the chapter the case of pure longitudinal waves, which is the easiest to deal with and is also the one that interacts most strongly with the optical field.

Optical fields cannot interact with sound waves whose wavelengths are much shorter than the wavelength of light. Hence, for the sound waves of interest in acousto-optics, there are many atoms per acoustic wavelength and we need not be concerned with the details of the sound wave on atomic scales. Instead, we discuss the continuum limit and consider the displacement of volume elements that are small on the scale of an acoustic wavelength and are large on atomic scales. The coordinates used to describe the sound wave are illustrated in Figure 11.3. The position of the volume element **r** replaces the position coordinate R_α

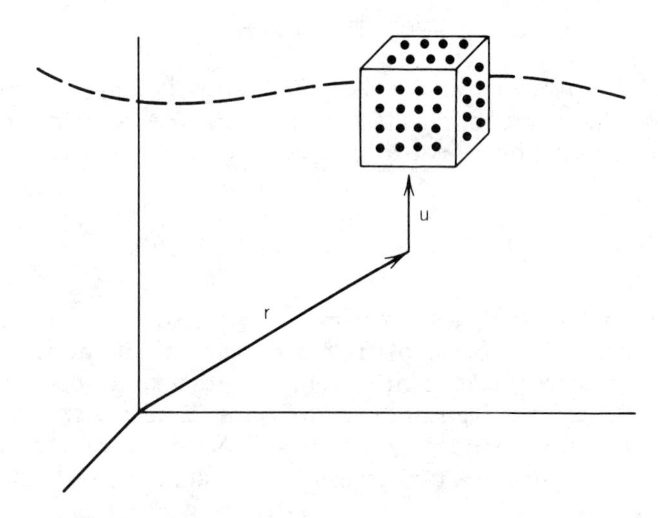

Figure 11.3. Displacement of medium in the continuum limit; **u** is displacement at position **r**.

in Eq. (2.32). The displacement of the volume element from its equilibrium position (**u** in Figure 11.3) replaces the parameter $\mathbf{r}_\alpha$ in Eq. (2.32). As a result, the displacement field for a single monochromatic acoustic normal mode can be written as

$$\mathbf{u}(\mathbf{r},t) = \frac{1}{2}\left(\boldsymbol{U}_{+}(\mathbf{K},\Omega)\, e^{i[\mathbf{K}\cdot\mathbf{r}-\Omega t]} + \boldsymbol{U}_{-}(\mathbf{K},\Omega)\, e^{i[-\mathbf{K}\cdot\mathbf{r}-\Omega t]} + cc\right), \quad (11.4)$$

where K and Ω have replaced k_β and ω_β in Eq. (2.32), and the amplitudes of the two oppositely propagating waves have been denoted as U_+ and U_-. If the sonic field contains multiple wave vector (**K**) and frequency components (Ω), the total field can be written either as a sum of amplitudes or as an integral over Fourier amplitudes (see, e.g., Eqs. (1.19) and (1.20)) of normal modes of appropriate **K** and Ω.

As noted above, in a three-dimensional solid there are three orthogonal modes for a given propagation direction, and in the general case, two are quasi-shear and one is quasi-longitudinal. In general, the displacement fields can be relatively complicated. For most cases of interest in acousto-optics, either the materials are isotropic (glasses, etc.) or the acoustic propagation takes place in symmetry planes and symmetry directions. As a result, the acoustic modes are usually pure shear or pure longitudinal waves, defined by $\nabla\cdot\mathbf{u} = 0$ and $\nabla\times\mathbf{u} = 0$, respectively. As illustrated in Figure 11.2b, shear waves have $\mathbf{K}\cdot\mathbf{u} = 0$, and the displacements lie in a plane orthogonal to the acoustic wave vector **K**. There are two such orthogonally polarized modes. For longitudinal waves as illustrated in Figure 11.2a, $\mathbf{K}\times\mathbf{u} = 0$, and the displacements are polarized along the propagation direction. The most common acousto-optic devices utilize longitudinal waves.

For acousto-optic applications, a transducer is used to generate the acoustic field. This means that a unidirectional wave is generated. The oppositely directed component may also occur via reflections at the end of the sample, but one normally tries to minimize these. For pure shear waves propagating along the i'th axis, and polarized along the j'th direction, the acoustic velocity is given by

$$v_T^2 = \frac{C_{ijij}}{\rho} \equiv \frac{C_{\mu\mu}}{\rho}, \quad \mu = 4,5, \text{ or } 6, \tag{11.5a}$$

where ρ is the mass density and $C_{\mu,\mu}$ is the appropriate elastic constant (C_{44} for isotropic media). For longitudinal waves propagating along the i'th axis

$$v_L^2 = \frac{C_{iiii}}{\rho} \equiv \frac{C_{\mu\mu}}{\rho}, \quad \mu = 1,2, \text{ or } 3. \tag{11.5b}$$

The elastic constant is C_{11} is appropriate for isotropic media. Typical values of v used in acousto-optic devices are listed in Table 11.1.

Table 11.1. Typical Values of Acoustic Velocity and the Acousto-Optic Figure of Merit M_2 for a Number of Materials

Material	n	v(m/s)	M_2
Fused quartz	1.46	5.95×10^3	1.500×10^{-21}
$As_2 S_3$	2.61	2.60×10^3	433.000×10^{-21}
YAG	1.83	8.60×10^3	0.073×10^{-21}
GaP	3.31	6.30×10^3	44.600×10^{-21}
α-Al_2O_3	1.76	11.20×10^3	0.340×10^{-21}

Source: Damon, R.W., Maloney, W.T. and McMahon, D.H. Interaction of Light with Ultrasound, in Mason, W.P. and Thurston, R.N. eds., *Physical Acoustics*, Academic Press, New York, 1970. p.308.
Note: M_2 defined by Eq. (11.44). Longitudinal acoustic waves and He-Ne laser radiation are assumed.

The acoustic displacement field $\mathbf{u}(\mathbf{r},t)$ is an awkward quantity to work with when dealing with the acousto-optic effect. The quantity of interest is usually how much light deflection one obtains per unit of applied acoustic power. The kinetic energy per unit volume of a displacement volume element is

$$T = \frac{1}{2} \rho \left| \frac{\partial \mathbf{u}(\mathbf{r},t)}{\partial t} \right|^2,$$

whose time average reads

$$T = \frac{1}{4} \Omega^2 \rho \left(|U_+|^2 + |U_-|^2 \right). \tag{11.6}$$

Since the displacement motion is essentially that of a simple harmonic oscillator, the time average potential and kinetic energies are equal and the total acoustic energy density is

$$U_{sound} = \frac{1}{2} \rho \Omega^2 \left(|U_+|^2 + |U_-|^2 \right). \tag{11.7}$$

We now consider a wave traveling in one direction with a velocity v so that the amplitude of the acoustic Poynting vector is

$$S_{sound} = \frac{\rho \Omega^2 v}{2} |U_+|^2, \tag{11.8}$$

which is the required relationship between amplitude and sonic flux (in units of power/unit area).

As indicated in Eq. (11.1), it is the acoustic strain field which enters into acousto-optic coupling. The stress tensor is defined as

$$S_{kl} = \frac{1}{2} \left(\frac{\partial u_k}{\partial x_l} + \frac{\partial u_l}{\partial x_k} \right). \tag{11.9}$$

This is a symmetric tensor and the indices k,l are equivalent and can be contracted via the Voigt notation so that we can denote $\mathbf{S}$ as S_μ. For scattering from longitudinal waves, the pertinent strain terms are S_1, S_2, and S_3. These terms are simply the three orthogonal contributions to $\nabla \cdot \mathbf{u}$ which is shown below to give electrostrictive coupling of the optical and sound fields. If shear waves couple at all, this coupling takes place through S_4, S_5, and S_6, which are related to $\nabla \times \mathbf{u}$.

11.1.2 Electrostrictive Coupling--Longitudinal Waves

The simplest mechanism for coupling of sound to light is to use the changes in density that occur as a sound wave traverses a medium. The term electrostriction refers to the change in refractive index with density. This case can be expressed in terms of the appropriate elasto-optic coefficients via Eq. (11.1). In this section we analyze this case from first principles and actually evaluate the pertinent elasto-optic coefficients for isotropic media.

We start by obtaining an expression for changes in density in terms of the acoustic displacement fields. We consider a volume element defined by three orthogonal vectors **a**, **b**, and **c** as shown in Figure 11.4. The volume is given by

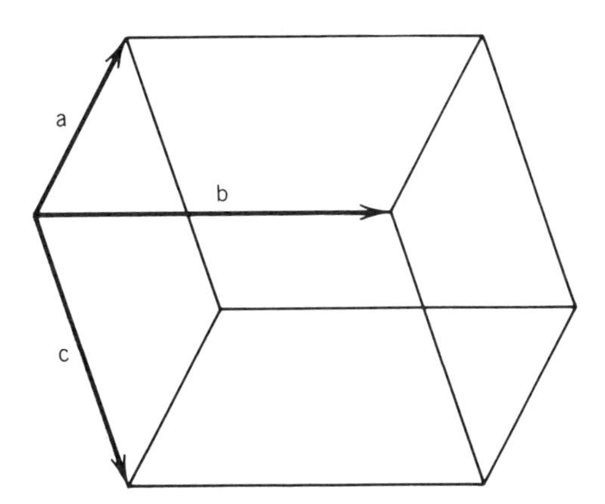

Figure 11.4. Vectors describing deformed volume element.

$$V = \mathbf{a}\cdot(\mathbf{b}\times\mathbf{c}).$$

The sound wave changes the vectors **a**, **b**, and **c** because it displaces the volume element nonuniformly as indicated in Figure 11.2a. Concentrating on a first,

$$\mathbf{a}(\mathbf{r}) = \mathbf{a} + [\mathrm{u}(\mathbf{r}+\mathrm{a}) - \mathrm{u}(\mathbf{r})], \tag{11.10}$$

where **a** is the vector describing the equilibrium length of the a'th edge of the volume and the vector $\mathrm{u}(\mathbf{r}+\mathbf{a})-\mathbf{u}(\mathbf{r})$ describes the amount by which **a** is changed by the sound wave. The displacement can itself be expanded in a Taylor's series as

$$\mathbf{u}(\mathbf{r}+\mathbf{r}') = \mathbf{u}(\mathbf{r}) + \mathrm{r}'\cdot\nabla\mathbf{u}(\mathbf{r},t). \tag{11.11}$$

Therefore

$$\mathbf{a}(\mathbf{r}) = \mathbf{a} + (\mathbf{a}\cdot\nabla)\,\mathbf{u}(\mathbf{r},t). \tag{11.12}$$

Including effects of the acoustic wave on **b** and **c** as well gives

$$V(\mathbf{r}) = [\mathbf{a} + (\mathbf{a}\cdot\nabla)\mathbf{u}(\mathbf{r},t)]\cdot[\mathrm{b} + (\mathbf{b}\cdot\nabla)\mathrm{u}(\mathbf{r},t)]\times[\mathrm{c} + (\mathbf{c}\cdot\nabla)\mathrm{u}(\mathbf{r},t)]$$

$$V(\mathbf{r}) \simeq \mathbf{a}\cdot(\mathbf{b}\times\mathbf{c}) + (\mathbf{a}\cdot\nabla)\mathrm{u}\cdot(\mathbf{b}\times\mathbf{c}) + (\mathbf{b}\cdot\nabla)\mathrm{u}\cdot(\mathbf{c}\times\mathbf{a}) + (\mathbf{c}\cdot\nabla)\mathrm{u}\cdot(\mathrm{a}\times\mathbf{b}). \tag{11.13}$$

The sum of the second, third and fourth terms in Eq. (1.13) describes the change in volume caused by the acoustic displacement **u**. For an isotropic liquid medium **a**, **b**, and **c** are orthogonal and of equal magnitude and

$$V(\mathbf{r}) = V(1+\nabla\cdot\mathrm{u}), \tag{11.14}$$

which is the required result.

We can now calculate the effective susceptibility which gives rise to the weak polarization. The number density of a solid (or liquid) is the number of particles per unit volume, or 1/(volume per particle). Therefore, in the presence of the sound wave,

$$N(\mathbf{r}) = \frac{1}{V(\mathbf{r})} \simeq N(1-\nabla\cdot\mathbf{u}). \tag{11.15}$$

The term $\nabla\cdot\mathbf{u}$ constitutes a "weak" perturbation on the number density. Let us write the total susceptibility as

$$\chi(\mathbf{r}) = N(\mathbf{r})\alpha,$$

where the scalar α (which we should denote as $\langle\alpha\rangle$, but do not for notational simplicity) is the mean molecular polarizability. Then

$$\chi_w = -N\alpha\nabla\cdot\mathbf{u} = -\chi\nabla\cdot\mathbf{u}. \tag{11.16}$$

The corresponding weak polarization is

$$\boldsymbol{P}_w = -\varepsilon_0\,\chi\nabla\cdot\mathbf{u}\,\boldsymbol{E}. \tag{11.17}$$

Since $\nabla\cdot\mathbf{u} = 0$ for shear waves, these sound waves do not couple to light via density fluctuations, but they can couple in other ways. Note that the sound wave written in Eq. (11.4) leads through Eqs. (11.16) and (11.17) to a phased array of susceptability as defined in Eq. (10.38). For the longitudinal wave U_+, this phased array is a scalar and reads

$$\chi_w = -\frac{i}{2}N\alpha\left(KU_+(K,\Omega)e^{i(Kz-\Omega t)} - iKU_+(K,\Omega)e^{i(-Kz+\Omega t)} + cc\right). \tag{11.18}$$

Note that for each sound wave there are two amplitudes for the phased array of susceptibility. It is left as an exercise to show that in the Bragg limit, only one of these is phase matched in any one particular geometry. Using $N\alpha = \chi = n^2-1$ allows us to write the amplitudes as

$$\chi_w(K,\Omega)_\pm = \mp\, i\,(n^2-1)\,K\,U_+(K,\Omega). \tag{11.19}$$

This is the final result for the electrostrictive contribution to the weak polarization. From a standpoint of a qualitative understanding of acousto-optic applications, one can work with this contribution alone. In real cases, there may be important terms originating from other contributions to the elasto-optical effect (i.e., effects that do not require density changes, but rather arise directly from the strain fields in the medium), which are discussed next.

11.1.3 Elasto-Optic Effect

In the preceding section we discussed the electrostrictive effect which leads to scattering by longitudinal sound waves in isotropic liquids via density fluctuations. In this section we look at all of the couplings allowed by Eq. (11.1) in both liquids and solids. To simplify the analysis it is assumed that the acoustic waves propagate along the z-axis. Other geometries follow by interchanges of axes.

Consider first scattering from longitudinal waves. In this case, only S_1 is nonzero and substituting into Eq. (11.1) gives

$$\chi_W = -i\ \varepsilon_0\ n^4\ p_{13}\ Ku_3. \tag{11.20}$$

Here n is the refractive index associated with the incident and scattered fields. Note that Eq. (11.19) implies the symmetries stated explicitly in Eq. (11.3), and one can use any of the nonzero p's to define χ_W. In anisotropic crystals there is, in general, a difference between the refractive indices of the incident and scattered light. However, since the scattering angles are usually small, the refractive index differences are also small and a single average value is a good approximation in Eq. (11.20). For a liquid, $p_{44} = 0$ and from Eqs. (11.19) and (11.20)

$$p_{13} = \frac{n^2-1}{n^4}. \tag{11.21}$$

Typically, $n \simeq 1.5$ and $p_{13} \simeq 0.25$.

The case of solids is slightly different. Compressional stress applied in one direction is not uniformly distributed in all directions as it is in a liquid. As a result, $p_{12} = p_{11} - 2p_{44}$ and $p_{11} \neq p_{12}$. Since p_{44} is usually small, p_{11} and p_{12} are comparable in magnitude and Eq. (11.21) provides a very useful approximation for their magnitudes.

Scattering from shear waves occurs only in solids. For acoustic propagation along the z axis, there are two possible polarizations for the shear wave, i.e., along the x and y axes. The pertinent stress fields are $S_5 = iKu_1$ and $S_4 = iKu_2$, and the possible couplings are summarized by the following nonzero elements:

$$\chi_{W,13} = -\ i\ \varepsilon_0\ p_{55}\ Ku_1 \tag{11.22}$$

and

$$\chi_{W,23} = -\ i\ \varepsilon_0\ p_{44}\ Ku_2. \tag{11.23}$$

Note that the indices on χ_W are not in Voigt notation, and that the Voigt indices on the p's imply symmetry of $\boldsymbol{\chi}_W$. It is clear from Eqs. (11.22) and (11.23) that scattering from shear waves involves a change in the polarization of the optical fields. This is called depolarized scattering.

11.2 DEFLECTED OPTICAL FIELDS

We now consider how to calculate the fields scattered by the sound wave. In the previous section we obtained expressions for the phased array of susceptibility, and we now need to use Eq. (10.40) to develop the corresponding weak polarization. If one attempts to develop expressions that are completely general, one needs a notation that deals with electric field polarizations, multiple scattered fields, shear and longitudinal sound waves, and the two susceptibility amplitudes in Eq. (11.19). Instead we sketch out the elements of the construction of $\mathbf{P}_W$, but confine actual constructions to a few interesting cases. We then use the SVEA to evaluate the scattered fields. First we examine the case of weak scattering in order to obtain the Raman-Nath parameter, which separates the Raman-Nath and Bragg regimes. The deflection of light by sound waves is discussed in these two limits, but strong scattering is discussed only in the Bragg limit.

11.2.1 General Treatment--Scattering Regimes

As noted in Eqs. (11.18) and (11.19), a sound wave of the form of Eq. (11.4) gives two contributions to $\mathbf{P}_W$, even when $U_- = 0$, which we take to be the case here. In general it is necessary to keep both terms. Using Eqs. (10.38) and (11.1) one finds that all contributions to the amplitude are of the form

$$\chi_{ij,k3,\pm} = \mp i \frac{\varepsilon_{ii}\varepsilon_{jj}}{\varepsilon_0^2} p_{ij,k3} K\, U_{+,k}(\mathbf{K},\Omega), \tag{11.24}$$

where the overall weak susceptibility is a sum over k, which denotes the three possible orthogonal components of the sound wave $\mathbf{U}_{+,k}(\mathbf{K},\Omega)$ whose components are denoted $U_{+,k}(\mathbf{K},\Omega)$. Note that one must be careful in this summation, since the components of the sound wave are not necessarily degenerate (i.e., for one K there may be more than one Ω), and hence there may be more than one weak susceptibility amplitude in Eq. (11.24).

We denote the incident and scattered fields as

$$\mathbf{E}_I = \frac{1}{2}\, \mathcal{E}_I\, e^{i(\mathbf{k}_I\cdot\mathbf{r} - \omega_I t)} + cc \tag{11.25}$$

and

$$\mathbf{E}_S = \frac{1}{2}\, \mathcal{E}_S\, e^{i(\mathbf{k}_S\cdot\mathbf{r} - \omega_S t)} + cc \;. \tag{11.26}$$

The configuration for the interacting geometry is sketched in Figure 11.5a. In this configuration we have an incident wave vector

$$\mathbf{k}_I = k_I(\cos\theta, 0, -\sin\theta). \tag{11.27}$$

The vector properties of the weak polarization depend on the polarization of the incident light field. It is conventional to use the term s-polarized in the case when the electric field is polarized normal to the $\mathbf{k}_I,\mathbf{K}$ plane, and the term p-polarized in the case when the electric field vector lies in the $\mathbf{k}_I$,K plane. These configurations are illustrated in Figure 11.5b. Thus

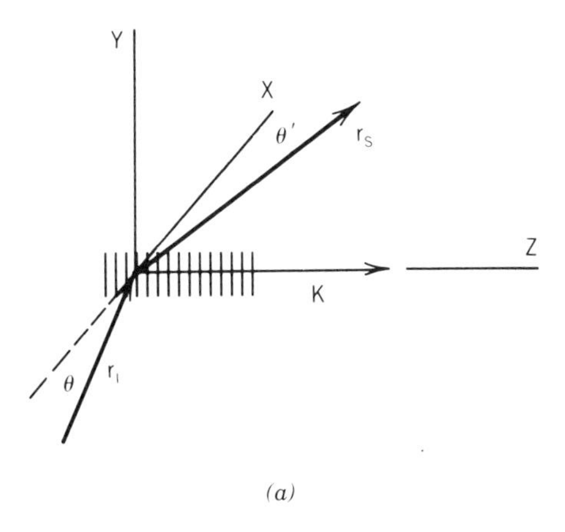

(a)

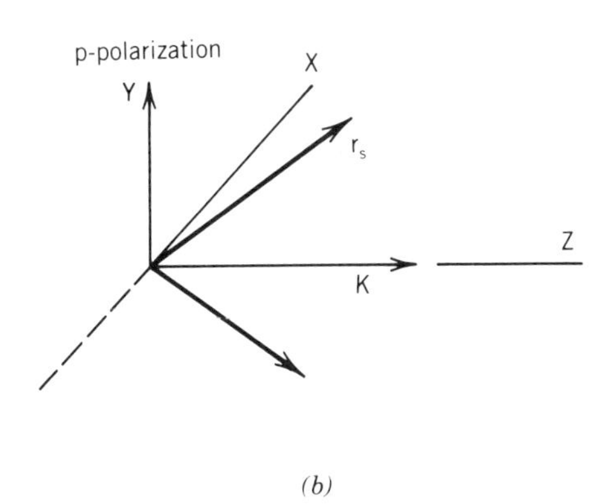

(b)

Figure 11.5. a) Scattering geometry for acousto-optics. b) Illustration of s- and p-polarization.

$$\boldsymbol{E}_I = E_I(0,1,0) \tag{11.28}$$

for s-polarized incident light, and

$$\boldsymbol{E}_I = E_I\,(\sin\theta,0,\cos\theta) \tag{11.29}$$

for p-polarized light. Similarly, the scattered fields are characterized by

$$\mathbf{k}_S = k_S(\cos\theta',0,\sin\theta') \tag{11.30}$$

and by

$$\boldsymbol{E}_S = E_S\ (0,1,0) \tag{11.31}$$

and

$$\boldsymbol{E}_S = E_S\ (-\sin\theta',0,\cos\theta') \tag{11.32}$$

for s- and p-polarized waves, respectively. The vector components of $\mathbf{k}_S$ are related to $\mathbf{k}_I$ and $\mathbf{K}$ through Eq. (10.7), which determines the direction of scattering. In the limit of small angles, it is the z component of $\mathbf{k}_p = \mathbf{k}_I \pm \mathbf{K}$ that determines the direction of scattering. It is left as an exercise to show that

$$k_S \sin\theta' = -k_I \sin\theta \pm K, \tag{11.33}$$

where the $\pm K$ terms refer to the frequencies $\omega_I \pm \Omega$. The SVEA in the form of Eqs. (10.17) and (10.18) reads

$$\frac{\partial E_S}{\partial x} = \frac{i\omega}{2n\varepsilon_0 \cos\theta'}\ \hat{\mathbf{e}} \cdot \mathbf{O}(\mathbf{k}) \cdot \boldsymbol{P}_W\ e^{i\Delta kx}, \tag{11.34}$$

where the polarization under consideration is left unspecified for the moment. The term $\cos\theta'$ in the denominator comes from the fact that the light wave is propagating at an angle with respect to the x direction, and is discussed in detail in Section 11.2.3. It is left as an exercise to make the indicated substitutions to obtain

$$\frac{\partial E_S}{\partial x} = i\,H\,E_I e^{i\Delta kx} \tag{11.35}$$

with

$$H = \frac{\omega n^3\, p_{eff}\, KU_+}{4c\cos\theta'}. \tag{11.36}$$

The term p_{eff} stands for the sum over all p's and the associated angular terms that arise from the sums in Eq. (11.1). Assuming that the refractive index is almost equal for all of the possible incident and scattered light polarizations, we have tabulated p_{eff} in Table 11.2 for a variety of acousto-optic interactions (in the limit of small scattering angles). It is evident from this table that scattering by longitudinal waves is polarized, and scattering from shear waves involves a change in the state of polarization.

Whether a useful fraction of the light is deflected by the sound wave depends on the magnitude of the phase match term $\Delta\mathbf{k}$. The z component of the phase match condition, Δk_z, is zero by virtue of Eq. (11.33), which determines the direction of scattering. Hence the phase match is governed by the x component of the phase match, i.e., $\Delta k = k_I \cos\theta - k_S \cos\theta'$. To evaluate this term, we note that Doppler shifts are small, and hence $k_S \simeq k_I = k$. This allows us to write

$$k\Delta k(\cos\theta + \cos\theta') = k^2(\cos\theta^2 - \cos\theta'^2),$$

Table 11.2. Formulae for p_{eff}.

Mode	s→s	p→p	p→s	s→p
Longitudinal	p_{12}	$-p_{12}\sin\theta\sin\theta' + p_{11}\cos\theta\cos\theta'$	0	0
Shear, u_x	0	$p_{44}(\cos\theta\sin\theta' - \cos\theta'\sin\theta)$	0	0
Shear, u_y	0	0	$p_{44}\cos\theta$	$p_{44}\cos\theta'$

Note: The sound wave propagates along the z axis.

which can be rewritten using Eq. (11.33) as

$$k_I^2\cos^2\theta - k_S^2\cos^2\theta' = k_I^2 - k_I^2\sin^2\theta\, k_S^2 + (-k_I\sin(\theta \pm K))^2. \tag{11.37}$$

This gives

$$\Delta k = + \frac{2K}{\cos\theta + \cos\theta'}\left(\mp\sin\theta + \frac{K}{2k_I}\right). \tag{11.38}$$

Integrating Eq. (11.35) over the width (L) of the acoustic beam, and assuming that $E(z=0) = 0$,

$$E(L) = \frac{iHE_I}{i\Delta k}\left(e^{i\Delta kL} - 1\right). \tag{11.39}$$

This can be rewritten as

$$\frac{E(L)}{E_I} = i\,HL\, e^{i\Delta kL/2}\, \mathrm{sinc}(\Delta kL/2), \tag{11.40}$$

which is the desired result.

The deflection efficiency for small ΔkL can be obtained from Eq. (11.40). Noting that the maximum ratio of deflected to incident light intensity is

$$\frac{S(L)}{S_I} = \left|\frac{E(L)}{E_I}\right|^2,$$

then

$$\frac{S(L)}{S_I} = H^2L^2 \tag{11.41}$$

or

$$\frac{S(L)}{S_I} = \frac{1}{16}\left(\frac{\omega}{c}\right)^2 \frac{n^6(KL)^2}{\cos^2\theta'} |U_+|^2 p_{eff}^2, \tag{11.42}$$

where p_{eff}, the effective elasto-optic coefficient, is tabulated in Table 11.2. Since we are usually interested in the amount of acoustic flux required to produce a deflection, substituting for $|U_+|^2$ from Eq. (11.8)

$$\frac{S(L)}{S_I} = \frac{1}{8}\frac{(k_vL)^2}{\cos^2\theta'}\frac{p_{eff}^2n^6}{\rho v^3} S_{sound} \tag{11.43}$$

where k_v is the vacuum wave vector of light of frequency ω, i.e., $k_v = \omega/c$. The quantity

$$M_2 = \frac{p_{eff}^2n^6}{\rho v^3} \tag{11.44}$$

contains only material properties and is usually defined as the figure of merit for the material. Values for M_2 are listed in Table 11.1.

11.2.2 Raman-Nath Regime

The variation in scattered light intensity with angle of incidence depends on the parameters of the scattering geometry and the ratio of acoustic to optical wave vectors. The pertinent parameter is the term

$$\frac{\sin^2(\Delta kL/2)}{(\Delta kL/2)^2}$$

in Eq. (11.40), which determines the maximum amount of power in the scattered field under non-phase matched operation. From Eq. (11.38) one sees that maximum deflection efficiency occurs when $\sin\theta = \pm K/2k_I$ where the $\pm$ refer explicitly to the scattered light frequencies $\omega_I \pm \Omega$. The pertinent question here is the dependence of the scattering efficiency on angle θ. If the scattered efficiency depends weakly on angle, i.e., if all fields are scattered under non-phase matched conditions, then one is working in the Raman-Nath regime. If, on the other hand, the phase match condition is the limiting factor, in which case at most one scattered field is phase matched, then one is working in the Bragg regime. A test for which regime one is in can be found from examining the phase match condition near $\theta = 0$ (normal incidence onto the acoustic beam). From Eq. (11.38) the phase match condition $\Delta kL/2 < 1$ can be written as

$$Q \leqslant 1, \tag{11.45}$$

where the Raman-Nath parameter Q is

$$Q = \frac{LK^2}{4k_I}. \tag{11.46}$$

For $Q \ll 1$, the scattered light efficiency is essentially independent of the angle of incidence, which is the key feature of the Raman-Nath limit. The Bragg (i.e., phase match) condition does not have to be satisfied in order to obtain diffracted light.

The angle at which the diffracted light appears can be obtained easily from Eq. (11.33). Recollecting that the small Doppler shift demands $k_S \simeq k_I$, one gets

$$\sin\theta' = -\sin\theta \pm \frac{K}{k_I}. \tag{11.47}$$

This is identical to the grating equation which describes the deflection of light by a ruled grating.

Since the diffracted light efficiency depends only weakly on the angle of incidence, second order diffraction can also occur via the scattering of light successively by two elastic phonons. One now lets the field in Eq. (11.40), with L replaced by z, act as an incident field for second order scattering. This problem is too complex to be discussed here, and we only give the result in the Raman-Nath regime. Light can be scattered into the ℓ'th diffraction order whose deflection angle is given by

$$\sin\theta'_{\pm\ell} = -\sin\theta \pm \frac{\ell K}{k_I}. \tag{11.48}$$

The intensities of the various orders are given approximately by

$$\frac{S_\ell}{S_I} = J_\ell^2(HL) \tag{11.49}$$

where J_ℓ is the ℓ'th order Bessel function. This solution works well for light diffraction by large amplitude, low frequency sound waves. For $\ell = \pm 1$, $J_\ell(HL) = \pm HL$ and Eq. (11.49) reduces to Eq. (11.41).

11.2.3 Bragg Regime

This scattering regime is characterized by a very strong dependence of the deflected light efficiency on the angle of incidence. In this limit, $Q \gg 1$ where Q is the Raman-Nath parameter defined in the preceding section. To fulfill the Bragg condition it is necessary that the x component of the phase match condition vanish, which, from Eqs. (11.27) and (11.30) requires $\theta = \theta'$. The angular function implied by the phase match is strongly peaked when

$$\sin\theta = \frac{K}{2k_I} \tag{11.50}$$

at the frequency $\omega_I + \Omega$, and whe .

$$\sin\theta = -\frac{K}{2k_I} \tag{11.51}$$

at the frequency $\omega_I - \Omega$. Only one of these two conditions can be satisfied at a time. Hence one deflected spot is obtained, and which one it is depends on whether Eq. (11.50) or (11.51) is valid. These phase match conditions are illustrated in Figure 11.6. The peak diffraction efficiency is given by Eq. (11.42).

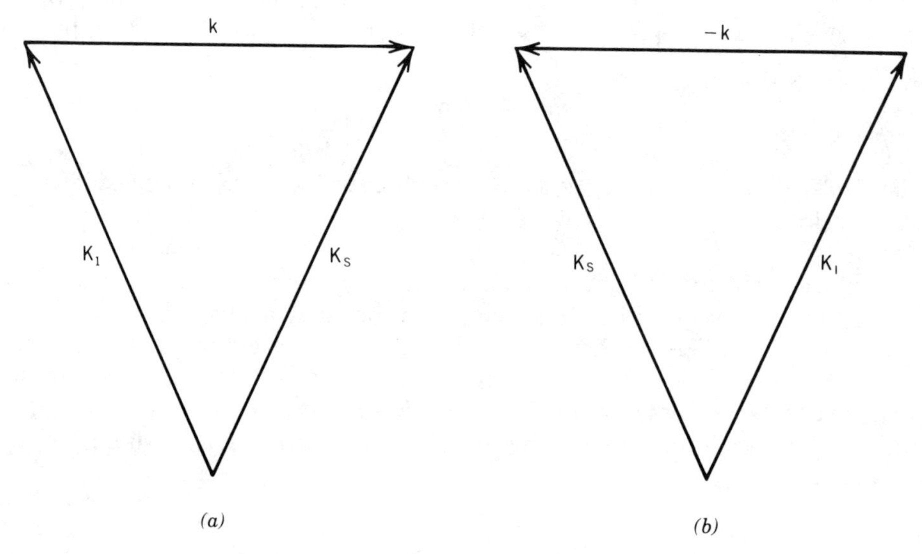

Figure 11.6. Phase match condition for Bragg limit. a) Term going as $\omega_I + \Omega$, b) term going as $\omega_I - \Omega$.

Large deflection efficiencies are possible in the Bragg limit because $Q >> 1$ usually implies wide (large L) acoustic beams, and $I \propto L^2$. The strong angular restrictions imposed by Eqs. (11.50) and (11.51) do not allow second order scattering. However, if the first order beam becomes strong, it can rescatter back into the incident light beam. This process satisfies the Bragg condition, which can be seen by simply reversing the wave vector arrows in Figure 11.6. Therefore, if large deflection efficiencies are important, both the incident and deflected beams must be treated as a function of propagation distance through the acoustic beam.

We now reformulate the scattering problem to include scattering by both the incident and reflected fields. Thus the weak polarization of Eq. (10.40) is extended to include both optical fields, i.e.,

$$\mathbf{P}_W = \frac{1}{4}\left(\boldsymbol{E}_I \cdot \boldsymbol{\chi}_W^*(\mathbf{K},\Omega)\, e^{i(\mathbf{k}_S\cdot\mathbf{r} - i\omega_S t)} + \boldsymbol{E}_S \cdot \boldsymbol{\chi}_W(\mathbf{K},\Omega)\, e^{i(\mathbf{k}_I\cdot\mathbf{r} - i\omega_I t)} + \mathrm{cc}\right) \quad (11.52)$$

where we have chosen the angle of incidence appropriate to the Bragg condition $\mathbf{k}_S = \mathbf{k}_I - K\hat{\mathbf{z}}$. Now use Eqs. (10.49) and (10.50) which give

$$\hat{\mathbf{k}}_S \cdot \nabla E_S = \frac{i\omega\chi_W^*}{4nc\varepsilon_0}\, \hat{\mathbf{e}}_S \cdot \mathbf{O}(\hat{\mathbf{k}}_I) \cdot \boldsymbol{E}_I \quad (11.53)$$

and

$$\hat{\mathbf{k}}_I \cdot \nabla E_I = \frac{i\omega\chi_W}{4nc\varepsilon_0}\, \hat{\mathbf{e}}_I \cdot \mathbf{O}(\hat{\mathbf{k}}_S) \cdot \boldsymbol{E}_S. \quad (11.54)$$

Equations (11.53) and (11.54) are obtained by dividing both sides by $\cos\theta$ (Eq. (11.53) is divided by $\cos\theta'$ in the general case), giving the angular factor in the denominator of Eq. (11.36). Substituting Eqs. (11.27), (11.28), (11.30), and (11.31) for s-polarized incident and scattered light

$$\left(-\tan\theta \frac{\partial}{\partial z} + \frac{\partial}{\partial x}\right) E_S = i\, H\, E_I \quad (11.55)$$

$$\left(\tan\theta \frac{\partial}{\partial z} + \frac{\partial}{\partial x}\right) E_I = i\, H\, E_S. \quad (11.56)$$

The solution to the coupled wave equations given in Eqs. (11.55) and (11.56) is relatively straightforward. We assume the usual boundary condition that $E_S(x=0) = 0$ and that $E_I(x=0) \neq 0$. Furthermore, since the solutions are plane waves, they must be independent of the z coordinate and $\partial/\partial z = 0$. Therefore,

$$\frac{\partial E_S}{\partial x} = i\, H\, E_I \quad (11.57)$$

and

$$\frac{\partial E_I}{\partial x} = i\, H\, E_S. \quad (11.58)$$

These can be solved by taking the derivative of Eq. (11.58) with respect to z, and then substituting Eq. (11.57) for $\partial E_S/\partial x$ to give

$$\frac{\partial^2 E_I}{\partial x^2} = -\, H^2\, E_I,$$

which has the solution (using the boundary conditions $E_S(x=0) = 0$, $E_I(x=0) \equiv E_I(0)$)

$$E_I(x) = E_I(0)\cos(xH). \tag{11.59}$$

Similarly,

$$E_S(x) = E_I(0)\sin(xH). \tag{11.60}$$

The scattered and incident fields vary sinusoidally with distance. This is a typical parametric mixing process. The sound waves mix with the incident light to produce the deflected wave, and the deflected wave mixes with the sound wave to produce a contribution to the incident wave. Note that the parameter H^{-1} has the units of distance and if $LH = \pi/2$, all of the incident beam is converted into the deflected wave. The acoustic flux required is

$$\bar{S}_{sound} = \frac{2\pi^2}{L^2}\left(\frac{c}{\omega}\right)^2\left(\frac{\cos\theta'}{n^3 p_{eff}}\right)^2 \rho v^3, \tag{11.61}$$

and for a square cross-section acoustic beam, the acoustic power is

$$P_s = 2\pi^2\left(\frac{c}{\omega}\right)^2\left(\frac{\cos\theta'}{n^3 p_{eff}}\right)^2 \rho v^3. \tag{11.62}$$

11.2.4 Manley-Rowe Relations

We now consider energy conservation. In this case, the acoustic beam energy removed by the scattering process is usually negligible and does not enter into the problem. However, as a matter of principle it is important to keep precise track of the energy of the system. To do this, recollect that, in Eqs. (11.53) and (11.54), the frequency ω that appears as a factor on the right-hand side of the equation should, in fact, be ω_S and ω_I respectively, which are not precisely equal. We start by multiplying Eqs. (11.53) and (11.54) by $E_S{}^*/\omega_S$ and $E_I{}^*/\omega_I$ respectively. This gives

$$\frac{1}{\omega_S} E_S{}^* \cdot \hat{k}_S \cdot \nabla E_S = \frac{i\chi_w{}^*}{4nc} \mathbf{E}_S{}^* \cdot \mathbf{O}(\hat{k}_S) \cdot \mathbf{E}_I \tag{11.63}$$

$$\frac{1}{\omega_I} E_I{}^* \cdot \hat{k}_I \cdot \nabla E_I = \frac{i\chi_w}{4nc} \mathbf{E}_I{}^* \cdot \mathbf{O}(\hat{k}_I) \cdot \mathbf{E}_S. \tag{11.64}$$

Since χ_w is pure imaginary, then $\chi_w{}^* = -\chi_w$ and the right-hand sides of Eqs. (11.63) and (11.64) are equal in magnitude but opposite in sign. Therefore

$$\frac{1}{\omega_S} E_S{}^* \hat{k}_S \cdot \nabla E_S + \frac{1}{\omega_I} E_I{}^* \hat{k}_I \cdot \nabla E_I = 0. \tag{11.65}$$

Noting that

$$\hat{\mathbf{k}} \cdot \nabla S = \frac{1}{2} \frac{\varepsilon_0}{\mu_0} n\, E^*(\hat{\mathbf{k}} \cdot \nabla) E + \text{cc} \tag{11.66}$$

where S is the optical flux, then

$$\frac{1}{\omega_S} \hat{\mathbf{k}}_S \cdot \nabla S_S + \frac{1}{\omega_I} \hat{\mathbf{k}}_I \cdot \nabla S_I = 0 \tag{11.67}$$

where S_S and S_I are the scattered and incident field fluxes, respectively. Equation (11.67) is called a Manley-Rowe relation. The Manley-Rowe relations do not state that the fluxes are conserved, which would read

$$\hat{\mathbf{k}}_S \cdot \nabla S_S + \hat{\mathbf{k}}_I \cdot \nabla S_I = 0. \tag{11.68}$$

Since ω_I and ω_S are very nearly equal, one can use Eq. (11.68) for any practical application. Let us now define

$$n = \frac{S}{\hbar\omega} \tag{11.69}$$

as the photon flux, i.e., the number of photons per unit area per second, then

$$\hat{\mathbf{k}}_S \cdot \nabla n_S + \hat{\mathbf{k}}_I \cdot \nabla n_I = 0. \tag{11.70}$$

Equation (11.70) can be interpreted as meaning that one photon of incident light is removed from the incident beam when one photon of scattered light is added to the scattered beam, or vice versa. This is a very powerful concept and is frequently used in the next volume, dealing with nonlinear optics. Note, however, that this photon interpretation has nothing to do with quantum mechanics. The conservation of photon flux is a classical concept that describes conservation of energy whenever one optical field scatters into another. The slight energy defect coming from the inequality between ω_I and ω_S, Eq. (11.70) is compensated by a change in the sound energy. The calculation of the Manley-Rowe relation between the field and sound wave, which gives the energy transfer to and from the sound wave, is deferred to the next volume, since it is just a special case of the Manley-Rowe relations developed for stimulated Brillouin scattering.

11.3 ACOUSTO-OPTIC DEVICES

There are a large number of applications of the acousto-optic effect. Here we discuss a few of current importance.

11.3.1 Acousto-Optic Modulators, Deflectors and Scanners

Acousto-optic devices are commercially available for the modulation, deflection and scanning of optical beams. They operate, typically, at acoustic frequencies in the 10 to 500 MHz frequency range and are configured so that the interaction takes place in the Bragg limit. This last condition is usually chosen in order to achieve large deflection efficiencies.

A typical Bragg cell is shown in Figure 11.7. A piezoelectric transducer generates a sound wave which propagates down the cell. The attenuation of sound waves is proportional to Ω^2, so that propagation distances fall off quickly with frequency. Hence the cells are seldom more than a few centimeters long. Depending on the end-face cell conditions, there may or may not be a reflected beam, and whether one wishes to have a reflected wave depends on the nature of the application. The direction of the incident light beam is fixed at the appropriate Bragg angle.

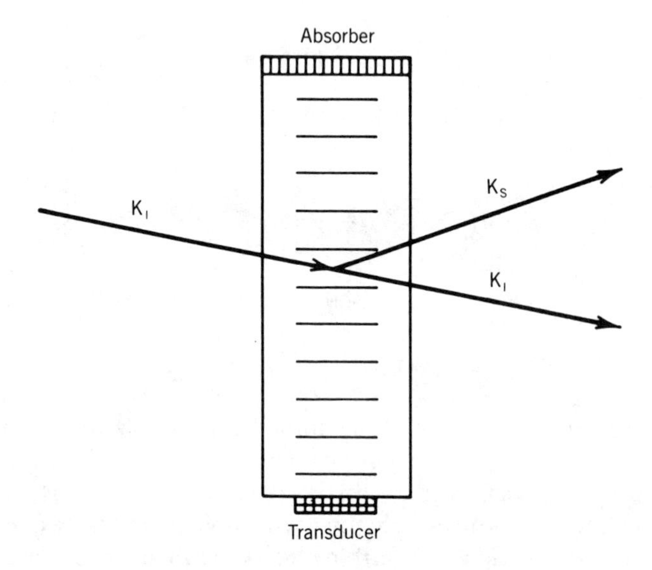

Figure 11.7. Illustration of Bragg cell.

Modulation of light is achieved by modulating the signal to the transducer. The speed at which the deflected beam can be modulated is normally limited by the transit time of the acoustic beam across the optical beam. For a light beam W wide, this transit time is $\tau = W/v$. For a 1 mm beam, $\tau \simeq 2 \times 10^{-7}$ sec, i.e., the maximum modulation frequency is about 5 MHz. This frequency can be increased by focusing the light, but at some point the spreading of the light beam spoils the Bragg angle condition and modulation efficiency decreases.

Scanning or deflection of the beam can be obtained by varying the acoustic frequency (and direction via transducer arrays in order to maintain the Bragg condition). The change in deflection angle $\Delta\theta$ is related to the change in acoustic frequency, $\Delta f = 2\pi \Delta\Omega$, by

$$\Delta\theta = \frac{\lambda}{v \cos\theta} \Delta f. \tag{11.71}$$

Because the optical beam has a finite width W, there is a characteristic divergence angle associated with the light, $\delta\theta = \lambda/W$. Thus the number of resolvable spots which varying the frequency can achieve is

$$N = \frac{\Delta\theta}{\delta\theta} = \frac{W\,\Delta f}{v\cos\theta}. \tag{11.72}$$

For example, for $\Delta f = 100$ MHz, $v \simeq 5 \times 10^5$ cm/sec and $W = 1$ cm, $N \simeq 200$ spots. However, if the acoustic beam direction is not varied as the frequency is swept, the limiting factor is not necessarily the frequency sweep but the divergence angle of the acoustic beam. In this case, the divergence of the sound beam allows the Bragg condition to be satisfied over a range of frequencies. Then $\Delta\theta = \Lambda/L$ and

$$N = \frac{\Lambda W}{\lambda L}. \tag{11.73}$$

A common figure of merit for a deflector is the number of separate positions to which a beam can be directed in unit time. Therefore this figure of merit is

$$\frac{N}{\tau} = \frac{\Delta f}{\cos\theta}, \tag{11.74}$$

which indicates why it is desirable to use high frequency modulators.

11.3.2 Devices in Laser Cavities

Acousto-optic deflectors are also used in laser cavities. One example is the acousto-optic Q-switch. To prevent a laser cavity from oscillating, a Bragg cell deflects some of the light out of the cavity so that losses exceed gain. When the acoustic power is cut off, the deflection loss goes to zero, the laser emission builds inside the resonator, and a Q-switched pulse, 5 to 25 nsec long, results.

Mode-locking can be achieved in a standing wave Bragg cell. An acousto-optic cell in which an acoustic standing wave is present is placed inside a laser cavity as illustrated in Figure 11.8. Since $U_+ = U_- = U \neq 0$, then for the total displacement,

$$u = U_0 \sin Kz \sin\Omega t. \tag{11.75}$$

This describes a grating which vanishes every $t = m\pi/\Omega$ where $m = 0,1,2,\cdots$. When located inside a laser resonator, the mode locker spoils the cavity except at the instant when the grating vanishes. If the half-period of the sound wave is equal to the round trip time of the cavity, then an ultrashort pulse can pass through the acousto-optic device each time $t = m\pi/\Omega$. Hence a mode-locked pulse builds up in the cavity. The length of the pulse depends on the depth of modulation of the standing wave, i.e., the time period over which the cavity is not sufficiently spoiled to permit lasing action. Limiting pulse durations of about 50 psec have been achieved.

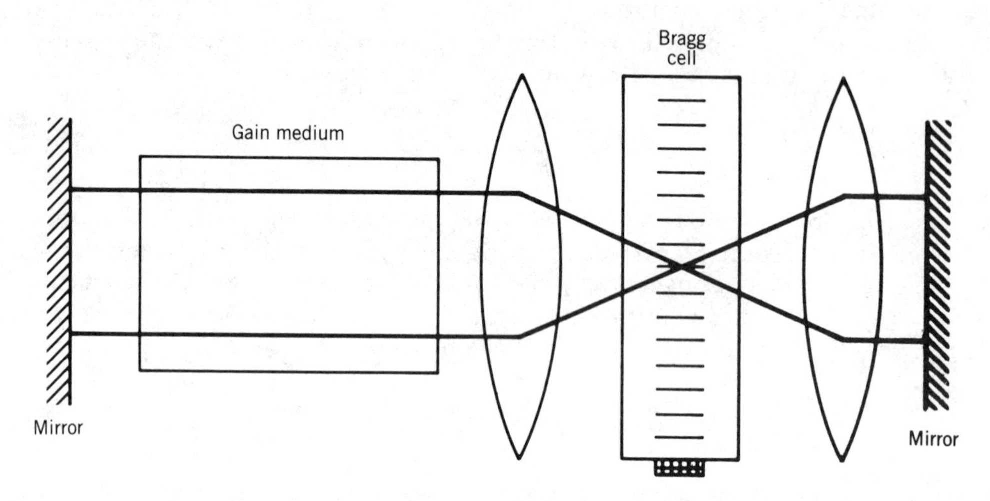

Figure 11.8. Acousto-optic mode locker.

11.3.3 Acousto-Optic Spectrum Analyzer

This is an extremely powerful device which has come into use in the past few years. Its function is to take a high frequency electrical signal and separate it into its frequency components. Such an operation carried out by an acousto-optic device can replace banks of filters.

The device is shown schematically in Figure 11.9. The electrical signal to be

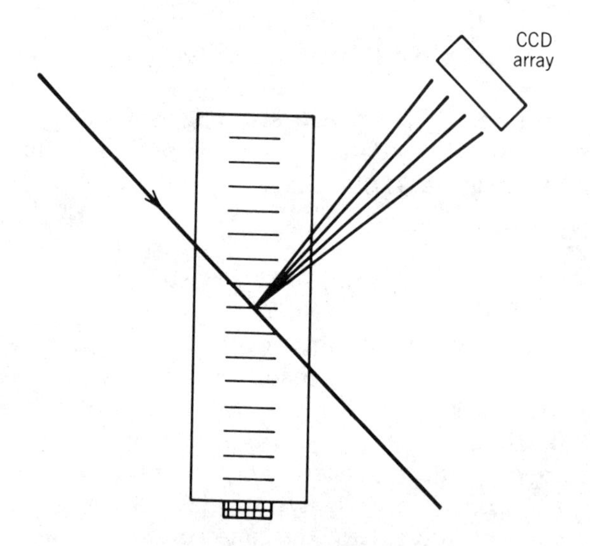

Figure 11.9. Acousto-optic spectrum analyzer.

analyzed excites sound waves via a broadband transducer. Light is deflected into a variety of directions, each direction corresponding to a different frequency component present in the electrical signal. The deflected light is gathered by a lens and focused onto a detector array placed in the focal plane of the lens. Because the lens acts as a Fourier transform element, there is a direct correspondence between the detector element illuminated and the frequency component in the signal. Hence the device acts as a real-time, parallel processing spectrum analyzer.

ADDITIONAL READING

Acoustics

Auld, B.A. *Acoustic Fields and Waves in Solids*, Wiley, New York, 1973. Volume 1.

Acousto-Optical Interactions with Coefficients

Yariv, A., *Quantum Electronics*, Wiley, New York, 1967. pp. 360-367.

Devices and Figures of Merit

Damon, R.W., Maloney, W.T. and McMahon, D.H. Interaction of Light with Ultrasound, in Mason, W.P. and Thurston, R.N. eds., *Physical Acoustics*, Academic Press, New York, 1970.

Coefficients

Presley, R.J., ed., *CRC Handbook of Lasers with selected data on Optical Technology*, Chemical Rubber Co., Cleveland, 1971. Chapter 17.

PROBLEMS

11.1. (a) Using the definitions of the inverse lattice vectors

$$\mathbf{A} = \mathbf{b} \times \mathbf{c}; \quad \mathbf{B} = \mathbf{c} \times \mathbf{a}; \quad \mathbf{C} = \mathbf{a} \times \mathbf{b}, \tag{11.76}$$

show that Eq. (11.13) can be written as

$$V(\mathbf{r}) = V + (\mathbf{aA} + \mathbf{bB} + \mathbf{cC}) \vdots \nabla \mathbf{u}(\mathbf{r}). \tag{11.77}$$

(b) Show that

$$\mathbf{aA} + \mathbf{bB} + \mathbf{cC} = \mathbf{a} \cdot (\mathbf{b} \times \mathbf{c})\, \mathbf{I} \tag{11.78}$$

so that the derivation in the text is valid for arbitrary symmetries.

11.2. Verify the contents of Table 11.2.

11.3. Verify that Eq. (11.33) follows from Eq. (10.7) in the limit of small angles.

11.4. Show that for realistic $\mathbf{K}$ and Ω, the condition $\mathbf{k}_S \simeq \mathbf{k}_I - \mathbf{K}$ precludes finding some $\mathbf{k}'_S \simeq \mathbf{k}_I + \mathbf{K}$ in the Bragg limit.

11.5. Derive the appropriate version of Eq. (11.19) for the case in which the sound wave is damped so that

$$u = \frac{1}{2} (U_+(k,\Omega)e^{i(Kz-\Omega t)} + cc)e^{-\gamma_s z/2}. \tag{11.79}$$

11.6. (a) Verify the correctness of the indices in Eqs. (11.22) and (11.23).

(b) Verify that Eq. (11.19) implies Eqs. (11.3).

(c) Verify that Eqs. (11.3) imply that χ_w is diagonal for all k_l and **K**. (This is just a variant on b.)

11.7. (a) Derive the appropriate version of Eqs. (11.35) and (11.36) for the case of a liquid in which the susceptibility amplitude is given by Eq. (11.19). Use the formulae in Section 10.5, bearing in mind that the propagation coordinate z in Chapter 10 is replaced with $x/\cos\theta$.

(b) Compare your result with Eqs. (11.35) and (11.36) using Table 11.2 and Eq. (11.21).

11.8. Evaluate Eq. (11.44) for the case of a liquid. Express the result in terms of n, $C_{\mu\mu}$ and N only.

11.9. Verify Eqs. (11.55) and (11.56).

11.10. Verify that a sonic power that optimizes Bragg scattering as given in Eq. (11.62) causes the deflection efficiency $\eta = S(L)/S(0)$ in the Raman-Nath limit (Eq. (11.43)) to be of the order unity.

11.11. Derive Eq. (11.52) from Eq. (10.40). Use the results of Problem 11.4 to eliminate non-phase matched interactions.

11.12. Define θ_h as the angle in the Bragg limit for which $\Delta kL = 1$, and define $\Delta\theta$ as the difference between θ_h and the angle at which exact phase matching takes place. Show that

$$\Delta\theta = \frac{\lambda}{8nL}, \tag{11.80}$$

where λ is the wavelength in vacuum.

12 Scattering

In the preceding chapters we ignored the optical effects of small-scale inhomogeneities in the optical susceptibilities. These inhomongeneities are produced by the random motion of the molecules and atoms whose position, orientation, etc. are changing from moment to moment. The microscopic inhomogeneities do not contribute to the average (macroscopic) fields, and hence can be ignored insofar as macroscopic electrodynamics is concerned, but they do scatter light. We deal with scattering in this chapter. The statistical fluctuations in the medium can be decomposed into the random excitation of normal modes. For example, acoustic, vibrational and rotational modes are all excited to some degree in a medium at a finite temperature T. In this chapter we examine three types of scattering processes: Rayleigh and Mie scattering, which arises from density fluctuations; Raman scattering, which arises from local modes of vibration and rotation; and Brillouin scattering, which arises from collective motions, i.e., sound waves (see Chapter 3 for discussion of local and collective motions).

We consider only those fluctuations which couple to optical fields. They can be described in terms of weak susceptibilities of the type discussed in Section 10.5. However, because the fluctuations are random, they are not properly described by a phased array of susceptibility in the sense of Chapter 10, in which the phased array is a field amplitude (e.g., Eq. (1.20)) and is itself an average quantity. Instead, in scattering from random disturbances, it is the Fourier amplitude (e.g., Eq. (1.19)) that is the phased array. Hence we write the susceptibility as

$$\boldsymbol{\chi} = \langle\boldsymbol{\chi}\rangle + \boldsymbol{\chi}_W(\mathbf{r},t) \tag{12.1}$$

where $\langle\boldsymbol{\chi}\rangle$ is the usual averaged dielectric tensor. The weak susceptibility is then written in terms of its Fourier amplitudes as

$$\boldsymbol{\chi}_W(\mathbf{r},t) = \frac{1}{2}\int d^3K \int d\Omega\, \tilde{\boldsymbol{\chi}}_W(\mathbf{K},\Omega)\, e^{i(\mathbf{K}\cdot\mathbf{r}-\Omega t)} + \text{cc}, \tag{12.2}$$

where $\tilde{\boldsymbol{\chi}}_W(\mathbf{K},\Omega)$ is the Fourier transform of $\boldsymbol{\chi}_W(\mathbf{r},t)$. In dealing with Fourier rather than field amplitudes it is important to bear in mind that they are physically different objects, having, for example, different units. The relations between Fourier and field amplitudes are covered in some detail in Problems 12.1 - 12.4.

The terms $\tilde{\chi}_W(\mathbf{K},\Omega)$ scatter light as discussed in Chapter 10. However, in computing intensities from electromagnetic fields, the proper technique is to construct the Poynting vector prior to taking statistical averages. In macroscopic electrodynamics, the averages of the fields are made prior to computing the Poynting vector. The difference between these two Poynting vectors gives the light scattered by the random fluctuations. The formal development used here may seem quite different from the methods used in Chapters 10 and 11. However it is actually just a variant on the methods of Chapter 10 that constructs the Poynting vector first, and hence allows the calculation of scattered fields.

While the mathematical methodology of the scattering formalism is related to the methods of Chapter 10, there are basic differences in approach. In dealing with macroscopic phased arrays, the assumptions are stated at the beginning. In contrast, scattering is treated using a formal solution of Maxwell's equations that follows from Fourier analysis of the fields. Except for the assumption that the susceptibility is weak, all basic assumptions are deferred until one attempts to extract real solutions from formal ones. Unless one is careful, it is easy to make mistakes in extracting real solutions, and for that reason we have avoided Fourier techniques up till now (many problems dealt with previously can be successfully solved with Fourier techniques; our point is that the beginner can easily get into trouble using them). As a matter of practice, the development in this chapter is restricted to one situation in which Fourier techniques are straightforward, namely one in which the incident field can be described by a field amplitude (e.g., a laser beam) that is not signifigantly attenuated in its passage through the scattering volume. Hence we assume that the scattering is weak. Moreover, we assume for convenience that the scattering volume is much larger than the wavelength of light.

Throughout this chapter we need to take the Fourier transform and its inverse. Since there are questions of notation involved, especially in the prefactor $(1/2\pi)^{3/2}$ or $(1/2\pi)^3$, we now define the versions to be used here. Using the weak susceptibility as an example,

$$\tilde{\chi}_W(\mathbf{K},t) = \frac{1}{2\pi^3} \int d^3r\, e^{-i\mathbf{K}\cdot\mathbf{r}}\, \chi_W(\mathbf{r},t) \tag{12.3a}$$

$$\tilde{\chi}_W(\mathbf{K},\Omega) = \frac{1}{2\pi} \int dt\, e^{+i\Omega t}\, \tilde{\chi}_W(\mathbf{K},t) \tag{12.3b}$$

$$\tilde{\chi}_W(\mathbf{r},\Omega) = \frac{1}{2} \int d^3K\, e^{i\mathbf{K}\cdot\mathbf{r}}\, \tilde{\chi}_W(\mathbf{K},\Omega) + \text{cc} \tag{12.3c}$$

$$\chi_W(\mathbf{r},t) = \frac{1}{2} \int d\Omega\, e^{-i\Omega t}\, \tilde{\chi}_W(\mathbf{r},\Omega) + \text{cc}. \tag{12.3d}$$

We follow the conventions of Eqs (1.19) and (1.20) for denoting field and Fourier amplitudes respectively in which script characters are field amplitudes and ordinary characters refer to total fields or Fourier amplitudes depending on the arguments. The susceptibility and the flux cannot be written with these conventions, and they are labeled with the symbol ~ to indicate their status as a Fourier amplitude rather than an ordinary amplitude. Also, for future reference

$$\int d^3r\, e^{i(\mathbf{K}-\mathbf{K}_\beta)\cdot\mathbf{r}} = (2\pi)^3\, \delta(\mathbf{K}-\mathbf{K}_\beta) \qquad (12.4a)$$

and

$$\int d^3K\, \delta(\mathbf{K}-\mathbf{K}_\beta) = 1. \qquad (12.4b)$$

12.1 FORMAL DEVELOPMENT OF THE SCATTERING PROBLEM

The goal of this section is to develop formulae for both the frequency spectrum of the scattered light and the total scattered intensity. At this stage we assume that $\tilde{\boldsymbol{\chi}}_w(\mathbf{K},\Omega)$ is given. The assumption of a spatially independent incident field allows us to use an analog of Eq. (10.40) to construct explicitly the weak polarization from the weak susceptibility at every point in the medium. Thus we can immediately extract real solutions that are valid as long as the assumption holds. If the assumption is false, then the field amplitude in Eq. (10.40) is an explicit function of position, and the interrelation between the weak polarization and the weak susceptibility must be treated with much greater care. Using the complex coherence function, the concept of the frequency spectrum is developed and the pertinent formulae derived. The properties of uncorrelated statistical fluctuations in the normal modes are used to simplify the results. Finally, we discuss depolarization effects in a general way.

12.1.1 The Scattered Fields

The calculation of a field radiated by a weak polarization term is relatively straightforward. In terms of procedure, we write the weak polarization as an effective current source (Eq. (1.14)) and use the standard results for the vector potential for the radiated field, i.e., Eqs. (1.40). The fields are found using Eq. (A.59), which is a variation on the results of Problem 1.4.

We start with a general weak polarization term of the form

$$\mathbf{P}_w(\mathbf{r},t) = \frac{1}{2} \int d^3k \int d\omega\, e^{i(\mathbf{k}\cdot\mathbf{r}-\omega t)}\, \mathbf{P}_w(\mathbf{k},\omega) + \text{cc}. \qquad (12.5)$$

From Eq. (1.14),

$$\mathbf{J}_w(\mathbf{r},t) = -\frac{i}{2} \int d^3k \int d\omega\omega\, e^{i(\mathbf{k}\cdot\mathbf{r}-\omega t)}\, \mathbf{P}_w(\mathbf{k},\omega) + \text{cc}.$$

Substituting into Eq. (1.40) and evaluating the integral over t' gives

$$\mathbf{A}(\mathbf{r},t) = -\frac{i\mu_0}{8\pi} \int d^3k \int d\omega \int d^3r'\, \omega\, e^{i[\mathbf{k}\cdot\mathbf{r}'-\omega(t-R/c)]}\, \frac{\mathbf{P}_w(\mathbf{k},\omega)}{R} + \text{cc}. \qquad (12.6)$$

Just as in Chapter 1, we expand R as

$$R \cong r - \hat{\mathbf{r}}\cdot\mathbf{r}',$$

where $\mathbf{r}$ is the distance from the observation point to the center of the scattering volume. Keeping both terms in the exponential, and approximating R by r in the denominator,

$$\mathbf{A}(\mathbf{r},t) = -\frac{i\mu_0}{8\pi r}\int d^3r' \int d^3k \int d\omega\, \omega\, \mathbf{P}_w(\mathbf{k},\omega) e^{i(\mathbf{k}\cdot\mathbf{r}' - k\hat{\mathbf{r}}\cdot\mathbf{r}') + i(kr-\omega t)} + \text{cc}, \quad (12.7)$$

where $k = \omega/c$, and we have assumed that the field is observed in vacuum. Evaluating the integral over r' gives

$$\int d^3r'\, e^{i(\mathbf{k}-k\hat{\mathbf{r}})\cdot\mathbf{r}'} = (2\pi)^3\delta(\mathbf{k}-k\hat{\mathbf{r}}) \quad (12.8)$$

and integrating over k results in

$$\mathbf{A}(\mathbf{r},t) = -\frac{i}{2}\frac{(2\pi)^3\mu_0}{4\pi r}\int d\omega\omega\, \mathbf{P}_w(k\hat{\mathbf{r}},\omega)\, e^{i(kr-\omega t)} + \text{cc}. \quad (12.9)$$

At this point we have to consider the structure of $\mathbf{P}_w(k\hat{\mathbf{r}},\omega)$ and express it in terms of the susceptibility and incident field. For an incident field of the form

$$\mathbf{E}(\mathbf{r},t) = \frac{1}{2}\boldsymbol{\mathcal{E}}_I\, e^{i(\mathbf{k}_I\cdot\mathbf{r}-\omega_I t)} + \text{cc}, \quad (12.10)$$

it is left as an exercise to show that

$$\mathbf{P}_w(k\hat{\mathbf{r}},\omega) = \frac{\varepsilon_0}{2}\boldsymbol{\mathcal{E}}_I\cdot\tilde{\boldsymbol{\chi}}_w(k\hat{\mathbf{r}}-\mathbf{k}_I,\omega-\omega_I) + \frac{\varepsilon_0}{2}\boldsymbol{\mathcal{E}}_I\cdot\tilde{\boldsymbol{\chi}}_w^*(k\hat{\mathbf{r}}+\mathbf{k}_I,\omega+\omega_I). \quad (12.11)$$

Since for the cases of interest here $\omega = \omega_I \pm \Omega$ with $\omega_I >> \Omega$, the last term does not couple to low frequency fluctuations and only the first term is retained. Substituting (12.11) into (12.9),

$$\mathbf{A}(\mathbf{r},t) = -\frac{i}{4}\frac{(2\pi)^3\mu_0\varepsilon_0}{4\pi r}\int d\omega\omega\, e^{i(kr-\omega t)}\boldsymbol{\mathcal{E}}_I\cdot\tilde{\boldsymbol{\chi}}_w(k\hat{\mathbf{r}}-\mathbf{k}_I,\omega-\omega_I) + \text{cc}. \quad (12.12)$$

Clearly, whether scattering occurs in the direction $k\hat{\mathbf{r}}$ with a frequency ω depends on whether the fluctuation responsible has Fourier components of wave vector and frequency at $k\hat{\mathbf{r}}-\mathbf{k}_I$ and $\omega-\omega_I$ respectively.

The scattered field $\mathbf{E}(\mathbf{r},t)$ can now be calculated using Eq. (A.59), which gives

$$\mathbf{E}(\mathbf{r},t) = \frac{1}{2}\frac{(2\pi)^3}{8\pi rc^2}\mathbf{O}(\hat{\mathbf{r}})\cdot\int_{-\infty}^{\infty} d\omega\omega^2\boldsymbol{\mathcal{E}}_I\cdot\tilde{\boldsymbol{\chi}}_w(k\hat{\mathbf{r}}-\mathbf{k}_I,\omega-\omega_I)\, e^{i(kr-\omega t)} + \text{cc}. \quad (12.13)$$

Noting that the integral of Eq. (12.13) over $-\infty < \omega < 0$ is just the complex conjugate of the integral over $0 > \omega > \infty$, Eq. (12.13) can be written in the more familiar form

$$\mathbf{E}(\mathbf{r},t) = \frac{1}{2}\frac{(2\pi)^3}{4\pi rc^2}\mathbf{O}(\hat{\mathbf{r}})\cdot\int_0^{\infty} d\omega\omega^2\boldsymbol{\mathcal{E}}_I\cdot\tilde{\boldsymbol{\chi}}_w(k\hat{\mathbf{r}}-\mathbf{k}_I,\omega-\omega_I)\, e^{i(kr-\omega t)} + \text{cc}. \quad (12.14)$$

This is the form for the field that we have been working toward.

12.1.2 Scattered Light Spectrum

In this section we describe a formalism for calculating the frequency spectrum and intensity of the scattered light. We start by considering optical measurements which are quadratic in the fields, i.e., the usual case of photoelectric detectors. In the general case we have two fields $\mathbf{E}(\mathbf{r}_1,t_1)$ and $\mathbf{E}(\mathbf{r}_2,t_2)$ in the far field pattern and we can evaluate the power in these fields, as well as specific polarization properties of the fields. In addition, we can consider making interference patterns between the two beams, as illustrated in Figure 12.1.

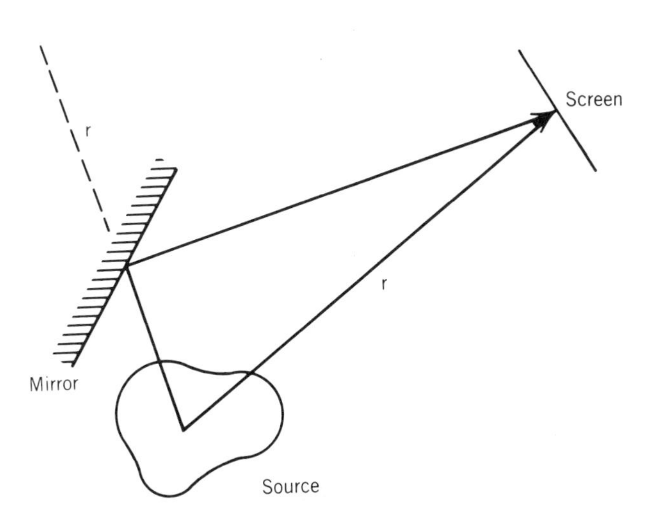

Figure 12.1. Schematic for measuring the complex coherence function. Polarizers can be inserted into the two scattered fields to obtain the components of the tensor.

One can observe the contrast in the fringes at the observation screen. On the way to the screen the polarization components can be isolated, and subsequently aligned so that the fringe contrast between different polarization components (labeled a and b) can be observed. The combination of these measurements is summarized in a complex coherence function which we denote as G, where

$$G(a,\mathbf{r},t;b,\mathbf{r}',t') = \frac{1}{c\mu_0} \langle E_a(\mathbf{r},t)E_b(\mathbf{r}',t')\rangle. \tag{12.15}$$

The bracket means a statistical average. If the statistical properties of the scattered light are invariant in time, which they are for the cases considered here, then it is also the time average. We can therefore formally define

$$\langle E_a(\mathbf{r},t)E_b(\mathbf{r}',t')\rangle = \lim_{T\to\infty} \frac{1}{T} \int_0^T dt' E_a(\mathbf{r}',t')E_b(\mathbf{r},t'+\tau) \tag{12.16}$$

where $t = t'+\tau$. This time average eliminates the rapidly oscillating terms.

The case of interest in light scattering requires that $r = r'$. For cases in which a and b refer to orthogonal senses of polarization, Eq. (12.16) can be rewritten as a tensor which is suitably defined to yield the Poynting vector in certain limits, i.e.,

$$\mathbf{G}(\mathbf{r},\tau) = \frac{1}{c\mu_0} \langle \mathbf{E}(\mathbf{r},t+\tau)\mathbf{E}(\mathbf{r},t)\rangle. \quad (12.17)$$

In component form, Eq. (12.17) reads

$$\mathbf{G}(\mathbf{r},\tau) = \frac{1}{c\mu_0}\begin{pmatrix} \langle E_a(\mathbf{r},t+\tau)E_a(\mathbf{r},t)\rangle & \langle E_a(\mathbf{r},t+\tau)E_b(\mathbf{r},t)\rangle \\ \langle E_b(\mathbf{r},t+\tau)E_a(\mathbf{r},t)\rangle & \langle E_b(\mathbf{r},t+\tau)E_b(\mathbf{r},t)\rangle \end{pmatrix}. \quad (12.18)$$

The total flux, i.e., total scattered intensity per unit area, is

$$S(\mathbf{r}) = \mathrm{Tr}[\mathbf{G}(\mathbf{r},\tau=0)], \quad (12.19)$$

where Tr is the trace, i.e., the sum of the diagonal components (note that Eq. (12.19) is invalid if a and b refer to coordinates that are not orthogonal).

The frequency spectrum is the Fourier transform of the complex coherence function. We define a generalized spectrum as

$$\tilde{\mathbf{S}}(\mathbf{r},\omega) = \frac{1}{4\pi} \int_{-\infty}^{\infty} d\tau\, e^{i\omega\tau}\, \mathbf{G}(\mathbf{r},\tau) + \mathrm{cc}, \quad (12.20)$$

and the conventional spectrum as

$$\tilde{S}(\mathbf{r},\omega) = \mathrm{Tr}\left(\tilde{\mathbf{S}}(\mathbf{r},\omega)\right). \quad (12.21)$$

The Fourier off-diagonal element is called the cross spectral density.

The calculation of the coherence function is simple but messy. Substituting Eq. (12.14) into (12.17) and discarding the rapidly varying terms gives

$$\mathbf{G}(\mathbf{r},\tau) = \frac{(2\pi)^6}{4(4\pi r)^2\mu_0 c^5} \left\langle \int_0^{\infty} d\omega'' \int_0^{\infty} d\omega'\omega'^2\omega''^2\, e^{-i\omega'(t+\tau)+i\omega'' t} \right.$$

$$\left. \times \mathbf{O}(\hat{\mathbf{r}})\cdot \boldsymbol{E}_I \cdot \tilde{\boldsymbol{\chi}}_W(k\hat{\mathbf{r}}-\mathbf{k}_I,\omega'-\omega_I)\cdot \tilde{\boldsymbol{\chi}}_W{}^*(k\hat{\mathbf{r}}-\mathbf{k}_I,\omega''-\omega_I)\cdot \boldsymbol{E}_I{}^* \cdot \mathbf{O}(\hat{\mathbf{r}}) + \mathrm{cc}\right\rangle. \quad (12.22)$$

The two terms affected by the statistical averaging are $\langle \exp[i(\omega''-\omega')t]\rangle$, which corresponds to a time average, and $\langle \tilde{\boldsymbol{\chi}}_W \tilde{\boldsymbol{\chi}}_W{}^*\rangle$, which corresponds to the statistical average. From Eq. (12.4a)

$$\langle e^{i(\omega''-\omega')t}\rangle = \lim_{T\to\infty} \frac{1}{T} \int_0^T e^{i(\omega''-\omega')t}\, dt = \frac{2\pi}{T}\, \delta(\omega'-\omega''). \quad (12.23)$$

The infinite time T is removed later. We now integrate over $d\omega''$ to take advantage of this δ-function and obtain

$$\mathbf{G}(\mathbf{r},\tau) = \frac{(2\pi)^7}{4(4\pi r)\quad c} \int_0^\infty d\omega'\omega'^4 e^{-i\omega'\tau} \frac{1}{T} \mathbf{O}(\hat{\mathbf{r}}) \cdot$$

$$\boldsymbol{E}_l \cdot \langle \tilde{\boldsymbol{\chi}}_W(k\hat{\mathbf{r}} - \mathbf{k}_l, \omega' - \omega_l)\, \tilde{\boldsymbol{\chi}}_W^*(k\hat{\mathbf{r}} - \mathbf{k}_l, \omega' - \omega_l) \rangle \cdot \boldsymbol{E}_l^* \cdot \mathbf{O}(\hat{\mathbf{r}}) + \mathrm{cc}. \tag{12.24}$$

Noting that

$$\mathbf{G}(\mathbf{r},\tau) = \frac{1}{2} \int_{-\infty}^{\infty} e^{-i\omega'\tau}\, \tilde{\mathbf{S}}(\mathbf{r},\omega')d\omega' + \mathrm{cc}$$

$$= \int_0^\infty e^{-i\omega'\tau}\, \tilde{\mathbf{S}}(\mathbf{r},\omega')d\omega' + \mathrm{cc},$$

then

$$\tilde{\mathbf{S}}(\mathbf{r},\omega) = \frac{(2\pi)^7}{4(4\pi r)^2 \mu_0 c^5}\, \omega^4 \mathbf{O}(\hat{\mathbf{r}}) \cdot \boldsymbol{E}_l$$

$$\cdot \frac{1}{T} \langle \tilde{\boldsymbol{\chi}}_W(k\hat{\mathbf{r}} - \mathbf{k}_l, \omega - \omega_l)\, \tilde{\boldsymbol{\chi}}_W^*(k\hat{\mathbf{r}} - \mathbf{k}_l, \omega - \omega_l) \rangle \cdot \boldsymbol{E}_l^* \cdot \mathbf{O}(\hat{\mathbf{r}}). \tag{12.25}$$

Usually one is interested in the power scattered per unit solid angle, which we define by $\tilde{\boldsymbol{\mathcal{S}}}(\mathbf{r},\omega) = r^2 \Delta\Omega \tilde{\mathbf{S}}(\mathbf{r},\omega)$ where $\Delta\Omega$ is the solid angle subtended at the sample. Thus

$$\frac{\tilde{\boldsymbol{\mathcal{S}}}(\mathbf{r},\omega)}{\Delta\Omega S_0} = \frac{(2\pi)^5}{8c^4}\, \omega^4\, \mathbf{O}(\hat{\mathbf{r}}) \cdot \hat{\mathbf{e}}_l \cdot \boldsymbol{\Delta}(k\hat{\mathbf{r}} - \mathbf{k}_l, \omega - \omega_l) \cdot \hat{\mathbf{e}}_l^* \cdot \mathbf{O}(\hat{\mathbf{r}}) \tag{12.26}$$

where S_0 is the incident power per unit area and

$$\boldsymbol{\Delta}(k\hat{\mathbf{r}} - \mathbf{k}_l, \omega - \omega_l) = \frac{1}{T} \langle \tilde{\boldsymbol{\chi}}_W(k\hat{\mathbf{r}} - \mathbf{k}_l, \omega - \omega_l) \tilde{\boldsymbol{\chi}}_W^*(k\hat{\mathbf{r}} - \mathbf{k}_l, \omega - \omega_l) \rangle. \tag{12.27}$$

All of the details of the fluctuations which lead to scattering are contained in the fourth rank tensor $\boldsymbol{\Delta}(k\hat{\mathbf{r}} - \mathbf{k}_l, \omega - \omega_l)$. The quantity on the left-hand side of Eq. (12.26) is the tensor generalization of a differential cross section which is conventionally denoted as

$$\frac{d^2\sigma}{d\omega d\Omega} = \mathrm{Tr}\left(\frac{\tilde{\boldsymbol{\mathcal{S}}}(\mathbf{r},\omega)}{\Delta\Omega S_0}\right). \tag{12.28}$$

12.1.3 Statistical Fluctuations

We now consider in more detail the fourth rank tensor defined by Eq. (12.28) with regard to eliminating the awkward term $1/T$. The weak susceptibility is caused by statistical fluctuations in the medium which lead directly to the excitation of the normal modes. That is, there are vibrational, electronic, rotational, and translational local normal modes, and acoustic collective modes which are oscillating. In the case of the local modes, the motions are uncorrelated from molecule to molecule, and also from one normal coordinate to the next. For the collective modes, each normal mode (i.e., each solution to the dispersion relation) is excited independently. From classical statistical mechanics, each mode has an energy KT. We discuss this point in detail later. For the present, the important feature is that the motions can be described in terms of normal modes which are randomly excited and hence are uncorrelated.

Based on the preceding discussion, we now simplify Eq. (12.27). Thus

$$\tilde{\boldsymbol{\chi}}_W = \sum_\beta \sum_\alpha \tilde{\boldsymbol{\chi}}_{\alpha,\beta}$$

where the summations over α and β are over the fluctuations that occur in the time T (the subscript α identifies a time interval) and over the normal modes (subscript β) respectively. Since we are dealing with statistical fluctuations

$$\sum_\beta \sum_{\beta'} \langle \tilde{\boldsymbol{\chi}}_{\alpha\beta} \tilde{\boldsymbol{\chi}}_{\alpha'\beta'}{}^* \rangle = \sum_\beta \langle \tilde{\boldsymbol{\chi}}_{\alpha\beta} \tilde{\boldsymbol{\chi}}_{\alpha'\beta}{}^* \rangle. \tag{12.29}$$

Furthermore, because there is no correlation between successive fluctuations

$$\sum_\alpha \sum_{\alpha'} \langle \tilde{\boldsymbol{\chi}}_{\alpha\beta} \tilde{\boldsymbol{\chi}}_{\alpha'\beta}{}^* \rangle = \sum_\alpha \langle \tilde{\boldsymbol{\chi}}_{\alpha\beta}, \tilde{\boldsymbol{\chi}}_{\alpha\beta}{}^* \rangle, \tag{12.30}$$

i.e., the product $\chi\chi^*$ must refer to a single fluctuation. Therefore Eq. (12.27) can be rewritten as

$$\boldsymbol{\Delta}(k\hat{\mathbf{r}} - \mathbf{k}_I, \omega - \omega_I) = \sum_\beta \frac{1}{T} \sum_\alpha \langle \tilde{\boldsymbol{\chi}}_{\alpha,\beta}(k\hat{\mathbf{r}} - \mathbf{k}_I, \omega - \omega_I) \tilde{\boldsymbol{\chi}}_{\alpha,\beta}{}^*(k\hat{\mathbf{r}} - \mathbf{k}_I, \omega - \omega_I) \rangle. \tag{12.31}$$

However, the term $\langle \chi\chi^* \rangle$ is independent of α because of the assumption that the statistical properties of the fluctuations are independent of time (i.e., our use of Eq. (12.16)). As a result

$$\frac{1}{T} \sum_\alpha \langle \tilde{\boldsymbol{\chi}}_{\alpha,\beta} \tilde{\boldsymbol{\chi}}_{\alpha,\beta}{}^* \rangle = \langle \tilde{\boldsymbol{\chi}}_\beta \tilde{\boldsymbol{\chi}}_\beta{}^* \rangle \frac{1}{T} \sum_\alpha. \tag{12.32}$$

The quantity $T^{-1} \sum_\alpha$ has a simple interpretation: it is the total number of fluctuations produced in the time T, divided by T, which is the rate at which fluctuations occur. Hence defining this rate for the β'th normal mode as

$$\Lambda_\beta = \frac{1}{T} \sum_\alpha$$

gives

$$\boldsymbol{\Delta}(k\hat{\mathbf{r}} - \mathbf{k}_I, \omega - \omega_I) = \sum_\beta \Lambda_\beta \langle \tilde{\boldsymbol{\chi}}_\beta(k\hat{\mathbf{r}} - \mathbf{k}_I, \omega - \omega_I) \tilde{\boldsymbol{\chi}}_\beta{}^*(k\hat{\mathbf{r}} - \mathbf{k}_I, \omega - \omega_I) \rangle. \tag{12.33}$$

Therefore, defining $\tilde{\mathcal{S}}_\beta(\mathbf{r},\omega)$ as the contribution to the spectrum of the β'th normal mode, we have

$$\frac{\tilde{\mathcal{S}}_\beta(\mathbf{r},\omega)}{\Delta\Omega S_0} = \frac{(2\pi)^5 \omega^4}{8c^4}$$

$$\times \Lambda_\beta \, \mathbf{O}(\hat{\mathbf{r}}) \cdot \hat{\mathbf{e}}_I \cdot \langle \tilde{\boldsymbol{\chi}}_\beta(k\hat{\mathbf{r}} - \mathbf{k}_I, \omega - \omega_I) \tilde{\boldsymbol{\chi}}_\beta{}^*(k\hat{\mathbf{r}} - \mathbf{k}_I, \omega - \omega_I) \rangle \cdot \hat{\mathbf{e}}_I \cdot \mathbf{O}(\hat{\mathbf{r}}), \tag{12.34}$$

which is the final result. Note that this result is intuitively clear, in spite of the tedium necessary to derive it. The operation $\hat{e}_l \cdot \boldsymbol{\chi}$ acting from left and right gives the dipole that is radiating. The tensor $\mathbf{O}(\hat{r})$ projects the dipole onto the observation plane. The remaining complication, i.e., the dyad product of the susceptibilities, comes from the fact that the dipoles produced from different fluctuations destructively interfere as often as they constructively interfere. Hence there can be no contribution to the answer from cross terms referring to different fluctuations. If, in contrast, we were to take all scattering terms to be completely coherent with respect to each other, then the dyad would become $\boldsymbol{\chi} \cdot \boldsymbol{\chi}$ and the expression would reduce to the structural form $|\mathbf{O}(\hat{r}) \cdot \mathbf{p}_0|^2$ as in Eq. (1.47).

12.1.4 Polarization Properties

The polarization properties of the light scattered by a given normal mode can reveal information about the scatterers. This is discussed in some detail later in sections dealing with Raman and Rayleigh scattering and is the subject of a large number of exercises. The point of the discussion is illustrated in Figure 12.2. If one considers the polarization parallel to an incident field, then there is a large net dipole contribution parallel to the field, which radiates light that has the same

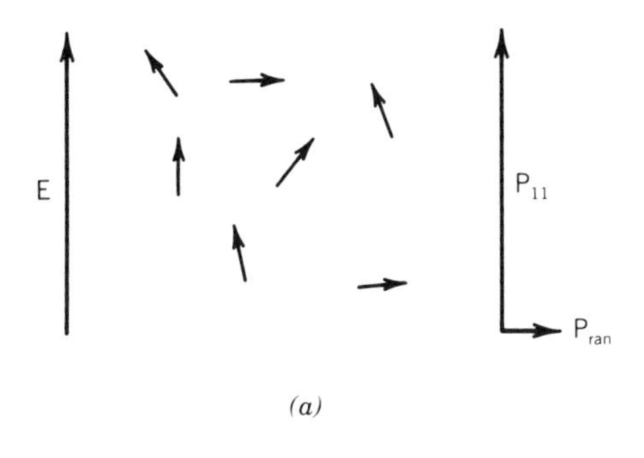

(a)

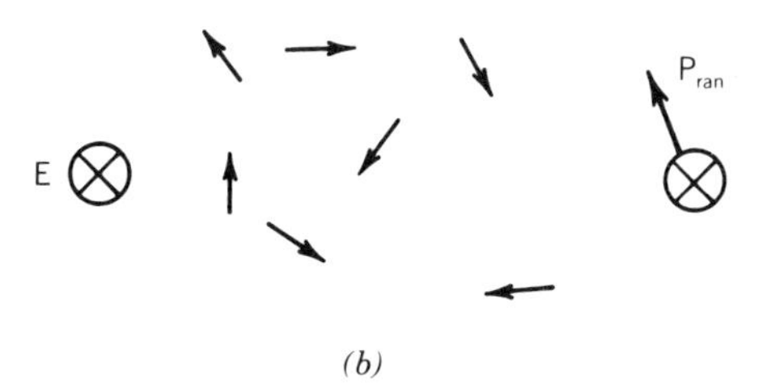

(b)

Figure 12.2. Depolarized scattering. a) View normal to the incident field; large polarization parallel to the field, small component perpendicular, scattering largely polarized. b) View in direction of incident field; no coherent component; light completely depolarized (i.e., direction of electric field vector completely statistical).

polarization properties as the incident field. However there is a small random polarization component perpendicular to the field due to random orientation of anisotropic molecules. If one observes along the direction of polarization of the incident field, then one only observes light emitted by this random polarization. Such fields, which have no systematic direction or sense of polarization, are said to be depolarized. Here we establish a formalism for calculating the degree of polarization. (Mechanisms for depolarized Brillouin scattering can be much more complex than a simple reorientation.)

Let us begin by considering what we mean by light that is in a definite state of polarization. In general, a field amplitude is written as

$$\boldsymbol{E} = (\hat{\mathbf{e}}_a a + \hat{\mathbf{e}}_b b)E$$

such that $\hat{\mathbf{e}}_a \cdot \hat{\mathbf{e}}_b = 0$, $\hat{\mathbf{e}}_a \cdot \hat{\mathbf{k}} = 0$, $\hat{\mathbf{e}}_b \cdot \hat{\mathbf{k}} = 0$ and

$$|a|^2 + |b|^2 = 1.$$

We now define a tensor $\boldsymbol{\rho}$ such that

$$\boldsymbol{\rho} = \frac{\tilde{\boldsymbol{S}}_\beta(\mathbf{r},\omega)}{\mathrm{Tr}\{\tilde{\boldsymbol{S}}_\beta(\mathbf{r},\omega)\}} \tag{12.35}$$

with the obvious property that

$$\mathrm{Tr}\{\boldsymbol{\rho}\} = 1. \tag{12.36}$$

For polarized light

$$\boldsymbol{\rho} = \begin{pmatrix} |a|^2 & ab^* \\ a^*b & |b|^2 \end{pmatrix} \tag{12.37}$$

and it is left as an exercise to show that

$$\mathrm{Tr}\{\boldsymbol{\rho}\cdot\boldsymbol{\rho}\} = 1. \tag{12.38}$$

The fact that the light is polarized is expressed in the off-diagonal elements of $\boldsymbol{\rho}$. For completely depolarized light,

$$\boldsymbol{\rho} = \frac{1}{2}\begin{pmatrix} 1 & 0 \\ 0 & 1 \end{pmatrix} \tag{12.39}$$

for which $\mathrm{Tr}\{\boldsymbol{\rho}\} = 1$, as required, but

$$\mathrm{Tr}\{\boldsymbol{\rho}\cdot\boldsymbol{\rho}\} = \frac{1}{2}\,. \tag{12.40}$$

The degree of polarization can be expressed by the parameter γ such that

$$\gamma^2 = 2\,\mathrm{Tr}\{\boldsymbol{\rho}\cdot\boldsymbol{\rho}\} - 1. \tag{12.41}$$

For $\gamma = 0$, the light is depolarized and $\gamma = 1$ describes polarized scattering. These definitions are more transparent in the context of the specific cases discussed in subsequent sections.

12.2 RAYLEIGH SCATTERING

Rayleigh and Mie scattering are caused by local fluctuations that exhibit no special dynamical structure. The example generally used to illustrate this case is a single atom existing as a fluctuation in a space that is otherwise empty. An equally good example is a perfectly uniform lattice with one atom missing. These two cases are illustrated in Figure 12.3. By Babinet's principle, the scattering patterns of the hole plus that of the single atom give the scattering pattern of a homogeneous lattice, which is zero. Hence the powers scattered in both (a) and (b) are identical. The details of this case were dealt with in Chapters 1 and 4 and are not repeated here.

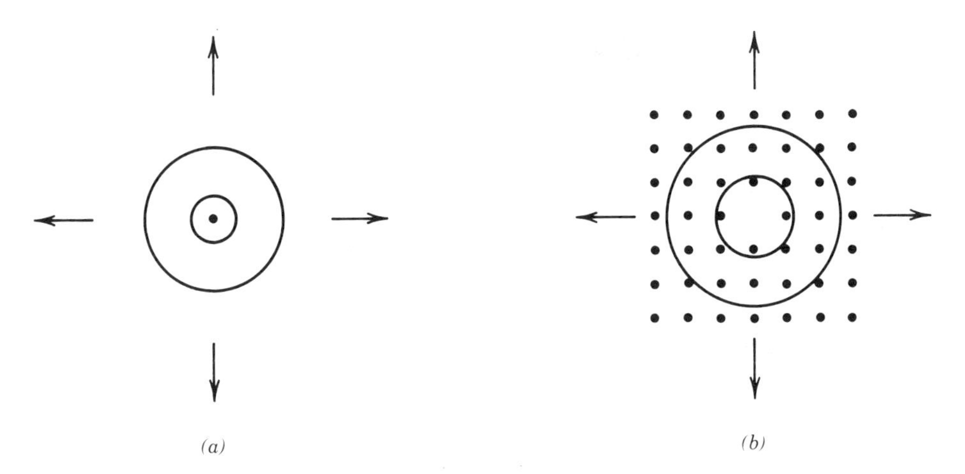

Figure 12.3. Rayleigh scattering a) from a single molecule b) from a vacancy in a lattice. If the molecules in a) and b) are identical, then Babinet's principle requires that the field in a) cancel the field in b).

12.2.1 WEAK SUSCEPTIBILITY

We consider a general description of this problem, one in which there is a space-time fluctuation in the local density. The susceptibility is written as

$$\boldsymbol{\chi} = \boldsymbol{\chi} + \frac{\delta N \boldsymbol{\alpha}}{\varepsilon_0} \tag{12.42}$$

where α is the polarizability of the atoms (or molecules). As a relatively manageable example of the fluctuation δN we take

$$\delta N = \begin{cases} \dfrac{\delta n_0}{(\pi\rho_0)^{3/2}} e^{-(r-r_0)^2/\rho_0^2} e^{-\Gamma(t-t_0)} & t > t_0 \\ 0 & t < t_0. \end{cases} \tag{12.43}$$

This fluctuation has an effective radius ρ_0 and lifetime Γ^{-1}. We now apply Eqs. (12.3), in the form

$$\tilde{\chi}_W(\mathbf{K},\Omega) = \frac{1}{2\pi^4} \int d^3r \int dt\, e^{-i\mathbf{K}\cdot\mathbf{r}+i\Omega t}\, \chi_W(\mathbf{r},t),$$

which gives

$$\tilde{\chi}_W(\mathbf{K},\Omega) = \frac{\delta n_0 \alpha}{(2\pi)^4 \varepsilon_0} e^{-i\mathbf{K}\cdot\mathbf{r}_0} e^{i\Omega t_0} e^{-K^2\rho_0^2/4} \frac{1}{\Gamma - i\Omega}. \tag{12.44}$$

From Eq. (12.44) we note that

$$\tilde{\chi}_W(\mathbf{K},\Omega) = \tilde{\chi}_W^*(-\mathbf{K},-\Omega), \tag{12.45}$$

which is a general relation for all $\tilde{\chi}_W(\mathbf{K},\Omega)$.

The nature of the scattering can be deduced from Eq. (12.44). Since Γ is usually less than 10^3 Hz, then Ω is small and Rayleigh scattering occurs, for all practical purposes, at the incident light frequency, i.e., $\omega = \omega_l \pm \Omega \cong \omega_l$. Furthermore, the magnitudes of the wave vector $\mathbf{K}$ available for the scattering process are determined by the $\exp(-K^2\rho_0^2/4)$ term. Since $\mathbf{k}_s = \mathbf{k}_l \pm \mathbf{K}$, the range of the scattered wave vector is determined by the range of $\mathbf{K}$ over which the Fourier coefficients $\tilde{\chi}_W(\mathbf{K},\Omega)$ are reasonably large. It is useful to distinguish between two regimes:

(a) If $\rho_0 k \ll 1$, i.e., $\rho_0 \ll \lambda$, the phase match condition constrains K to vary from 0 to 2k. In this regime $\exp(-K^2\rho_0^2/4) \simeq 1$ and there is no spatial variation due to this term. The angular variation of scattering is determined only by the angular terms implicit in the tensor dot products. This limit is called Rayleigh scattering, which is caused by inhomogeneities whose dimensions are small on the scale of an optical wavelength.

(b) If $\rho_0 k \gg 1$ (i.e., $\rho_0 \gg \lambda$), the range of allowed K's is small compared to $\mathbf{k}_s$ and $\mathbf{k}_l$. Since $|\mathbf{k}_s - \mathbf{k}_l| = |\mathbf{K}| \ll |\mathbf{k}|$, scattering occurs primarily in the forward direction. This is called Mie scattering.

The name Tyndall scattering is sometimes applied to describe scattering by dust on surfaces, for example. It is just a special case of Mie scattering. Note that $\tilde{\chi}_W$ is dependent on α, which is frequently a random variable with respect to the orientation of a molecule. This in turn can lead to depolarized Rayleigh scattering.

12.2.2 Frequency Spectrum

In order to evaluate the frequency spectrum of the scattered light, it is first necessary to deal with the susceptibility correlation function. In the case of single atoms (or molecules) $\delta N = 1$ and $\exp(-K^2\rho_0^2/4) \cong 1$ in Eq. (12.44). Then

$$\langle \tilde{\boldsymbol{\chi}}_\beta(k\hat{\mathbf{r}}-\mathbf{k}_I)\tilde{\boldsymbol{\chi}}_\beta^*(k\hat{\mathbf{r}}-\mathbf{k}_I)\rangle = \frac{\langle \boldsymbol{\alpha}_\beta \boldsymbol{\alpha}_\beta^* \rangle}{(2\pi)^8 \varepsilon_0^2(\Gamma_\beta^2 + (\omega-\omega_I)^2)}. \qquad (12.46)$$

If we further assume that the atom (molecule) is isotropic,

$$\langle \boldsymbol{\alpha}_\beta \boldsymbol{\alpha}_\beta^* \rangle = \alpha_\beta^2 \mathbf{II}$$

and

$$\frac{\tilde{\boldsymbol{S}}_\beta(\mathbf{r},\omega)}{\Delta\Omega S_0} = \frac{\omega^4}{8c^4(2\pi)^3\varepsilon_0^2}(\mathbf{O}(\hat{\mathbf{r}})\cdot\hat{\mathbf{e}}_I)(\mathbf{O}(\hat{\mathbf{r}})\cdot\hat{\mathbf{e}}_I)\sum_\beta \frac{\Lambda_\beta \alpha_\beta^2}{\Gamma_\beta^2 + (\omega-\omega_I)^2}. \qquad (12.47)$$

Rayleigh scattering is characterized by a Lorentzian line shape. By measuring the half-width of the Lorentzian, the fluctuation decay rate can be deduced. Furthermore, if the scattering molecules are isotropic, the scattering is called polarized and the radiation pattern is that of an oscillating dipole. By integrating $\tilde{S}(\mathbf{r},\omega)$ over ω one obtains the total scattered intensity into the solid angle $\Delta\Omega$ as

$$\frac{\boldsymbol{S}_\beta(\mathbf{r})}{\Delta\Omega S_0} = \frac{\omega^4}{4c^4(4\pi)^2\varepsilon_0^2}(\mathbf{O}(\hat{\mathbf{r}})\cdot\hat{\mathbf{e}}_I)(\mathbf{O}(\hat{\mathbf{r}})\cdot\hat{\mathbf{e}}_I)\sum_\beta N_\beta \, \alpha_\beta^2, \qquad (12.48)$$

where N_β gives the steady state excitation of the β'th normal mode. Following the convention in Eq. (12.28) the trace of the right-hand side of Eq. (12.48) is also a differential cross section and is denoted $d\sigma/d\Omega$. If one integrates Eq. (12.47) over solid angle instead of frequency the result is written $d\sigma/d\omega$, and if one integrates over both variables one obtains the scattering cross section discussed in Section 4.2.

12.2.3 Depolarization Ratios

We now consider in more detail the polarization properties of Rayleigh scattering for general $\boldsymbol{\alpha}$. We define a fourth rank tensor $\mathbf{f}$ such that

$$\mathbf{f} = \langle \alpha\alpha \rangle$$

where we have assumed that all of the coefficients are real numbers (i.e., that we are far from resonances). In Section 8.2.2 we expressed the polarizability in terms of symmetry axes and planes as

$$\boldsymbol{\alpha} = \alpha_\perp \mathbf{O}(\hat{\mathbf{s}}_a) + \alpha_\parallel \hat{\mathbf{s}}_a\hat{\mathbf{s}}_a$$

where $\hat{\mathbf{s}}_a$ is a unit vector which points along the symmetry axis of a linear molecule or lies in the plane of a plane of symmetry. Thus

$$\mathbf{f} = \langle(\alpha_\perp \mathbf{O}(\hat{\mathbf{s}}_a) + \alpha_\parallel \hat{\mathbf{s}}_a\hat{\mathbf{s}}_a)(\alpha_\perp \mathbf{O}(\hat{\mathbf{s}}_a) + \alpha_\parallel \hat{\mathbf{s}}_a\hat{\mathbf{s}}_a)\rangle. \tag{12.49}$$

The computing of averages of fourth rank tensors is tedious, and we simplify matters by noting that there are only three nonzero components for the cases of interest, i.e., Eq. (12.49). These are the same nonzero components we saw for the electronic Kerr effect tensor $\mathbf{S}$ discussed in Section 8.2.2 . Using the Voigt notation, they can be written as

$$f_{11} = \alpha_\perp^2 + \frac{2}{3}\alpha_\perp(\alpha_\parallel - \alpha_\perp) + \frac{1}{5}(\alpha_\parallel - \alpha_\perp)^2, \tag{12.50a}$$

$$f_{12} = \alpha_\perp^2 + \frac{2}{3}\alpha_\perp(\alpha_\parallel - \alpha_\perp) + \frac{1}{15}(\alpha_\parallel - \alpha_\perp)^2, \tag{12.50b}$$

$$f_{44} = \frac{1}{15}(\alpha_\parallel - \alpha_\perp)^2, \tag{12.50c}$$

where

$$f_{11} = f_{22} = f_{33}, \tag{12.50d}$$

$$f_{12} = f_{13} = f_{23} = f_{32} = f_{31} = f_{21}, \tag{12.50e}$$

and

$$f_{44} = f_{55} = f_{66}. \tag{12.50f}$$

As an example, we consider an incident field polarized along the x (denoted 1) axis. Then the only nonzero components of $\hat{e}_1 f_{1ij1} \hat{e}_1$ are

$$\hat{\mathbf{e}}_I \cdot \mathbf{f} \cdot \hat{\mathbf{e}}_I = \begin{pmatrix} f_{1111} & 0 & 0 \\ 0 & f_{1221} & 0 \\ 0 & 0 & f_{1331} \end{pmatrix}.$$

In terms of contracted indices we can write this as

$$f_{44}\mathbf{I} + (f_{11} - f_{44})\hat{\mathbf{x}}\hat{\mathbf{x}}. \tag{12.51}$$

When the scattered field direction is also taken as the x axis, i.e., we take a 90° scattering angle, $\mathbf{O}(\hat{\mathbf{x}}) = \mathbf{I} - \hat{\mathbf{x}}\hat{\mathbf{x}}$ and

$$\mathbf{O}(\hat{\mathbf{x}}) \equiv \begin{pmatrix} 0 & 0 & 0 \\ 0 & 1 & 0 \\ 0 & 0 & 1 \end{pmatrix}. \tag{12.52}$$

Therefore

$$\mathbf{O}(\hat{\mathbf{x}}) \cdot \{\hat{\mathbf{e}}_I \cdot \mathbf{f} \cdot \hat{\mathbf{e}}_I\} = f_{44}(\mathbf{I} - \hat{\mathbf{x}}\hat{\mathbf{x}}) = f_{44}\mathbf{O}(\hat{\mathbf{x}}). \tag{12.53}$$

Furthermore,

$$\mathbf{O}(\hat{\mathbf{x}}) \cdot \{\hat{\mathbf{e}}_I \cdot \mathbf{f} \cdot \hat{\mathbf{e}}_I\} \cdot \mathbf{O}(\hat{\mathbf{x}}) = f_{44}\mathbf{O}(\hat{\mathbf{x}}). \tag{12.54}$$

Since $\tilde{\boldsymbol{S}}_\beta(\mathbf{r},\omega) \propto f_{44}\mathbf{O}(\hat{\mathbf{x}})$ and $\mathrm{Tr}\{\mathbf{O}(\hat{\mathbf{x}})\} = 2$, then

$$\boldsymbol{\rho} = \frac{1}{2}\,\mathbf{O}(\hat{\mathbf{x}}). \tag{12.55}$$

As a result, $\mathrm{Tr}\{\boldsymbol{\rho}\cdot\boldsymbol{\rho}\} = 1/2$ and Eq. (12.41) gives $\gamma = 0$, i.e., the light is completely depolarized, which is the result anticipated in Figure 12.2b.

If one observes in the $\hat{\mathbf{y}}$ or $\hat{\mathbf{z}}$ direction (the two are equivalent) then it is possible to measure the degree of optical anisotropy of the molecule by measuring the depolarization. The details are left as an exercise and we outline here the major steps. Starting from Eq. (12.51),

$$\mathbf{O}(\hat{\mathbf{y}}) \cdot \{\hat{\mathbf{e}}_I \cdot \mathbf{f} \cdot \hat{\mathbf{e}}_I\} \cdot \mathbf{O}(\hat{\mathbf{y}}) = f_{44}\mathbf{O}(\hat{\mathbf{y}}) + (f_{11} - f_{44})\hat{\mathbf{x}}\hat{\mathbf{x}}. \tag{12.56}$$

The trace is $f_{11} + f_{44}$. Then

$$\boldsymbol{\rho} = \frac{f_{44}\mathbf{O}(\hat{\mathbf{y}}) + (f_{11}-f_{44})\hat{\mathbf{x}}\hat{\mathbf{x}}}{f_{11}+f_{44}},$$

from which one obtains

$$\boldsymbol{\rho}\cdot\boldsymbol{\rho} = \frac{f_{44}{}^2\mathbf{O}(\hat{\mathbf{y}}) + (f_{11}{}^2-f_{44}{}^2)\hat{\mathbf{x}}\hat{\mathbf{x}}}{(f_{11}+f_{44})^2}$$

and therefore

$$\mathrm{Tr}(\boldsymbol{\rho}\cdot\boldsymbol{\rho}) = \frac{f_{11}{}^2+f_{44}{}^2}{(f_{11}+f_{44})^2}$$

and

$$\gamma = \left|\frac{f_{11} - f_{44}}{f_{11} + f_{44}}\right|. \tag{12.57}$$

Combining this result with Eqs. (12.50a), (12.50b), and (12.50c),

$$\gamma = \left|\frac{1 + \frac{2}{3}\left\{\frac{\alpha_\parallel}{\alpha_\perp} - 1\right\} + \frac{2}{15}\left\{\frac{\alpha_\parallel}{\alpha_\perp} - 1\right\}^2}{1 + \frac{2}{3}\left\{\frac{\alpha_\parallel}{\alpha_\perp} - 1\right\} + \frac{4}{15}\left\{\frac{\alpha_\parallel}{\alpha_\perp} - 1\right\}^2}\right|$$

and the ratio $\alpha_\parallel/\alpha_\perp$ can be deduced from a measurement of γ. Note that for highly anisotropic molecules $\gamma = 1/2$, and for isotropic molecules $\gamma = 1$ and hence the scattering is polarized.

12.3 RAMAN SCATTERING

Raman scattering arises from the motion of those degrees of freedom of individual molecules that are not themselves dipole active but instead modulate the dipoles of the dipole active modes. These Raman active modes are discussed in Section 3.3, and include both rotational and vibrational modes. In this section we discuss how these modes scatter light.

12.3.1 Weak Susceptibilities

We start with the simple example of a spherical oscillating charge distribution (which is not dipole active and hence nonradiative). An example of this case is the methane molecule undergoing its symmetric mode of vibration as illustrated in Figure 12.4. We model this normal coordinate as a radially symmetric (i.e., isotropic) polarizable body that expands and contracts radially, inducing a change in polarizability.

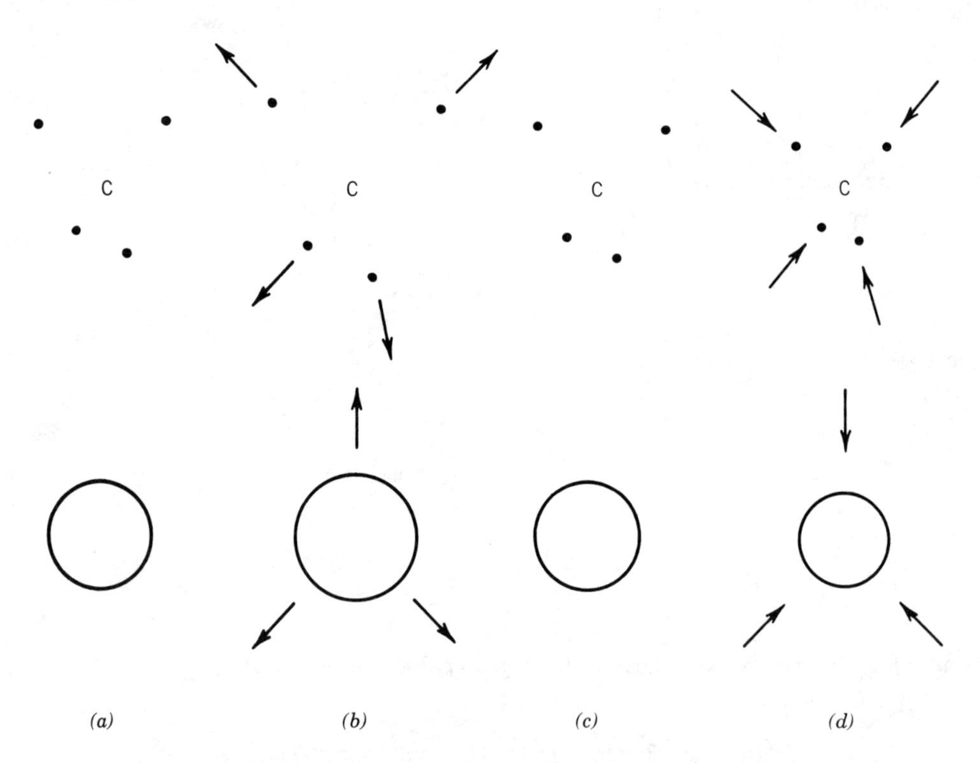

Figure 12.4. Breathing mode of methane (CH_4, hydrogens shown as dots); columns a)-d) illustrate complete cycle of oscillation. Below, model of motion; a radially pulsating polarizable body.

The time dependence of the normal coordinate can be written as

$$q_\beta = q_{\beta 0}\cos[\Omega_\beta(t-t_0) + \phi_\beta]\, e^{-\Gamma_\beta(t-t_0)}, \quad t > t_0, \tag{12.58}$$

where ϕ_β is the random phase of the oscillation. In the general case, we can write for one molecule

$$\boldsymbol{\chi}_w = \frac{\delta\boldsymbol{\alpha}}{\varepsilon_0}\, \frac{e^{-(\mathbf{r}-\mathbf{r}_0)^2/\rho_0^2}}{(\pi\rho_0)^{3/2}} \tag{12.59}$$

where ρ_0 is typically the radius of the molecule and the $\delta\boldsymbol{\alpha}$ is the fluctuation produced in the polarizability by a normal mode vibration. For the symmetric case $\delta\boldsymbol{\alpha} = \delta\alpha\mathbf{I}$. Furthermore,

$$\delta\boldsymbol{\alpha} = \sum_\beta \frac{\partial\boldsymbol{\alpha}}{\partial q_\beta}\, q_{\beta 0}\cos(\Omega_\beta(t-t_0)+\phi_\beta)\, e^{-\Gamma_\beta(t-t_0)}, \tag{12.60}$$

where $\partial\boldsymbol{\alpha}/\partial q_\beta$ determines the change in polarizability produced by the β'th Raman active mode. The influence of a polarizability described by Eq. (12.60) on the scattered field is illustrated in Figure 12.5a. The scattered field is amplitude modulated and a spectral resolution of the modulated field gives a carrier and two sidebands as illustrated in Figure 12.5b. We now Fourier transform $\boldsymbol{\chi}_w(\mathbf{r},t)$,

$$\tilde{\boldsymbol{\chi}}_w(\mathbf{K},\Omega) = \frac{1}{(2\pi)^4}\, e^{-i\mathbf{K}\cdot\mathbf{r}_0 + i\Omega t_0}\, e^{-K^2\rho_0^2/4}$$

$$\sum_\beta \partial\boldsymbol{\alpha}/\partial q_\beta\, q_{0\beta}\, \frac{1}{2\varepsilon_0}\left(\frac{e^{i\phi_\beta}}{\Gamma_\beta - i(\Omega+\Omega_\beta)} + \frac{e^{-i\phi_\beta}}{\Gamma_\beta - i(\Omega-\Omega_\beta)}\right). \tag{12.61}$$

For molecules, $\lambda \gg \rho_0$ and we can drop the term $\exp(-K^2\rho_0^2/4)$ (see Eq. (12.44)). From the structure of Eq. (12.61), we expect light to be scattered at the frequencies $\omega_I \pm \Omega_\beta$. The case $\omega_I - \Omega_\beta$ is called the Stokes line and the light at $\omega_I + \Omega_\beta$ is called the anti-Stokes line (see Figure 12.5b). If one chooses a molecule whose Raman polarizability $\partial\boldsymbol{\alpha}/\partial q_\beta$ is spatially anisotropic, for example, a linear molecule, this results in depolarized vibrational Raman scattering. In practice, the anti-Stokes line is often not observed. This is a quantum mechanical effect and is discussed in more detail later.

Rotating molecules also give rise to Raman scattering. The effect does not occur for molecules whose polarizability (at optical frequencies) is spherically symmetric since there is no variation in polarizability for a rotating isotropic charge. The rotational Raman effect always involves anisotropic molecules and therefore needs tensors for its analysis. In addition, the rotational Raman effect involves frequency shifts of twice the rotational frequency Ω_β (see Problem 3.5).

The details of the rotational Raman effect involve careful consideration of the angle which the molecular axis makes with the applied field. In Figure 12.6 we illustrate the coordinate system for a linear molecule rotating with the angular velocity Ω_β. The normal coordinate is orthogonal to the symmetry axis of the molecule denoted by $\hat{\mathbf{s}}_\beta$. The general angular velocity is given by

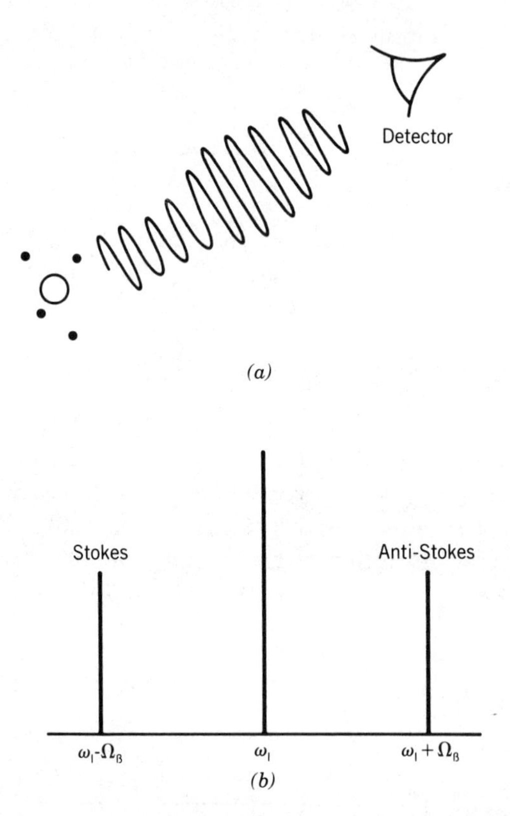

Figure 12.5. Scattered radiation from CH_4 (see Figure 12.4). a) Amplitude modulated light scattered due to sinusoidal variation in polarizability. b) Spectrum of scattered field showing Stokes and anti-Stokes sidebands. Note that the modulation frequency, and hence the sideband frequency, is independent of scattering angle. This is in contrast to Brillouin scattering shown in Figure 12.8.

$$\mathbf{\Omega}_\beta = (\sin\theta\ \cos\phi,\ -\sin\theta\sin\phi,\ \cos\theta)\Omega_\beta \tag{12.62}$$

and the vector $\hat{\mathbf{s}}_\beta$ is always orthogonal to $\mathbf{\Omega}_\beta$. We simplify matters by noting that the pair of vectors

$$\hat{\mathbf{e}}_1 = (\sin\phi,\ \cos\phi,\ 0)$$

$$\hat{\mathbf{e}}_2 = (\cos\theta\cos\phi,\ -\cos\theta\sin\phi,\ -\sin\theta)$$

is always orthogonal to $\mathbf{\Omega}_\beta$. Thus the rotation of the vector $\hat{\mathbf{s}}_\beta$ can be described by

$$\hat{\mathbf{e}}_+ = \frac{1}{\sqrt{2}}\,(\hat{\mathbf{e}}_1 + i\hat{\mathbf{e}}_2)$$

and

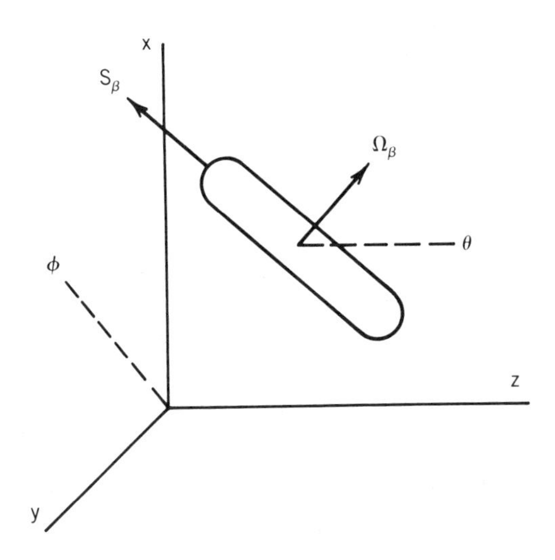

Figure 12.6. Schematic of rotating Raman active molecule. Angle θ defined with respect to z axis, ϕ with respect to x axis.

$$\hat{\mathbf{s}}_\beta = \frac{1}{2}\,[\hat{\mathbf{e}}_+\, e^{-i[\Omega_\beta(t-t_0)+\phi_\beta]} + cc]\; e^{-\Gamma_\beta(t-t_0)}. \tag{12.63}$$

Rotation in the opposite direction, given by $\hat{\mathbf{e}}_-$, can be obtained by $\boldsymbol{\Omega}_\beta \to -\boldsymbol{\Omega}_\beta$, which is already included in the general form of Eq. (12.62). From Eq. (8.22) for the polarizability (retaining only the time dependent terms)

$$\boldsymbol{\alpha}(t) = (\alpha_\| - \alpha_\perp)\hat{\mathbf{s}}_\beta\hat{\mathbf{s}}_\beta \tag{12.64}$$

where $\alpha_\|$ and $\alpha_\perp$ are optical frequency polarizabilities. Substituting Eq. (12.63) into (12.64),

$$\boldsymbol{\alpha}(t) = (\alpha_\| - \alpha_\perp)\frac{1}{4}\,[\hat{\mathbf{e}}_+\hat{\mathbf{e}}_+\, e^{-2i[\Omega_\beta(t-t_0)+\phi_\beta]} + cc]\, e^{-2\Gamma_\beta(t-t_0)} \tag{12.65}$$

Note that this equation shows explicitly the $2\Omega_\beta$ nature of the scattering.

We are now able to evaluate the weak susceptibility term. One writes

$$\boldsymbol{\chi}_W = \frac{\delta n_0 \boldsymbol{\alpha}(t)}{\varepsilon_0}\;\frac{e^{-(r-r_0)^2/\rho_0^2}}{\pi^{3/2}\rho_0^3}$$

and makes the same assumptions used before that $\rho_0 \ll \lambda$ and $\delta n_0 = 1$ to obtain

$$\tilde{\boldsymbol{\chi}}_W(\mathbf{K},\Omega) = \frac{(\alpha_\parallel - \alpha_\perp)}{(2\pi)^4 \varepsilon_0} e^{-iKr_0+i\Omega t_0}$$

$$\frac{1}{4}\Big(\hat{\mathbf{e}}_+\hat{\mathbf{e}}_+ \frac{e^{2i\phi_\beta}}{2\Gamma_\beta - i(\Omega+2\Omega_\beta)} + \hat{\mathbf{e}}_-\hat{\mathbf{e}}_- \frac{e^{-2i\phi_\beta}}{2\Gamma_\beta - i(\Omega-2\Omega_\beta)}\Big). \tag{12.66}$$

From this equation it is clear that there are Stokes and anti-Stokes components to the scattered light, both shifted by $2\Omega_\beta$ from the incident light frequency.

12.3.2 Frequency Spectrum

The key to calculating the frequency spectrum is the evaluation of the statistical quantity $\boldsymbol{\Delta}(k\hat{\mathbf{r}}-\mathbf{k}_I,\omega-\omega_I)$ given in Eq. (12.33). In this section we evaluate this parameter for both rotational and vibrational Raman scattering.

12.3.2.1 Vibrational Raman Scattering

First we treat the susceptibility correlation function. Thus

$$\langle \tilde{\boldsymbol{\chi}}_\beta(k\hat{\mathbf{r}}-\mathbf{k}_I,\omega-\omega_I)\tilde{\boldsymbol{\chi}}_\beta^*(k\hat{\mathbf{r}}-\mathbf{k}_I,\omega-\omega_I)\rangle = \frac{1}{4(2\pi)^8\varepsilon_0^2}$$

$$\sum_\beta \langle \frac{\partial\boldsymbol{\alpha}}{\partial q_\beta}\frac{\partial\boldsymbol{\alpha}^*}{\partial q_\beta}\rangle \langle q_{0\beta}q_{0\beta}^*\rangle \Big(\frac{1}{\Gamma_\beta^2+(\omega-\omega_I+\Omega_\beta)^2} + \frac{1}{\Gamma_\beta^2+(\omega-\omega_I-\Omega_\beta)^2}\Big). \tag{12.67}$$

Note that the cross terms between the Stokes and anti-Stokes components of Eq. (12.61) vanish on statistical averaging over the random phase ϕ_β. We also assume that fluctuations between different modes β and β' are uncorrelated and hence

$$\sum_\beta\sum_\beta \rightarrow \sum_\beta \delta_{\beta\beta'}.$$

Since the vibrations are thermally excited, the energy per mode, u_β, is given by twice the potential energy:

$$u_\beta = \langle q_{0\beta}q_{0\beta}^*\rangle m_\beta\Omega_\beta^2. \tag{12.68}$$

For KT >> $\hbar\Omega_\beta$, the energy per mode is just KT. Therefore

$$\langle q_{0\beta}q_{0\beta}^*\rangle = \frac{KT}{m_\beta\Omega_\beta^2}. \tag{12.69}$$

Furthermore, under steady state conditions $\Lambda_\beta = \Gamma_\beta$ and hence

$$\boldsymbol{\Delta}(k\hat{\mathbf{r}}-\mathbf{k}_I,\omega-\omega_I) = \frac{1}{4(2\pi)^7\varepsilon_0^2}\sum_\beta \langle\frac{\partial\boldsymbol{\alpha}}{\partial q_\beta}\frac{\partial\boldsymbol{\alpha}^*}{\partial q_\beta}\rangle \frac{kT}{m_\beta\Omega_\beta^2}$$

$$\times \left(\frac{\Gamma_\beta/2\pi}{(\omega-\omega_I+\Omega_\beta)^2+\Gamma_\beta{}^2} + \frac{\Gamma_\beta/2\pi}{(\omega-\omega_I-\Omega_\beta)^2+\Gamma_\beta{}^2} \right). \quad (12.70)$$

The result is written in this particular form because the integral over frequency required to evaluate the total scattered intensity just gives unity for the quantity inside the larger parenthases.

If $KT \gg \hbar\Omega_\beta$ is not valid, the intensities of the two frequency-shifted components are no longer equal. In that case, one must introduce the Boltzmann factor and ask whether the scattering process creates a vibrational quantum (the Stokes process) or destroys a vibrational quantum. The energy normalization is different for the two processes. The mode energy u_β is now given by the product of the optical phonon energy ($\hbar\Omega_\beta$) and the probability that the initial state (prior to scattering) is occupied. This gives

$$u_\beta = \hbar\Omega_\beta \frac{1}{1-e^{-\hbar\Omega_\beta/KT}}, \quad (12.71a)$$

for the Stokes line and

$$u_\beta = \hbar\Omega_\beta \frac{e^{-\hbar\Omega_\beta/KT}}{1-e^{-\hbar\Omega_\beta/KT}}, \quad (12.71b)$$

for the anti-Stokes line. In both cases, if $KT \gg \hbar\Omega_\beta$, Eq. (12.68), the classical limit, is recovered. Thus Eq. (12.70) now becomes

$$\mathbf{\Delta}(k\hat{\mathbf{r}}-\mathbf{k}_I,\omega-\omega_I) = \frac{1}{4(2\pi)^7\varepsilon_0} \sum_\beta \left\langle \frac{\partial \boldsymbol{\alpha}}{\partial q_\beta} \frac{\partial \boldsymbol{\alpha}^*}{\partial q_\beta} \right\rangle \frac{1}{m_\beta \Omega_\beta{}^2}$$

$$\times \frac{\hbar\Omega_\beta}{1-e^{-\hbar\Omega_\beta/KT}} \left(\frac{\Gamma_\beta/2\pi}{(\omega-\omega_I+\Omega_\beta)^2+\Gamma_\beta{}^2} + e^{-\hbar\Omega_\beta/KT} \frac{\Gamma_\beta/2\pi}{(\omega-\omega_I-\Omega_\beta)^2+\Gamma_\beta{}^2} \right). \quad (12.72)$$

If $\hbar\Omega_\beta > KT$, only the Stokes process contributes to the scattered light spectrum, as observed experimentally. Equation (12.72) now contains all of the information about the normal modes.

To obtain the detailed frequency spectrum of light scattered by vibrational normal modes, one substitutes Eq. (12.72) in (12.34). The result is

$$\frac{\tilde{\boldsymbol{S}}_\beta(\mathbf{r},\omega)}{\Delta\Omega S_0} = \frac{\omega^4}{8c^4(4\pi)^2\varepsilon_0{}^2} \mathbf{O}(\hat{\mathbf{r}})\cdot\hat{\mathbf{e}}_I \cdot \sum_\beta \left\langle \frac{\partial \boldsymbol{\alpha}}{\partial q_\beta} \frac{\partial \boldsymbol{\alpha}^*}{\partial q_\beta} \right\rangle \frac{\hbar}{m_\beta \Omega_\beta} \frac{1}{1-e^{-\hbar\Omega_\beta/KT}}$$

$$\times \left(\frac{\Gamma_\beta/2\pi}{(\omega-\omega_I+\Omega_\beta)^2+\Gamma_\beta{}^2} + e^{-\hbar\Omega_\beta/KT} \frac{\Gamma_\beta/2\pi}{(\omega-\omega_I-\Omega_\beta)^2+\Gamma_\beta{}^2} \right)\cdot\hat{\mathbf{e}}\cdot\mathbf{O}(\hat{\mathbf{r}}). \quad (12.73)$$

For spherically symmetrical normal modes far from resonances

$$\langle \frac{\partial \boldsymbol{\alpha}}{\partial q_\beta} \frac{\partial \boldsymbol{\alpha}^*}{\partial q_\beta} \rangle = \left| \frac{\partial \alpha}{\partial q_\beta} \right|^2 \blacksquare$$

and the results simplify to

$$\frac{\tilde{\boldsymbol{S}}_\beta(\mathbf{r},\omega)}{\Delta\Omega S_0} = \frac{\omega^4}{8c^4(4\pi)^2\varepsilon_0^2} (\mathbf{O}(\hat{\mathbf{r}})\cdot\hat{\mathbf{e}}_l)(\mathbf{O}(\hat{\mathbf{r}})\cdot\hat{\mathbf{e}}_l) \sum_\beta \left|\frac{\partial\alpha}{\partial q_\beta}\right|^2 \frac{1}{1-e^{-\hbar\Omega_\beta/KT}}$$

$$\times \frac{\hbar}{m_\beta\Omega_\beta} \left(\frac{\Gamma_\beta/2\pi}{(\omega-\omega_l+\Omega_\beta)^2+\Gamma_\beta^2} + e^{-\hbar\Omega_\beta/KT} \frac{\Gamma_\beta/2\pi}{(\omega-\omega_l-\Omega_\beta)^2+\Gamma_\beta^2} \right). \tag{12.74}$$

The terms $\mathbf{O}(\hat{\mathbf{r}})\cdot\hat{\mathbf{e}}_l$ imply that the scattered light is completely polarized and give the usual $\sin\theta$ term where θ is the angle between $\hat{\mathbf{r}}$ and $\hat{\mathbf{e}}_l$. The frequency spectrum is then given by

$$\frac{\tilde{S}(\mathbf{r},\omega)}{\Delta\Omega S_0} = \frac{\omega^4}{8c^4(4\pi)^2\varepsilon_0^2} \sin^2\theta \sum_\beta \left|\frac{\partial\alpha}{\partial q_\beta}\right|^2 \frac{\hbar}{m_\beta\Omega_\beta} \frac{1}{1-e^{-\hbar\Omega_\beta/KT}}$$

$$\times \left(\frac{\Gamma_\beta/2\pi}{(\omega-\omega_l+\Omega_\beta)^2+\Gamma_\beta^2} + e^{-\hbar\Omega_\beta/KT} \frac{\Gamma_\beta/2\pi}{(\omega-\omega_l-\Omega_\beta)^2+\Gamma_\beta^2} \right). \tag{12.75}$$

For anisotropic Raman polarizabilities, the discussion is essentially identical to that for Rayleigh scattering.

12.3.2.2 Rotational Raman Scattering

We obtained in Eq. (12.66) a form for the weak susceptibility for rotational Raman scattering. In this case the statistical fluctuations involve collisions, and there are subtle issues involving the interrelationship between the distributions of angular rotational velocities and the damping term Γ_β which are beyond the scope of this discussion. These are handled somewhat more easily in quantum mechanical approaches. The statistical average required in this case is taken over the angular orientation of the molecules, which changes from collision to collision.

We now evaluate the susceptibility correlation function. From Eq. (12.66),

$$\boldsymbol{\Delta}(k\hat{\mathbf{r}}-\mathbf{k}_l,\omega-\omega_l) = \sum_\beta \frac{(\alpha_\parallel-\alpha_\perp)^2}{(2\pi)^8\varepsilon_0^2} \frac{\Lambda_\beta}{16} \left(\langle \hat{\mathbf{e}}_+\hat{\mathbf{e}}_+\hat{\mathbf{e}}_+^*\hat{\mathbf{e}}_+^* \rangle \frac{1}{4\Gamma_\beta^2+(\omega-\omega_l+2\Omega_\beta)^2} \right.$$

$$\left. + \langle \hat{\mathbf{e}}_-\hat{\mathbf{e}}_-\hat{\mathbf{e}}_-^*\hat{\mathbf{e}}_-^* \rangle \frac{1}{4\Gamma_\beta^2+(\omega-\omega_l-2\Omega_\beta)^2} \right). \tag{12.76}$$

Substituting into Eq. (12.34),

$$\frac{\tilde{\boldsymbol{S}}_\beta(\mathbf{r},\omega)}{\Delta\Omega S_0} = \frac{\omega^4(\alpha_\parallel-\alpha_\perp)^2}{64(4\pi)^2c^4}\sum_\beta \frac{\Lambda_\beta}{\Gamma_\beta}$$

$$\mathbf{O}(\hat{\mathbf{r}})\cdot\hat{\mathbf{e}}_l\cdot\Big(\langle\hat{\mathbf{e}}_+\hat{\mathbf{e}}_+\hat{\mathbf{e}}_+{}^*\hat{\mathbf{e}}_+{}^*\rangle \frac{\Gamma_\beta/\pi}{4\Gamma_\beta{}^2+(\omega-\omega_l+2\Omega_\beta)^2}$$

$$+ \langle\hat{\mathbf{e}}_-\hat{\mathbf{e}}_-\hat{\mathbf{e}}_-{}^*\hat{\mathbf{e}}_-{}^*\rangle \frac{\Gamma_\beta/\pi}{4\Gamma_\beta{}^2+(\omega-\omega_l-2\Omega_\beta)^2}\Big)\cdot\hat{\mathbf{e}}_l\cdot\mathbf{O}(\hat{\mathbf{r}}). \tag{12.77}$$

The polarization properties of the scattered light are complicated to compute (and even worse to denote) and we just summarize the results. For the fourth rank **f** tensor we obtain

$$f_{11} = \frac{2}{15}(\alpha_\parallel-\alpha_\perp)^2, \tag{12.78a}$$

$$f_{12} = -\frac{1}{15}(\alpha_\parallel-\alpha_\perp)^2, \tag{12.78b}$$

$$f_{44} = \frac{1}{10}(\alpha_\parallel-\alpha_\perp)^2. \tag{12.78c}$$

Substituting into Eq. (12.57), $\gamma = 1/7$, which indicates that the degree of polarization is low, i.e., the scattered light is almost completely depolarized.

12.4 BRILLOUIN SCATTERING

Sound waves are the collective modes of a sample and they can be excited by thermal fluctuations. In the Chapter 11 we discussed the scattering of light by sound waves in some detail, and we are able to use some of that formalism here. Since the calculation can be complex for optically and acoustically anisotropic media, we restrict our discussion to media which are both acoustically and optically isotropic.

12.4.1 Weak Polarization

The weak polarization term for acoustic interactions is written in terms of the elasto-optic tensor $\mathbf{p}$ (Eq. (11.1)) as

$$\boldsymbol{\chi}_W = -\sum_\beta n^4 i\,\mathbf{p}\cdot\mathbf{K}_\beta\cdot\frac{1}{2}\Big(\boldsymbol{U}_+(\mathbf{K}_\beta,\Omega)\,e^{i(\mathbf{K}_\beta\cdot\mathbf{r}-\Omega_\beta(t-t_0))}$$

$$+\boldsymbol{U}_-(\mathbf{K}_\beta,\Omega)\,e^{i(\mathbf{K}_\beta\cdot\mathbf{r}+\Omega_\beta(t-t_0))}+cc\Big)\,e^{-(t-t_0)\Gamma_\beta}, \tag{12.79}$$

where $\boldsymbol{U}_+$ and $\boldsymbol{U}_-$ are the acoustic displacements for forward and backward propagating normal mode sound waves. In Eq. (12.79) we have left out the random phases which have the same consequences here as in Raman scattering, namely to eliminate all contributions to the final result from cross terms involving the two counter-propagating acoustic waves. What makes this calculation different from the preceding cases (Rayleigh and Raman scattering) is that $\mathbf{K}_\beta$ and

Ω_β are connected by the dispersion relation $\Omega_\beta/K_\beta = v_\beta$ where v_β is the acoustic velocity of the β'th mode.

We now Fourier analyze $\boldsymbol{\chi}_w$ as indicated by Eqs. (12.3a) and (12.3b). The calculation is straightforward and yields

$$\tilde{\boldsymbol{\chi}}_w(\mathbf{K},\Omega) = -\frac{i}{2\pi} e^{-i\Omega t_0} \sum_\beta n^4\, \mathbf{p}\cdot\mathbf{K}_\beta \cdot \Big(\boldsymbol{U}_+(\mathbf{K}_\beta,\Omega_\beta)\,\delta(\mathbf{K}-\mathbf{K}_\beta)\,\frac{1}{\Gamma - i(\Omega-\Omega_\beta)}$$

$$+ \boldsymbol{U}_-(\mathbf{K}_\beta,\Omega_\beta)\delta(\mathbf{K}-\mathbf{K}_\beta)\,\frac{1}{\Gamma - i(\Omega+\Omega_\beta)}\Big). \tag{12.80}$$

The summation over β is over the discrete solutions to the dispersion relation as discussed in Section 2.3.2. We also note that there are three different types of modes possible, two shear and one longitudinal. Thus Eq. (12.80) is not complete and should also include a summation over the three mutually orthogonal acoustic modes. We return to this point later.

12.4.2 Statistical Fluctuations

The calculation of the $\boldsymbol{\Delta}$ term (Eq. (12.33)) is complicated by the collective and propagating character of sound waves. For the steady state case we take $\Lambda_\beta = \Gamma_\beta$, i.e., the fluctuations which generate the sound waves are replenished as fast as the acoustic waves are damped (this assumes that all sound waves have the same amplitude). Proceeding with the calculation,

$$\frac{1}{T}\langle\tilde{\boldsymbol{\chi}}(k\hat{\mathbf{r}}-\mathbf{k}_I,\omega-\omega_I)\tilde{\boldsymbol{\chi}}^*(k\hat{\mathbf{r}}-\mathbf{k}_I,\omega-\omega_I)\rangle = \sum_\beta\sum_{\beta'}\frac{n^8}{4}$$

$$\times\Big((\mathbf{p}\cdot\mathbf{K}_\beta)\cdot\langle\boldsymbol{U}_+(\mathbf{K}_\beta,\Omega_\beta)\boldsymbol{U}_+^*(\mathbf{K}_\beta,\Omega_{\beta'})\rangle\cdot(\mathbf{p}\cdot\mathbf{K}_{\beta'})\delta(\mathbf{K}-\mathbf{K}_\beta)$$

$$\delta(\mathbf{K}-\mathbf{K}_{\beta'})\,\frac{\Gamma_\beta/\pi}{\Gamma_\beta - i(\Omega-\Omega_\beta)}\,\frac{1}{\Gamma_{\beta'}+i(\Omega-\Omega_{\beta'})}$$

$$+(\mathbf{p}\cdot\mathbf{K}_\beta)\cdot\langle\boldsymbol{U}_-(\mathbf{K}_\beta,\Omega_\beta)\boldsymbol{U}_-^*(\mathbf{K}_{\beta'},\Omega_{\beta'})\rangle\cdot(\mathbf{p}\cdot\mathbf{K}_{\beta'})\delta(\mathbf{K}-\mathbf{K}_\beta)$$

$$\delta(\mathbf{K}-\mathbf{K}_{\beta'})\,\frac{\Gamma_\beta/\pi}{\Gamma_\beta - i(\Omega+\Omega_\beta)}\,\frac{1}{\Gamma_{\beta'}+i(\Omega+\Omega_{\beta'})}\Big). \tag{12.81}$$

Since statistical fluctuations are uncorrelated between different modes,

$$\sum_\beta\sum_{\beta'} \rightarrow \sum_\beta\sum_{\beta'}\delta_{\beta\beta'}.$$

Furthermore,

$$\delta(\mathbf{K}-\mathbf{K}_\beta)\delta(\mathbf{K}-\mathbf{K}_\beta) = \frac{1}{(2\pi)^3}\,\delta(\mathbf{K}-\mathbf{K}_\beta)\int d^3r'\, e^{i(\mathbf{K}-\mathbf{K}_\beta)\cdot\mathbf{r}'}. \tag{12.82}$$

In the integral, $\mathbf{K} = \mathbf{K}_\beta$ and the integral just yields the sample volume V. Therefore,

$$\delta(\mathbf{K}-\mathbf{K}_\beta)\delta(\mathbf{K}-\mathbf{K}_\beta) = \frac{V}{(2\pi)^3}\,\delta(\mathbf{K}-\mathbf{K}_\beta) \tag{12.83}$$

and

$$\boldsymbol{\Delta}(k\hat{\mathbf{r}}-\mathbf{k}_l,\omega-\omega_l) = \sum_\beta \frac{n^8 V}{4(2\pi)^3}\Big((\mathbf{p}\cdot\mathbf{K}_\beta)\cdot\langle \boldsymbol{U}_+(K_\beta,\Omega_\beta)\boldsymbol{U}_+{}^*(K_\beta,\Omega_\beta)\rangle$$

$$\cdot(\mathbf{p}\cdot\mathbf{K}_\beta)\delta(k\hat{\mathbf{r}}-\mathbf{k}_l-\mathbf{K}_\beta)\frac{\Gamma_\beta/\pi}{\Gamma_\beta{}^2+(\omega-\omega_l-\Omega_\beta)^2} + (\mathbf{p}\cdot\mathbf{K}_\beta)\cdot\langle \boldsymbol{U}_-(K_\beta,\Omega_\beta)\boldsymbol{U}_-{}^*(K_\beta,\Omega_\beta)\rangle$$

$$\cdot(\mathbf{p}\cdot\mathbf{K}_\beta)\;\delta(k\hat{\mathbf{r}}-\mathbf{k}_l-\mathbf{K}_\beta)\;\frac{\Gamma_\beta/\pi}{\Gamma_\beta{}^2+(\omega-\omega_l+\Omega_\beta)^2}\Big). \tag{12.84}$$

We note that $\langle \boldsymbol{U}_+(K_\beta,\Omega_\beta)\boldsymbol{U}_+{}^*(K_\beta,\Omega_\beta)\rangle = \langle \boldsymbol{U}_-(K_\beta,\Omega_\beta)\boldsymbol{U}_-{}^*(K_\beta,\Omega_\beta)\rangle$ and Eq. (12.84) simplifies to

$$\boldsymbol{\Delta}(k\hat{\mathbf{r}}-\mathbf{k}_l,\omega-\omega_l) = \sum_\beta \frac{n^8 V}{(2\pi)^4}(\mathbf{p}\cdot\mathbf{K}_\beta)\cdot\langle \boldsymbol{U}_\beta\boldsymbol{U}_\beta{}^*)\rangle\cdot(\mathbf{p}\cdot\mathbf{K}_\beta)\delta(k\hat{\mathbf{r}}-\mathbf{k}_l-\mathbf{K}_\beta)$$

$$\Big(\frac{\Gamma_\beta/2\pi}{\Gamma_\beta{}^2+(\omega-\omega_l-\Omega_\beta)^2} + \frac{\Gamma_\beta/2\pi}{\Gamma_\beta{}^2+(\omega-\omega_l+\Omega_\beta)^2}\Big). \tag{12.85}$$

Since the acoustic modes are thermally excited, their amplitudes can be evaluated by equating their energy to KT. That is, noting that the total energy is twice the time-averaged kinetic energy,

$$KT = \rho\Omega_\beta{}^2\,\langle \boldsymbol{U}_\beta\boldsymbol{U}_\beta{}^*\rangle \int d^3r'.$$

This gives

$$\langle \boldsymbol{U}_\beta\boldsymbol{U}_\beta{}^*\rangle = \frac{KT}{\rho\Omega_\beta{}^2 V}$$

and therefore

$$\langle \boldsymbol{U}_\beta\boldsymbol{U}_\beta{}^*\rangle = \frac{KT}{\rho\Omega_\beta{}^2 V}\,\hat{\mathbf{u}}\hat{\mathbf{u}}. \tag{12.86}$$

The remaining problem is to express Σ_β, the summation over the normal acoustic modes, in terms of an integral. As discussed in Section 2.3.2, the normal modes satisfy periodic boundary conditions. We illustrate the allowed standing wave normal modes for the case of a rectangular sample in Figure 12.7. However, we have developed the formalism using running wave components, so we need the proper expression for the allowed components in this case. Periodic boundary conditions on the x component lead to

$$e^{iK_x(x+L_x)} = e^{iK_x x},$$

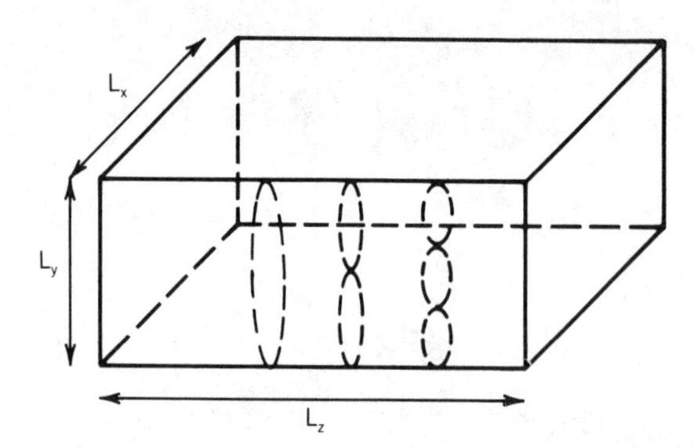

Figure 12.7. Illustration of the volume of the sample for Brillouin scattering. Three standing sound wave fields are illustrated on the front face of the sample.

so that

$$K_x = n_x \frac{2\pi}{L_x}. \tag{12.87a}$$

Identical arguments give the other two components, which read

$$K_y = n_y \frac{2\pi}{L_y} \tag{12.87b}$$

$$K_z = n_z \frac{2\pi}{L_z}, \tag{12.87c}$$

where n_x, n_y, and n_z are integers. Therefore the summations over the normal modes correspond to summations over the integers n_x, n_y, and n_z. The volume in **K** space occupied by one mode corresponds to

$$\Delta K_x \Delta K_y \Delta K_z = \frac{(2\pi)^3}{V}.$$

Therefore the summation over the modes is equivalent to an integral over **K** space where the mode density is $V/(2\pi)^3$, i.e., 1/(volume occupied in **K** space by one mode). Hence

$$\sum_{\beta} \rightarrow \frac{V}{(2\pi)^3} \int d^3K. \tag{12.88}$$

In terms of Eq. (12.85,)

$$\int \delta(k\hat{\mathbf{r}} - \mathbf{k}_l - \mathbf{K}) d^3K = 1.$$

Therefore, finally

$$\mathbf{\Delta}(k\hat{\mathbf{r}}-\mathbf{k}_I,\omega-\omega_I)=\frac{n^8V}{(2\pi)^7}\frac{KT}{\rho\Omega_\beta^{\,2}}[\mathbf{p}\cdot(k\hat{\mathbf{r}}-\mathbf{k}_I)]\cdot\hat{\mathbf{u}}\hat{\mathbf{u}}\cdot[\mathbf{p}\cdot(k\hat{\mathbf{r}}-\mathbf{k}_I)]$$

$$\left(\frac{\Gamma_\beta/2\pi}{\Gamma_\beta^{\,2}+(\omega-\omega_I-\Omega_\beta(k\hat{\mathbf{r}}-\mathbf{k}_I))^2}+\frac{\Gamma_\beta/2\pi}{\Gamma_\beta^{\,2}+(\omega-\omega_I+\Omega_\beta(k\hat{\mathbf{r}}-\mathbf{k}_I))^2}\right). \tag{12.89}$$

The meaning of the term $\Omega_\beta(k\hat{\mathbf{r}}-\mathbf{k}_I)$ is given below.

12.4.3 Frequency Spectrum of Scattered Light

It is now a simple matter to calculate the frequency spectrum. We substitute Eq. (12.89) into (12.33) and (12.34) to get

$$\frac{\tilde{\boldsymbol{S}}_\beta(\mathbf{r},\omega)}{\Delta\Omega S_0}=\sum_\beta\frac{n^8V\omega^4KT}{8(2\pi)^2\rho\Omega_\beta^{\,2}}$$

$$\times\ [\mathbf{O}(\hat{\mathbf{r}})\cdot\hat{\mathbf{e}}_I\cdot(\mathbf{p}\cdot(k\hat{\mathbf{r}}-\mathbf{k}_I))\cdot\hat{\mathbf{u}}_\beta\hat{\mathbf{u}}_\beta\cdot(\mathbf{p}\cdot(k\hat{\mathbf{r}}-\mathbf{k}_I))\cdot\hat{\mathbf{e}}_I\cdot\mathbf{O}(\hat{\mathbf{r}})]$$

$$\left(\frac{\Gamma_\beta/2\pi}{\Gamma_\beta^{\,2}+(\omega-\omega_I-\Omega_\beta(k\hat{\mathbf{r}}-\mathbf{k}_I))^2}+\frac{\Gamma_\beta/2\pi}{\Gamma_\beta^{\,2}+(\omega-\omega_I+\Omega_\beta(k\hat{\mathbf{r}}-\mathbf{k}_I))^2}\right), \tag{12.90}$$

where we have summed over β to include scattering from all three possible acoustic modes, two shear and one longitudinal.

Now let us deal with the frequency $\Omega_\beta^{\,2}(k\hat{\mathbf{r}}-\mathbf{k}_I)$ in the denominator in Eq. (12.90). This term comes from the delta function in Eq. (12.85), which expresses the phase match condition that comes from the collective character of the sound wave. This phase match condition is precisely the same as obtained in Chapter 11 for scattering off coherently driven sound waves. The only difference is that the oppositely-directed thermally-excited sound waves have equal powers (see Eq. (12.79)) and hence there is both an up- and downshifted component in the scattered spectrum. The argument $k\hat{\mathbf{r}}-\mathbf{k}_I$ in the frequency $\Omega_\beta^{\,2}(k\hat{\mathbf{r}}-\mathbf{k}_I)$ is just the acoustic wave vector that phase matches the scattering of the incident field ($\mathbf{k}_I$) into the scattered field ($\mathbf{k}_S = k\hat{\mathbf{r}}$). The frequency shift Ω and the acoustic wave vector $\mathbf{K}$ are linked by the acoustic dispersion $\Omega = Kv$. Since $\mathbf{K} = \mathbf{k}_S-\mathbf{k}_I$, and $\Omega \ll \omega_I$, then

$$\Omega = 2n\,\frac{v}{c}\,\sin\frac{\theta}{2}\,, \tag{12.91}$$

where θ is the angle between $\mathbf{k}_I$ and $\mathbf{k}_S$, the incident and scattered wave vectors respectively. Equation (12.91) is called the Brillouin equation. Because $\Omega \sim \sin(\theta/2)$, the Brillouin spectrum changes with scattering angle, as indicated in Figure 12.8.

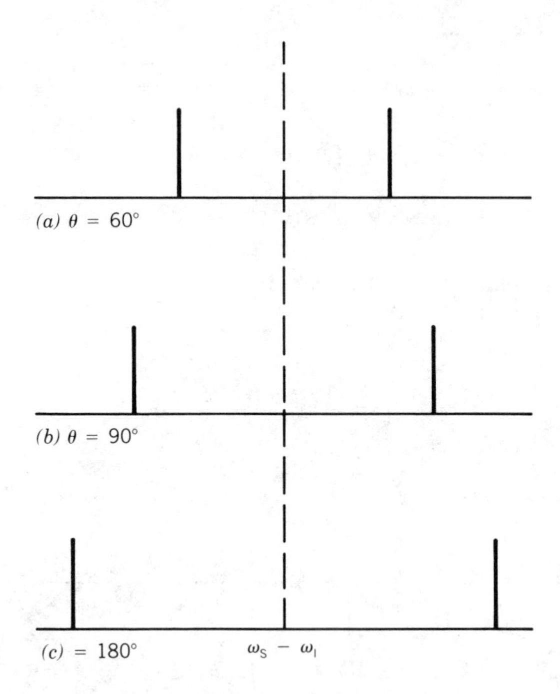

Figure 12.8. Brillouin sidebands as a function of angle. Contrast with Raman scattering in which the sidebands are independent of angle. The angular dependence is due to collective character of the normal coordinate.

12.5 CHARACTERISTICS OF THE OPTICAL SCATTERING SPECTRA

We saw in the preceding sections that statistical fluctuations in a medium can excite a variety of normal modes which scatter light by Raman-type processes. The strength of the scattering is usually proportional to the temperature: the higher the temperature the stronger the scattered signal. The characteristics of the frequency spectrum and of the spatial dependence of the scattered light reveal information about the dynamics of the normal modes, which makes light scattering a powerful tool for probing normal modes.

Since the normal modes are dynamic (versus static) phenomena, the scattered light is always shifted in frequency from the incident light frequency. For the cases characterized by normal mode frequencies, the scattered light is centered at frequencies shifted from the incident light by the normal mode frequencies. This occurs for the rotational, vibrational, and collective (acoustic) modes. (In the rotational case, the shift occurs at twice the rotational frequency.) In all cases, the damping of the normal modes is manifest by a spectral broadening of the scattered light. Hence by measuring the frequency shift and line width, the normal mode frequency and relaxation time are evaluated. For Rayleigh scattering, there is no normal mode frequency and the scattered light spectrum is centered at the incident light frequency. However, in this case the decay time of the fluctuations can also be measured from the spectral line width.

Fortunately the ranges of frequency shifts involved for the normal modes are such that the spectral features do not overlap (except for low frequency sound waves, which may be difficult to resolve in the presence of strong Rayleigh lines). Typical frequency shifts are shown on a logarithmic scale in Figure 12.9. The frequency shifts and line widths are essentially independent of the scattering angle for all cases except the Brillouin spectrum.

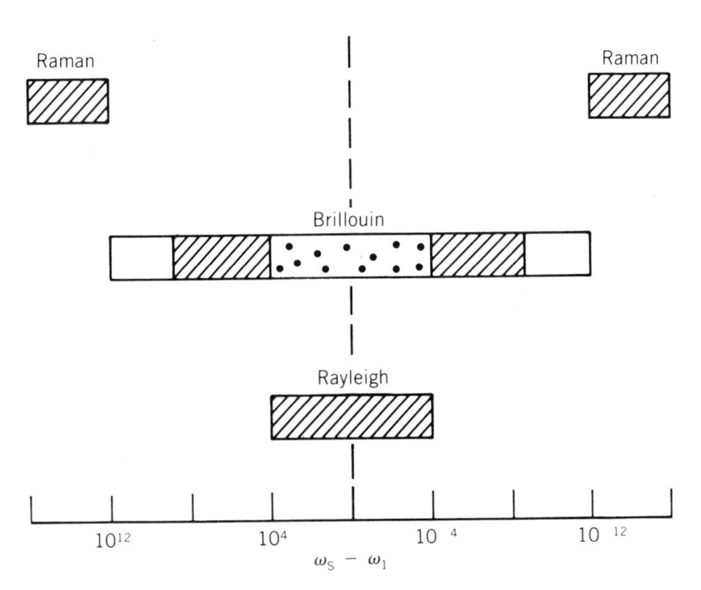

Figure 12.9. Experimentally accessible regions of the scattering mechanisms. The Brillouin spectrum near zero frequency is unresolvable due to strong Rayleigh scattering.

Information can also be obtained from the variation in scattering intensity with polarization and scattering geometry. This is especially true for Rayleigh and Raman scattering, for which cases the molecular anisotropy of the polarizability can be measured.

ADDITIONAL READING

Strong, J.M. *Radiation and Optics*, McGraw-Hill, New York, 1963. Chapter 14.

Cumins, H.Z. and Schoen, P.E. Linear Scattering from Thermal Fluctuations, *Laser Handbook*, Vol II, Arecchi, F.J. and Schultz-Dubois, E.O. eds., North Holland, Amsterdam, 1972. pp. 1029-1075.

PROBLEMS

12.1. Use Eq. (12.45) to show that Eq. (12.2) is equivalent to the usual definition of Fourier transform

$$\boldsymbol{\chi}_W(r,t) = \int d^3K \int d\Omega\ \tilde{\boldsymbol{\chi}}_W(k,\Omega)e^{i(\mathbf{K}\cdot\mathbf{r}-\Omega t)}.$$

12.2. (a) Develop an expression for the Fourier amplitude of $\mathbf{A}(\mathbf{r},t)$ starting with Eq. (12.12). Show explicitly that the expression $A(k_s,\omega_s)$ has the same phase match condition developed in Sections 10.4 and 10.5.

(b) Derive Eq. (12.11).

12.3. (a) Show that the Fourier amplitude of $\boldsymbol{\chi}$ (e.g., as used in Eq. (12.12)) is related to the conventional amplitude $\boldsymbol{\chi}(\mathbf{K},\Omega)$ (e.g., in Eq. (10.38) as

$$\boldsymbol{\chi}_W(\mathbf{K},\Omega) = \tilde{\boldsymbol{\chi}}'_W(\mathbf{K}',\Omega')\delta^{(3)}(\mathbf{K} - \mathbf{K}')\delta(\Omega - \Omega'). \tag{12.92}$$

(b) Show that the spectrum of a monochromatic field (e.g., as defined by Eq. (1.30)) can be written in terms of the flux in Eq. (12.21) (denoted $\underset{\sim}{S}$) as

$$S(\omega) = \tilde{S}(\omega')\delta(\omega - \omega'). \tag{12.93}$$

(c) Show that for a general monochromatic field, the flux defined in Eq. (1.30) can be obtained as

$$S = \mathrm{Tr}(\mathbf{G}(\tau=0)).$$

12.4. The electric field amplitude is related to the Fourier amplitude as

$$\mathbf{E}(\mathbf{k},\omega) = \mathbf{E}(\mathbf{k}',\omega')\delta(\omega-\omega')\delta^{(3)}(\mathbf{k} - \mathbf{k}').$$

Use Eqs. (12.82) and (12.83) to rederive Eq. (12.93) from Eq. (1.30).

12.5. Show that an electric field having the form

$$\mathbf{E} = \frac{1}{2}\left(a\hat{\mathbf{e}}_a e^{i\phi_a(t)} + b\hat{\mathbf{e}}_b e^{i\phi_b(t)}\right) \mathcal{E}\, e^{i(\mathbf{k}\cdot\mathbf{r}-\omega t)} + \mathrm{cc}$$

is depolarized if ϕ_a and ϕ_b are random with respect to each other, i.e., if

$$\langle e^{i\phi_a(t)}\, e^{-i\phi_b(t)}\rangle = 0.$$

Hint: Show that $\boldsymbol{\rho} = \frac{1}{2}\,\mathbf{I}$.

12.6. Show that if $\mathbf{S}(r) = f(\omega)\mathbf{VV}$ where $\mathbf{V}$ is an arbitrary vector, then the light is completely polarized. Use this to show that the light scattered by the fluctuation described by Eq. (12.43) is polarized.

12.7. Consider the fluctuation described by Eq. (12.43) where the incoming field is depolarized. We proceed formally as follows. The tensor in this case is

$$\boldsymbol{\rho} = \mathbf{O}(\hat{\mathbf{r}}) \cdot \langle \hat{\mathbf{e}}_I \hat{\mathbf{e}}_I \rangle \cdot \mathbf{O}(\hat{\mathbf{r}}).$$

Let us suppose $\mathbf{k}_I$ points in the $\hat{z}$ direction, then

$$\langle e_I e_I \rangle = \frac{1}{2} \begin{pmatrix} 1 & 0 & 0 \\ 0 & 1 & 0 \\ 0 & 0 & 0 \end{pmatrix} .$$

Use this to compute γ from Eq. (12.48). For a short version of this problem, compute the degree of polarization only for the cases $\hat{\mathbf{r}} = \hat{x}$, $\hat{\mathbf{r}} = \hat{z}$. Show that for $\hat{\mathbf{r}} = \hat{\mathbf{x}}$ the light is completely polarized and for $\hat{\mathbf{r}} = \hat{z}$ the light is completely depolarized.

12.8. Show that if in any case one has a $\boldsymbol{\rho}$ that can be written as

$$\boldsymbol{\rho} = \frac{1}{2}\,\mathbf{O}(\hat{\mathbf{r}}),$$

then the light is depolarized. One can often use the formula

$$\mathbf{O}(\hat{\mathbf{r}}) \cdot \mathbf{O}(\hat{\mathbf{r}}) = \mathbf{O}(\hat{\mathbf{r}})$$

to quickly assess whether the light observed in a direction $\hat{\mathbf{r}}$ is depolarized.

12.9. Show that Eq. (12.41) can be rewritten as

$$\gamma = (1-4\,\det\boldsymbol{\rho})^{1/2},$$

where det signifies the determinant.

12.10. Many different expressions can be found in the literature to describe depolarized scattering. One involves assuming a geometry we have used repeatedly in which the incoming field is linearly polarized, and the observation direction is perpendicular to both the polarization direction and $\mathbf{k}_I$. For specificity take $\hat{\mathbf{e}}_I = \hat{\mathbf{x}}$, $\hat{\mathbf{k}}_I = \hat{z}$, and $\hat{\mathbf{r}} = \hat{\mathbf{y}}$.

(a) Define R as the ratio between the flux observed through a polarizer set to pass $\hat{z}$ polarization to the flux observed through a polarizer set to pass $\hat{\mathbf{x}}$ polarization. Show that

$$R = \frac{f_{44}}{f_{11}}$$

and

$$\gamma = \frac{1-R}{1+R}.$$

(b) Use Eqs. (12.50a) and (12.50c) to show

$$f_{11} = \left(\frac{2\alpha_{\parallel} + \alpha_{\perp}}{3}\right)^2 + \frac{4}{45}\,(\alpha_{\parallel} - \alpha_{\perp})^2$$

and hence

$$R = \frac{3(\alpha_{\parallel} - \alpha_{\perp})^2}{45\left(\frac{2\alpha_{\parallel} + \alpha_{\perp}}{3}\right)^2 + 4\left(\frac{2\alpha_{\parallel} - \alpha_{\perp}}{3}\right)^2}$$

12.11. (a) Develop Eq. (12.44) explicitly to obtain the properties of Rayleigh and Mie scattering.

(b) Use Eq. (12.91) to obtain the range of frequencies for Brillouin scattering shown in Figure 12.9.

(c) Use Table 11.1 and $\lambda = 0.5 \times 10^{-6}$m to obtain the upper bound on the frequency of Brillouin scattering for fused quartz.

12.12. (a) How does Figure 12.9 appear if one has Raman active optical phonons that are decoupled (see the end of Section 2.3.2 and Figure 2.6) from the acoustical phonons? How is this result affected if the optical phonons are (b) weakly coupled and (c) strongly coupled?

12.13. Verify Eqs. (12.50a) to (12.50f). Note that this is a difficult exercise, but one can make good use of the discussion in Section 8.2.2.

12.14. Work out the all intermediate steps in detail the derivation of Eq. (12.57) from Eq. (12.56).

APPENDIX **A**

Useful Formulae

In this appendix we summarize a few useful formulae for reference in the second volume of the book. In addition we use this opportunity to discuss some notational questions and provide some exercises and discussion of tensors for students who are not familiar with them.

A.1 VECTORS AND TENSORS

This discussion of tensor notation is brief, and follows closely the discussion and notation in Portis, which is much more comprehensive and should be consulted as additional reading as well as for exercises for those who need practice. Vectors and tensors are defined by their transformation properties. This aspect of their properties is beyond the scope of this appendix. However it is important to remember that not all sets of numbers can be treated as vectors and tensors, and that our treatment is consequently oversimplified. With this caution, we define a three vector **A** [**B**, **C**, etc.] as an ordered triplet of numbers (a_1,a_2,a_3) [(b_1,b_2,b_3), (c_1,c_2,c_3), etc.].

The dot product of one vector with another is a scalar, which reads

$$\mathbf{A} \cdot \mathbf{B} = a_1b_1 + a_2b_2 + a_3b_3. \tag{A.1}$$

The cross product of one vector with another leads to a vector, which reads

$$\mathbf{A} \times \mathbf{B} = (a_2b_3 - a_3b_2, a_3b_2 - a_2b_3, a_1b_2 - a_2b_1). \tag{A.2}$$

Tensors are denoted with a stronger bolding than vectors. A second rank tensor **T** is a 3 x 3 array of numbers denoted t_{ij}. In detail the tensor reads

$$\mathbf{T} = \begin{pmatrix} t_{11} & t_{12} & t_{13} \\ t_{21} & t_{22} & t_{23} \\ t_{31} & t_{32} & t_{33} \end{pmatrix} . \tag{A.3}$$

In many cases we refer to tensors by denoting their elements, e.g., **T** is referred to as t_{ij}. This is not strictly proper, but it simplifies the notational problems in tensors of higher rank (rank refers to the number of subscripts) such as r_{ijk} or s_{ijkl}.

One can make higher rank tensors from lower rank tensors (a vector is a tensor of rank 1) through an outer product. The product is often denoted with a cross centered in an o, which prints rather badly on our machine. Instead we follow a fairly standard procedure by using no symbol. For example, the outer or dyad product **AB** is defined by

$$\mathbf{AB} = \begin{pmatrix} a_1b_1 & a_1b_2 & a_1b_3 \\ a_2b_1 & a_2b_2 & a_2b_3 \\ a_3b_1 & a_3b_2 & a_3b_3 \end{pmatrix} . \tag{A.4}$$

A shorthand notation for this is a_ib_j. The trace of a second rank tensor is the sum of its diagonal elements. One can observe by inspection that the trace of Eq. (A.4) is the dot product defined in Eq. (A.1). One can construct tensors of arbitrarily high rank by making outer products. For example, $t_{ij}a_k$ is a third rank tensor; $t_{ij}t_{kl}$ and $a_it_{jk}b_l$ are fourth rank tensors.

The operation that occurs repeatedly in this book is the dot product of tensors. This operation is simple in the case in which tensors are dyad products of vectors. For example,

$$\mathbf{AB} \cdot \mathbf{C} = \mathbf{A}\ (\mathbf{B} \cdot \mathbf{C}). \tag{A.5}$$

The double dot symbol means to take two dot products:

$$\mathrm{AB} : \mathbf{CDE} = \mathbf{A}\ (\mathbf{B} \cdot \mathbf{C}) \cdot \mathbf{DE} = (\mathbf{B} \cdot \mathbf{C})\ (\mathbf{A} \cdot \mathbf{D})\ \mathbf{E}. \tag{A.6}$$

In cases in which the tensors cannot be written as dyad products of vectors, these symbols refer to a contraction of nearest indices, e.g., if **R** and **U** are second rank tensors, then

$$\mathbf{T} = \mathbf{R} \cdot \mathbf{U}, \tag{A.7}$$

which means

$$t_{ij} = \sum_k r_{ik}u_{kj}. \tag{A.8}$$

and the double dot gives a scalar t, i.e.,

$$t = \mathbf{R} : \mathbf{U} \tag{A.9}$$

means

$$t = \sum_{ij} r_{ij}u_{ji} \tag{A.10}$$

For third rank tensors **U** and **W**

$$\mathbf{W} = \mathbf{T} \cdot \mathbf{U} \tag{A.11}$$

is equivalent to

$$w_{ijk} = \sum_{l} t_{il} u_{ljk} \tag{A.12}$$

and

$$\mathbf{W} = \mathbf{T} \vdots \mathbf{U} \tag{A.13}$$

is equivalent to

$$w_i = \sum_{jk} t_{jk} u_{kji}. \tag{A.14}$$

A.2 IDENTITIES

The following set of vector and differential operator identities can be useful in working out problems. These follow the notation in Portis as far as possible, so one can consult that work for proofs and discussion.

A.2.1 Vector Identities

$$\mathbf{A} \cdot (\mathbf{B} \times \mathbf{C}) = \mathbf{C} \cdot (\mathbf{A} \times \mathbf{B}) \tag{A.15}$$

$$\mathbf{A} \times (\mathbf{B} \times \mathbf{C}) = \mathbf{B}(\mathbf{A} \cdot \mathbf{C}) - \mathbf{C}(\mathbf{A} \cdot \mathbf{B}) \tag{A.16}$$

$$(\mathbf{A} \times \mathbf{B}) \cdot (\mathbf{C} \times \mathbf{D}) = (\mathbf{A} \cdot \mathbf{C})(\mathbf{B} \cdot \mathbf{D}) - (\mathbf{A} \cdot \mathbf{D})(\mathbf{B} \cdot \mathbf{C}) \tag{A.17}$$

$$\mathbf{A} \cdot [\mathbf{B} \times (\mathbf{C} \times \mathbf{D})] = [(\mathbf{A} \times \mathbf{B}) \times \mathbf{C}] \cdot \mathbf{D} \tag{A.18}$$

$$(\mathbf{A} \times \mathbf{B}) \times (\mathbf{C} \times \mathbf{D}) = [\mathbf{A} \cdot (\mathbf{B} \times \mathbf{D})]\mathbf{C} - [\mathbf{A} \cdot (\mathbf{B} \times \mathbf{C})]\mathbf{D} \tag{A.19}$$

$$(\mathbf{A} \times \mathbf{B}) \times (\mathbf{C} \times \mathbf{D}) = (\mathbf{A} \times \mathbf{C})(\mathbf{B} \cdot \mathbf{D}) - (\mathbf{A} \times \mathbf{D})(\mathbf{B} \cdot \mathbf{C}) \tag{A.20}$$

$$[\mathbf{A}(\mathbf{B} \times \mathbf{C})] \cdot \mathbf{D} = (\mathbf{A}\mathbf{B}) \cdot (\mathbf{C} \times \mathbf{D}) \tag{A.21}$$

A.2.2 Vector Calculus

$$\nabla(\phi\psi) = \phi\nabla\psi + \psi\nabla\phi \tag{A.22}$$

$$\nabla \cdot (\phi\mathbf{A}) = (\nabla\phi) \cdot \mathbf{A} + \phi(\nabla \cdot \mathbf{A}) \tag{A.23}$$

$$\nabla \times (\phi\mathbf{A}) = (\nabla\phi) \times \mathbf{A} + \phi(\nabla \times \mathbf{A}) \tag{A.24}$$

$$\nabla \cdot (\mathbf{A} \times \mathbf{B}) = \mathbf{B} \cdot (\nabla \times \mathbf{A}) - \mathbf{A} \cdot (\nabla \times \mathbf{B}) \tag{A.25}$$

$$\nabla \times (\mathbf{A} \times \mathbf{B}) = \mathbf{A}(\nabla \cdot \mathbf{B}) - \mathbf{A} \cdot \nabla)\mathbf{B} - \mathbf{B}(\nabla \cdot \mathbf{A}) + (\mathbf{B} \cdot \nabla)\mathbf{A} \tag{A.26}$$

$$\nabla(\mathbf{A} \cdot \mathbf{B}) = (\mathbf{A} \cdot \nabla)\mathbf{B} + \mathbf{A} \times (\nabla \times \mathbf{B}) + (\mathbf{B} \cdot \nabla)\mathbf{A} + \mathbf{B} \times (\nabla \times \mathbf{A}) \tag{A.27}$$

$$\nabla(\mathbf{A}\cdot\mathbf{B}) = \mathrm{A}(\nabla\cdot\mathrm{B}) + (\mathbf{A}\times\nabla)\times\mathbf{B} + \mathrm{B}(\nabla\cdot\mathbf{A}) + (\mathrm{B}\times\nabla)\times\mathrm{A} \tag{A.28}$$

$$\nabla\times(\nabla\phi) = 0 \tag{A.29}$$

$$\nabla\cdot(\nabla\times\mathbf{A}) = 0 \tag{A.30}$$

$$\nabla\times(\nabla\times\mathbf{A}) = \nabla(\nabla\cdot\mathbf{A}) - \nabla^2\mathrm{A} \tag{A.31}$$

A.2.3 Vector Integrals

In the following, dV (also denoted d^3r and **dr**) is the volume element dxdydz. The differential **dS** (also denoted $d\sigma$) is a differential element of area that points in a direction normal to the surface, and **dr** is a differential element along a one dimensional curve that points along the tangent to the curve. The symbols V, S and C refer to a volume, surface and curve respectively. When appearing on the right and left-hand sides of an expression, S is the boundary of V and C is the boundary of S.

$$\int_V (\nabla\cdot\mathbf{A})dV = \int_S d\mathbf{S}\cdot\mathbf{A} \tag{A.32}$$

$$\int_S (\nabla\times\mathrm{A})dV = \int_S d\mathbf{S}\times\mathrm{A} \tag{A.33}$$

$$\int_S d\mathbf{S}\cdot(\nabla\times\mathbf{A}) = \int_C d\mathbf{r}\cdot\mathbf{A} \tag{A.34}$$

$$\int_S d\mathbf{S}\times\nabla\phi = \int_C \phi\, d\mathbf{r} \tag{A.35}$$

$$\int_S (d\mathbf{S}\times\nabla)\times\mathrm{A} = \int_C d\mathbf{r}\times\mathbf{A} \tag{A.36}$$

A.2.4 Dyad Identities and Calculus

$$\nabla(\mathrm{A}\cdot\mathbf{B}) = (\nabla\mathrm{A})\cdot\mathrm{B} + (\nabla\mathbf{B})\cdot\mathrm{A} \tag{A.37}$$

$$\nabla\cdot(\mathbf{AB}) = (\nabla\cdot\mathrm{A})\mathbf{B} + (\mathrm{A}\cdot\nabla)\mathrm{B} \tag{A.38}$$

$$\nabla\times(\mathbf{AB}) = (\nabla\times\mathrm{A})\mathbf{B} - (\mathrm{A}\times\nabla)\mathrm{B} \tag{A.39}$$

$$\nabla(\mathrm{A}\times\mathrm{B}) = (\nabla\mathrm{A})\times\mathrm{B} - (\nabla\mathrm{B})\times\mathrm{A} \tag{A.40}$$

$$\nabla\times(\mathrm{A}\times\mathbf{B}) = \nabla\cdot(\mathrm{BA} - \mathbf{AB}) \tag{A.41}$$

$$\nabla\cdot(\mathbf{ABC}) = \mathbf{BC}(\nabla\cdot\mathrm{A}) + [(\mathrm{A}\cdot\nabla)\mathrm{B}]\mathrm{C} + \mathrm{B}(\mathrm{A}\cdot\nabla)\mathrm{C} \tag{A.42}$$

A.3 NOTATION

Optics is a notational nightmare, and much of the apparent (and sometimes real) notational confusion in the text is unavoidable. In this section we discuss the conventions followed in the text with respect to coordinates and indices of summation.

A.3.1 Conventions for Subscripts

One convention for subscripts is that the latin characters i,j,k,l always refer to triplets. Whenever a summation refers to these indices a summation from 1 to 3 is implied. Greek symbols are used whenever a summation other than 1 to 3 is implied. The range of summation must then be inferred from context. For Voigt notation, 1 to 6 is the rule.

The greatest ambiguity occurs in summation over normal coordinates, since there are many special cases. The ambiguity in notation follows a real ambiguity in the way the nomenclature of normal coordinates is used in practice. One speaks of an absorption peak due to a dipole active normal coordinate. In real experiments, many molecules are involved in the absorption, and the normal coordinates of one molecule are different coordinates (in a notational sense) from those of another molecule. For example, consider the nuclear motions of a diatomic. The first molecule has six normal coordinates, and one might let α = 1, 2 and 3 label the translations, α = 4 label the vibration and α = 5 and 6 label the rotations. In the second molecule we start with α = 7, and we might let α = 7, 8 and 9 label translations, α = 10 labels vibration and α = 11 and 12 label rotations. A sum over vibrational motions thus runs α = 4, 10, 16, ···, which is clumsy. Instead we label the molecules by a = 1, 2, ···, N and let the symbol a,4 stand for the vibrational coordinate of molecule a. In addition, the act of summation over many molecules is also an average. Thus the following notations for summation over the vibrational motion are equivalent provided that the medium is spatially homogeneous (brackets are averages, Σ_α runs α = 4, 10, etc.):

$$\sum_\alpha \mathbf{p}_\alpha \equiv \sum_a \mathbf{p}_{a,4} \equiv \sum_a \langle \mathbf{p}_{a,4} \rangle \equiv N \langle \mathbf{p}_4 \rangle. \tag{A.43}$$

In constructing a total polarization, one sums over all normal coordinates, in which case the following are equivalent:

$$\sum_\alpha \mathbf{p}_\alpha \equiv \sum_{a,\alpha} \mathbf{p}_{a,\alpha} \equiv \sum_{a,\alpha} \langle \mathbf{p}_{a,\alpha} \rangle \equiv N \sum_\alpha \langle \mathbf{p}_\alpha \rangle. \tag{A.44}$$

When Raman active modes are of interest, we typically let α stand for the dipole active modes and let β stand for the Raman active modes. In a sum $\Sigma_{\alpha\beta}$ we mean that one runs α over all indices labeling dipole active modes and β over all Raman active modes in a double summation. The summation over β is implicit in the averaging process of Eqs. (A.43) and (A.44), but since the Raman active coordinates may be altered by the optical interaction, we may be interested in the details of the process.

A.3.2 Coordinate Notation

We hold to the following notation for coordinates insofar as convention allows (we always try to adopt conventional notation for the problem under consideration, even if it results in denoting the same variable more than one way). The vector $\mathbf{r}$ = (x,y,z) denotes a coordinate system chosen for its utility in solving the problem of wave propagation. This is sometimes referred to as a propagation axis system. The system $\mathbf{x} = (x_1,x_2,x_3)$ refers to a system fixed in the optical medium or to a system fixed by an external magnetic or DC electric field; in crystals $\mathbf{x}$ always denotes a principal axis system. The system $(r_1,r_2,\cdots)$ refers to coordinates describing the location of particles, and $(q_1,q_2,\cdots)$ refer to normal coordinates (except in acousto-optics where the conventional system is denoted by u's).

The plane-wave eigenvectors, which are the normal coordinates of the electromagnetic fields, are denoted differently in most subdisciplines of optics. In electro-optics, the standard usage is to let $\mathbf{E}_o$ and $\mathbf{E}_e$ stand for the two orthogonal modes. In nonlinear optics it is not possible to use this notation, and the standard is to label the fields a, b, and c for three wave mixing and a, b, c, and d for four wave mixing. Unfortunately even this usage is difficult to follow consistently (see, e.g., second harmonic generation). In scattering, stimulated scattering and acousto-optics, the conventional labels are $\mathbf{E}_I$ for the incident field and $\mathbf{E}_S$ for scattered fields, in spite of the resultant ambiguity when the scattered fields are Stokes and anti-Stokes waves. Since $\mathbf{E}_I$ stands for a field incident on a scatterer, we need a separate notation $\mathbf{E}_{in}$ for a wave incident on a medium.

A.4 AMPLITUDE NOTATION

Field amplitudes are exploited throughout the two volumes as a means of solving problems. Whenever practicable, the field amplitude is denoted with the capital italic character that corresponds to the capital latin character denoting the field. The following are all field amplitudes:

$$E(z,t) = \frac{1}{2}\,\mathcal{E}(\mathbf{k},\omega)\,e^{i(\mathbf{k}\cdot\mathbf{r}-\omega t)} + cc, \tag{A.45}$$

$$E(z,t) = \frac{1}{2}\,\mathcal{E}(z,\omega)\,e^{-i\omega t} + cc, \tag{A.46}$$

$$E(z,t) = \frac{1}{2}\,\mathcal{E}(\mathbf{k},t)\,e^{i\mathbf{k}\cdot\mathbf{r}} + cc. \tag{A.47}$$

In the case of spherically diverging waves, the field amplitudes (nearly always the vector potential) are written as

$$A(z,t) = \frac{1}{2}\,\frac{\mathcal{A}(\mathbf{k},\omega)}{r}\,e^{i(kr-\omega t)} + cc. \tag{A.48}$$

Field amplitudes are different from Fourier amplitudes (among other things, the units are different). Since Fourier amplitudes are used only in Chapter 12, there is no reason for a summary here.

It is frequently necessary to convert a result derived in terms of field amplitudes to an expression involving fluxes. A convenient formula is

$$S = \frac{n\varepsilon_0 c}{2}\,|\mathcal{E}|^2. \tag{A.49}$$

Here we see several conventions adopted in the text. An unbolded symbol for a quantity that is normally a vector is the scalar amplitude defined by

$$\mathbf{E} = \hat{\mathbf{e}}\, E, \tag{A.50}$$

where $\hat{\mathbf{e}}$ is the unit vector pointing in the direction of $\mathbf{E}$. The vertical bars around a scalar stand for absolute value. Note that there are several notational ambiguities inherent in denoting the flux. For clarity in presentation, a bar over the symbol is used in the first four chapters to denote the time-averaged flux. Later this notation is dropped, since the customary usage of the symbol in the literature refers to the time-averaged quantity. Since there are no phasors in the flux, it has no field amplitude as such. Note that in order for a field amplitude to exist, there must be a well-defined index of refraction n, and the index used in Eq. (A.49) is the one found from the dispersion relation that defines the amplitude $\boldsymbol{\mathcal{E}}$ in the equation.

A.5 RELATIONSHIPS INVOLVING PLANE WAVES

The following relations apply to a general plane wave field $\mathbf{F}$. Note that we follow the procedure of the text in dropping the arguments in cases in which they can be determined by context. While these relations are stated for fields of the form of Eq. (A.47), they apply also to fields of the form of Eq. (A.45).

$$\nabla \cdot \mathbf{F}(\mathbf{r}) = \frac{1}{2}\, i\mathbf{k} \cdot \boldsymbol{\mathcal{F}}\, e^{i\mathbf{k}\cdot\mathbf{r}} + cc \tag{A.51}$$

$$\nabla \times \mathbf{F}(\mathbf{r}) = \frac{1}{2}\, i\mathbf{k} \times \boldsymbol{\mathcal{F}}\, e^{i\mathbf{k}\cdot\mathbf{r}} + cc \tag{A.52}$$

$$\nabla(\nabla \cdot \mathbf{F}(\mathbf{r})] = -\frac{1}{2}\, \mathbf{k}(\mathbf{k} \cdot \boldsymbol{\mathcal{F}})\, e^{i\mathbf{k}\cdot\mathbf{r}} + cc \tag{A.53}$$

$$\nabla^2 \mathbf{F}(\mathbf{r}) = -\frac{1}{2}\, k^2\, \boldsymbol{\mathcal{F}}\, e^{i\mathbf{k}\cdot\mathbf{r}} + cc \tag{A.54}$$

$$\nabla \times [\nabla \times \mathbf{F}(\mathbf{r})] = -\frac{1}{2}\, \mathbf{k} \times (\mathbf{k} \times \boldsymbol{\mathcal{F}})\, e^{i\mathbf{k}\cdot\mathbf{r}} + cc. \tag{A.55}$$

Equation (A.55) occurs repeatedly in the two volumes. It can be reduced to a conceptually and computationally simpler form through the orthogonal projection operator $\mathbf{O}(\hat{\mathbf{e}})$ defined by

$$\mathbf{O}(\hat{\mathbf{e}}) = \mathbf{I} - \hat{\mathbf{e}}\hat{\mathbf{e}} \tag{A.56}$$

such that

$$\mathbf{k} \times (\mathbf{k} \times \boldsymbol{\mathcal{F}}) = -\, k^2\, \mathbf{O}(\hat{\mathbf{k}}) \cdot \boldsymbol{\mathcal{F}}, \tag{A.57}$$

or

$$\nabla \times (\nabla \times \mathbf{F}) = \frac{1}{2}\, k^2\, \mathbf{O}(\hat{\mathbf{k}}) \cdot \boldsymbol{\mathcal{F}}\;\; e^{i\mathbf{k}\cdot\mathbf{r}} + cc. \tag{A.58}$$

These relations are particularly valuable in scattering problems since one can obtain a gauge-independent answer for **E**. In particular, if **A** is a plane-wave field, then

$$\boldsymbol{\mathcal{E}} = - i\omega\, \mathbf{O}(\hat{\mathbf{k}}) \cdot \boldsymbol{\mathcal{A}}, \tag{A.59}$$

and if **A** is a spherically diverging wave in the far field of a radiator, then

$$\boldsymbol{\mathcal{E}} = - i\omega\, \mathbf{O}(\hat{\mathbf{r}}) \cdot \frac{\boldsymbol{\mathcal{A}}}{r}, \tag{A.60}$$

where $\hat{\mathbf{r}}$ is a vector pointing from the radiating source to the point of observation.

A.6 SLOWLY VARYING FIELDS

In the second volume virtually all problems are handled with slowly varying fields. These fields are denoted

$$\mathbf{E}(\mathbf{z},t) = \frac{1}{2}\, \boldsymbol{\mathcal{E}}(\mathbf{z},\omega)\, e^{i(\mathbf{k}\cdot\mathbf{r}-\omega t)} + \text{cc}, \tag{A.61}$$

or more rarely

$$\mathbf{E}(\mathbf{z},t) = \frac{1}{2}\, \boldsymbol{\mathcal{E}}(\mathbf{z},t)\, e^{i(\mathbf{k}\cdot\mathbf{r}-\omega t)} + \text{cc}. \tag{A.62}$$

The slowly varying amplitudes make no sense unless they vary slowly in space over the scale of a wavelength and vary slowly in time over an optical period. We continue discussing only the case defined in Eq. (A.61). It is assumed that there exists a weak polarization field denoted

$$\mathbf{P}_W = \frac{1}{2}\, \boldsymbol{\mathcal{P}}_W\, e^{i(\mathbf{k}_p\cdot\mathbf{r}-\omega t)} + \text{cc}. \tag{A.63}$$

Then the equation of motion for any one field (in nonlinear optics, two or more fields are typically coupled) reads

$$\frac{\partial \mathcal{E}}{\partial z} = i\, \frac{\omega}{2n\varepsilon_0 c}\, \hat{\mathbf{e}}\cdot\mathbf{O}(\hat{\mathbf{k}}) \cdot \boldsymbol{\mathcal{P}}_W\, e^{i\Delta kz}, \tag{A.64}$$

where $\hat{\mathbf{e}}$ is a unit vector pointing in the direction $\mathbf{O}(\hat{\mathbf{k}})\cdot\mathbf{E}$, and

$$\Delta k\, \hat{\mathbf{z}} = \mathbf{k}_p - \mathbf{k}. \tag{A.65}$$

In Eq. (A.65) we follow the customary convention in the text in taking $\mathbf{k}_p$ to point in the $\hat{\mathbf{z}}$ direction, thus defining the propagation axis.

A.7 HARMONIC OSCILLATORS

In both volumes we are primarily concerned with the response of a harmonic oscillator to a driving field. Assume fields of the form given in Eq. (A.45) or Eq. (A.46). Denote the force field as F for generality. The oscillator equation reads

$$m_\alpha(\ddot{q}_\alpha + \Gamma_\alpha \dot{q}_\alpha + \omega_\alpha^2 q_\alpha) = F, \tag{A.66}$$

which is solved by

$$q_\alpha = \frac{1}{2} Q_\alpha e^{-i\omega t} + cc, \tag{A.67}$$

$$Q_\alpha = \frac{F}{m_\alpha D_\alpha(\omega)}, \tag{A.68}$$

$$D_\alpha(\omega) = \omega_\alpha^2 - \omega^2 - i\Gamma_\alpha \omega. \tag{A.69}$$

Here we have ignored Doppler broadening since it is unimportant for most applications. In various regimes of interest, D_α can be approximated. Near resonance

$$D_\alpha(\omega) \simeq 2\omega\,(\omega_\alpha - \omega - \frac{i\Gamma_\alpha}{2}), \tag{A.70}$$

far from resonance

$$D_\alpha(\omega) \simeq (\omega_\alpha^2 - \omega^2), \tag{A.71}$$

which can be further simplified below resonance as

$$D_\alpha(\omega) \simeq \omega_\alpha^2, \tag{A.72}$$

and above resonance as

$$D_\alpha(\omega) \simeq -\omega^2. \tag{A.73}$$

A.8 DIPOLES AND FORCES

We are nearly always interested in dipole active modes interacting with electric fields. A normal mode is associated with a charge separation vector $\boldsymbol{\ell}_\alpha$, which plays the same role in the classical problem as a dipole matrix element plays in the quantum problem. The dipole is defined as

$$p_\alpha = q_\alpha \, \ell_\alpha. \tag{A.74}$$

The amplitude ℓ_α describes an effective charge if q_α denotes a position. If q_α denotes an angle, ℓ_α has units of charge times length, and m_α must be replaced by I_α, the moment of inertia, in Eqs. (A.66) and (A.68). Note that in the special case in which the normal coordinates correspond to three spatially orthogonal motions, they are subscripted i = 1,2,3, and the three vectors $\boldsymbol{\ell}_i$ are orthogonal to each other.

A.8.1 Dipole active modes

The forces are computed from potentials V as

$$F_\alpha = -\frac{\partial V}{\partial q_\alpha}. \tag{A.75}$$

The interaction potential which determines the forces on a dipole is denoted V_{int}, and reads

$$Vint = -\mathbf{p}_\alpha \cdot \mathbf{E}. \tag{A.76}$$

In this case Eq. (A.68) reads

$$Q_\alpha = \frac{\boldsymbol{\ell}_\alpha \cdot \boldsymbol{E}}{m_\alpha D_\alpha(\omega)}. \tag{A.77}$$

From the definition of the polarizability tensor $\boldsymbol{\alpha}$, as

$$\mathbf{p}_\alpha = \frac{1}{2}\,\boldsymbol{\alpha}_\alpha \cdot \boldsymbol{E}\, e^{-i\omega t} + cc, \tag{A.78}$$

which is usually written in shorthand version as

$$\mathbf{p}_\alpha = \boldsymbol{\alpha}_\alpha \cdot \mathbf{E}, \tag{A.79}$$

we find

$$\boldsymbol{\alpha}_\alpha = \frac{\boldsymbol{\ell}_\alpha \boldsymbol{\ell}_\alpha}{m_\alpha D_\alpha(\omega)}. \tag{A.80}$$

A.8.2 Notation of Optical Phenomenology

The interactions in optics are summarized by the susceptibility, dielectric and polarizability tensors. Each follows from the other by definition, so there is no science involved in their use, but it is easy to make algebraic errors when going from one to the other. Unfortunately it has recently been decided to officially adopt MKS units which makes these transformations intractable. The pundits have spoken, so we are stuck with it.

The susceptability is related to the polarizability by

$$\boldsymbol{\chi} = \frac{1}{\varepsilon_0} N \sum_\alpha \langle \boldsymbol{\alpha}_\alpha \rangle. \tag{A.81}$$

The susceptibility is related to the dielectric tensor through

$$\boldsymbol{\varepsilon} = \varepsilon_0 (\mathbf{I} + \boldsymbol{\chi}). \quad \text{(A.82)}$$

We specialize to the case in which these quantities are all real, and diagonalize these tensors in a principal axis system, in which case there exist plane waves with indices of refraction

$$n_i = \left(\frac{\varepsilon_i}{\varepsilon_0}\right)^{1/2}. \quad \text{(A.83)}$$

Observing that we have multiplied and divided each equation by a useless but large coefficient that any sensible person would set to unity, we now hope that we do not make large error in obtaining the only answer containing science, namely

$$n_i^2 = 1 + \frac{1}{\varepsilon_0} N \frac{\ell_i^2}{m_i D_i(\omega)}, \quad \text{(A.84)}$$

where we specialize the notation to the case of crystals, which is the form we need most often in the next volume. In isotropic media, the expression for the index reads

$$n^2 = 1 + \frac{1}{\varepsilon_0} N \sum_\alpha \left\langle \frac{\boldsymbol{\ell}_\alpha \boldsymbol{\ell}_\alpha}{m_\alpha D_\alpha(\omega)} \right\rangle, \quad \text{(A.85)}$$

where the average makes the tensor isotropic, and one replaces the unit tensor by one. Equations (A.84) and (A.85) are needed in many versions in nonlinear optics. In particular,

$$\frac{\ell_i^2}{m_i D_i(\omega)} = \frac{\varepsilon_0}{N} (n_i^2 - 1), \quad \text{(A.86)}$$

$$\frac{\ell_i^2}{m_i D_i(\omega)} = \frac{1}{N} (\varepsilon_i - \varepsilon_0), \quad \text{(A.87)}$$

$$\frac{\ell_i^2}{m_i D_i(\omega)} = \frac{\varepsilon_0}{N} \chi_i. \quad \text{(A.88)}$$

A.8.3 Raman Active Modes

The influence of Raman active modes is buried in the coefficients given above through the averaging process. To render them explicit, the averages must be dropped (e.g., use Eq. (A.43)). We summarize the methodology only for the commonly-encountered case of optically isotropic media. The interaction potential reads

$$V_{int} = -\frac{1}{2} \mathbf{p}_\alpha \cdot \mathbf{E}. \quad \text{(A.89)}$$

This interaction potential does not conflict with Eq. (A.76) (note that here and in the next volume all interaction potentials are denoted the same, since this is how they are encountered in the literature). To avoid errors, the potential in Eq. (A.89) must be developed using any of the formulae from Eq. (A.79) to Eq. (A.88) or variants thereof. The use of Eq. (A.89) assumes that the solution for $\mathbf{p}_\alpha$ has already been found and is expressed as an explicit function of the Raman active coordinate.

Local modes are most frequently analyzed beginning with

$$V_{int} = -\frac{1}{2}\,\boldsymbol{\alpha}_\alpha : \mathbf{EE} \tag{A.90}$$

or

$$V_{int} = -\frac{1}{2}\sum_\alpha \left\langle \frac{\boldsymbol{\ell}_\alpha \boldsymbol{\ell}_\alpha}{m_\alpha D_\alpha(\omega)} \right\rangle : \mathbf{EE}. \tag{A.91}$$

Collective Raman active modes are most frequently analyzed using

$$V_{int} = -\frac{1}{2}\,\varepsilon_0\,\boldsymbol{\chi}_\alpha : \mathbf{EE}, \tag{A.92}$$

where $\boldsymbol{\chi}_\alpha$ is the contribution to the susceptibility from the collective mode.

ADDITIONAL READING

Portis, A.M. *Electromagnetic Fields: Sources and Media*, Wiley, New York, 1978. Appendices B,C.

PROBLEMS

A.1. (a) Let **A** = (1,2,3), **B** = (−1,1,2) show that **A** • **B** = 7.
(b) Show that

$$\mathbf{AB} = \begin{pmatrix} -1 & 1 & 2 \\ -2 & 2 & 4 \\ -3 & 3 & 6 \end{pmatrix} \quad \mathbf{AB} = \begin{pmatrix} -1 & -2 & -3 \\ 1 & 2 & 3 \\ 2 & 4 & 6 \end{pmatrix}.$$

A.2. Let **A** = (1,2,1) and **B** = (3,−1,−2). Show that **A** × **B** = (3,5,7). Verify numerically that **A** • (**A** × **B**) = **B** • (**A** × **B**) = 0.

A.3. (a) The unit dyad reads

$$\mathbf{I} = \begin{pmatrix} 1 & 0 & 0 \\ 0 & 1 & 0 \\ 0 & 0 & 1 \end{pmatrix} . \tag{A.93}$$

show explicitly that for an arbitrary vector $A = (a_1, a_2, a_3)$, $\mathbf{I} \cdot \mathbf{A} = \mathbf{A}$.

(b)Let $\mathbf{A} = (1,2,1)$ and

$$\mathbf{T} = \begin{pmatrix} -1 & 2 & 6 \\ -1 & 1 & -1 \\ 3 & 4 & 4 \end{pmatrix} .$$

Show that $\mathbf{T} \cdot \mathbf{A} = (9,0,5)$ and $\mathbf{A} \cdot \mathbf{T} = (0,8,8)$.

A.4. (a) Let $\hat{\mathbf{e}} = (1,1,2)/\sqrt{6}$. Show that

$$\mathbf{O}(\hat{\mathbf{e}}) = \frac{1}{6} \begin{pmatrix} 5 & -1 & -2 \\ -1 & 5 & -2 \\ -2 & -2 & 2 \end{pmatrix} .$$

(b) Verify numerically that $\mathbf{O}(\hat{\mathbf{e}}) \cdot \hat{\mathbf{e}} = (0,0,0)$.

(c) Let $\mathbf{A} = (1,2,1)$. Show that $\mathbf{O}(\hat{\mathbf{e}}) \cdot \mathbf{A} = (1,7,-4)/6$, and verify numerically that $\mathbf{O}(\hat{\mathbf{e}}) \mathbin{\dot{:}} \mathbf{A}\hat{\mathbf{e}} = 0$.

A.5. Using $\mathbf{A}$ and $\mathbf{B}$ in Problem A.2, verify numerically that $(\mathbf{A} \times \mathbf{B}) \cdot \mathbf{O}(\hat{\mathbf{A}}) \cdot \mathbf{B} = 0$.

A.6. Let

$$\mathbf{U} = \begin{pmatrix} -1 & 2 & 1 \\ 3 & 1 & 1 \\ 1 & 2 & 1 \end{pmatrix} \quad \mathbf{V} = \begin{pmatrix} -1 & 1 & -1 \\ 1 & 1 & -1 \\ 2 & 1 & 0 \end{pmatrix} .$$

(a) Verify that

$$\mathbf{U} \cdot \mathbf{V} = \begin{pmatrix} 5 & 2 & -1 \\ 0 & 5 & -4 \\ 3 & 4 & -3 \end{pmatrix} .$$

(b) Verify that $\mathbf{U} \mathbin{\dot{:}} \mathbf{V} = 7$. Verify numerically that $\mathbf{U} \mathbin{\dot{:}} \mathbf{V} = \mathbf{V} \mathbin{\dot{:}} \mathbf{U}$.

A.7. Show that the trace of a matrix can be written $\mathbf{I} \mathbin{\dot{:}} \mathbf{T}$, where $\mathbf{I}$ is the unit matrix.

A.8. (a) Let $\mathbf{A} = (1,2,3)$. Show that in Voigt (Table 8.1) notation that the dyad $\mathbf{AA}$ can be written as a six vector (1,4,9,6,3,2).

A.9. Let A = (7,1,2), **B** = (1,2,1) and C = (1,2,3). Show that the third rank tensor (**AB**+**BA**)**C** can be written in Voigt notation as

$$\begin{pmatrix} 4 & 2 & 4 \\ 8 & 4 & 8 \\ 12 & 6 & 12 \\ 5 & 4 & 5 \\ 10 & 8 & 10 \\ 15 & 12 & 15 \end{pmatrix} .$$

A.10. Let $\hat{\mathbf{e}} = (1,0,1)/\sqrt{2}$, A = (2,2,1), **B** = (1,2,2), C = (1,2,3), **D** = (0,1,0) **E** = (1,0,0). Let $\mathbf{R}$ = (AB+**BA**)**C**. Show that

$$\mathbf{O}(\hat{e}) \cdot \mathbf{R} \vdots \mathbf{DE} = (-1,0,1).$$

APPENDIX **B**

Multipole Fields and Radiation Patterns

A central problem in this book is to determine the fields and power radiated by various moments of a charge distribution. Although we discuss charge distributions and currents in terms of individual charges, the formal discussion is given in terms of a charge distribution $\rho(\mathbf{r}',t')$ which varies continuously in space and time. The problem as sketched in Figure 1.2 is to calculate the fields $E(\mathbf{r},t)$ and $B(\mathbf{r},t)$ created by a source described by $\rho(\mathbf{r}',t')$. Note that since electromagnetic fields travel at the velocity c, the field at $\mathbf{r}$,t is generated by the source at $\mathbf{r}'$ at the time $t' = t - R/c$.

It turns out that it is easier to calculate multipole fields in terms of vector and scalar potentials $\mathbf{A}$ and ϕ respectively. Hence we start by briefly reviewing these potential fields.

B.1 VECTOR AND SCALAR POTENTIALS

All of the usual electromagnetic fields **E**, **D**, **B** and **H** can be expressed in terms of the vector potential A and scalar potential ϕ:

$$\mathbf{B} = \nabla \times \mathbf{A} \tag{B.1}$$

and

$$\mathbf{E} = -\nabla\phi - \frac{\partial \mathbf{A}}{\partial t}. \tag{B.2}$$

If we now substitute Eq. (B.1) into Maxwell's equations, or the wave equation, we obtain wave equations for **A**, i.e., Eq. (1.18), and ϕ, i.e.,

$$-\nabla^2\phi + \frac{1}{c^2}\frac{\partial^2\phi}{\partial t^2} = \frac{\rho}{\varepsilon_0}, \tag{B.3}$$

driven by real currents **J** and charges ρ. Since the sources **J** and ρ are linked by the equation of continuity

$$\nabla \cdot \mathbf{J} + \frac{\partial \rho}{\partial t} = 0, \tag{B.4}$$

the solutions to Eqs. (B.2) and (B.3) must also satisfy

$$\nabla \cdot \mathbf{A} + \frac{1}{c^2}\frac{\partial \phi}{\partial t} = 0. \tag{B.5}$$

There is an arbitrariness in the form of A and ϕ (see Problem B.9), which involves a discussion of gauges. We avoid discussing gauges since they are treated in standard texts on electromagnetic theory. Instead we have developed means by which answers can be obtained in a manner independent of gauge.

The solutions to Eqs. (1.18) and (B.3) can be found in standard texts. The vector potential is given in Eq. (1.40), and

$$\phi = \frac{1}{4\pi\varepsilon_0}\int d\mathbf{r}' \int \frac{\rho(\mathbf{r}',t')}{R}\,\delta(t'-t+R/c)dt'. \tag{B.6}$$

Before we calculate the fields created by various multipole sources, we examine the mathematical forms for each multipole in the DC limit (i.e., $\omega = 0$).

B.2 ELECTRIC DIPOLE AND QUADRAPOLE

In the DC limit the frequency is zero, the magnetic and electric terms are decoupled and $E = -\nabla\phi$. Hence it is sufficient to deal with the scalar potential ϕ, i.e., to solve

$$\phi(\mathbf{r}) = \frac{1}{4\pi\varepsilon_0}\int \frac{\rho(\mathbf{r}')}{|\mathbf{r}-\mathbf{r}'|}\,d\mathbf{r}'. \tag{B.7}$$

We are primarily interested in fields outside the source with field points sufficiently far away that $r \gg r'$. The term $1/R$ is expanded in a power series with $\mathbf{r}'$ as the "small" variable. In general,

$$f(a+r') = f(a) + \mathbf{r}'\cdot[\nabla_a f(a)] + \frac{1}{2}\,\mathbf{r}'\mathbf{r}':[\nabla_a\nabla_a\, f(a)] + \cdots \tag{B.8}$$

where the symbol $:$ means to take two dot products. Applying this to the present case,

$$\frac{1}{|\mathbf{r}-\mathbf{r}'|} = \frac{1}{r} + \mathbf{r}'\cdot\frac{\mathbf{r}}{r^3} + \frac{1}{2}\,\mathbf{r}'\mathbf{r}':\left(\frac{3\mathbf{r}\mathbf{r}}{r^5} - \frac{\mathbf{I}}{r^3}\right) + \cdots. \tag{B.9}$$

Substituting this into Eq. (B.7) gives

$$\phi(\mathbf{r}) = \frac{1}{4\pi\varepsilon_0 r} \int \rho(\mathbf{r}')\, d\mathbf{r}' + \frac{\mathbf{r}}{4\pi\varepsilon_0 r^3} \cdot \int \mathbf{r}'\rho(\mathbf{r}')\, d\mathbf{r}'$$

$$+ \frac{1}{2}\left(\frac{3\mathbf{r}\mathbf{r}}{r^5} - \frac{\mathbf{I}}{r^3}\right) : \int \mathbf{r}'\mathbf{r}'\rho(\mathbf{r}')\, d\mathbf{r}' + \cdots. \tag{B.10}$$

At this point we identify the total charge Q (monopole) as

$$Q = \int \rho(\mathbf{r}')\, d\mathbf{r}'. \tag{B.11}$$

The dipole moment **p** is given by

$$\mathbf{p} = \int \mathbf{r}'\rho(\mathbf{r}')\, d\mathbf{r}'. \tag{B.12}$$

Discussion of the quadrapole term is deferred for the moment.

The term "dipole moment" is used frequently and we now examine what it means. Consider the dipole moment **p** calculated in the **r**' and **r**" coordinate systems illustrated in Figure B.1:

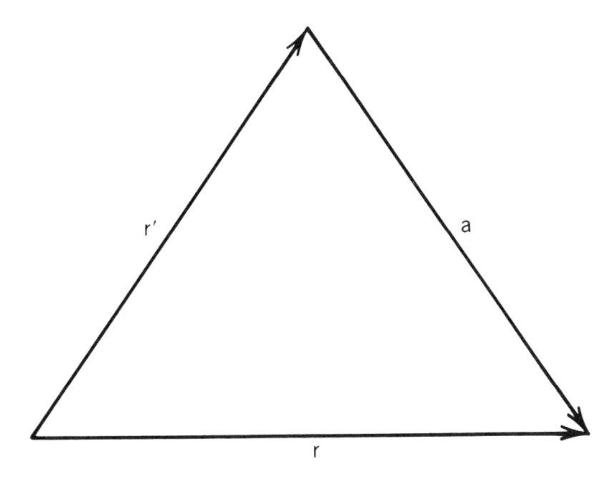

Figure B.1. Coordinates for computing a dipole moment.

$$\mathbf{p}' = \int \mathbf{r}''\rho(\mathbf{r}'')\, d\mathbf{r}'' = \int \mathbf{r}'\rho(\mathbf{r}')\, d\mathbf{r}' - \mathbf{a}\int \rho(\mathbf{r}')\, d\mathbf{r}'$$

or

$$\mathbf{p}' = \mathbf{p} - \mathbf{a}Q. \tag{B.13}$$

Hence there is no unique value of **p** and we may refer to "the dipole moment" only if the total charge $Q = 0$.

The potential due to the quadrapole term can be written as

$$\phi(\mathbf{r}) = \frac{1}{4\pi\varepsilon_0}\,\mathbf{T}:\mathbf{q}, \tag{B.14}$$

where

$$\mathbf{T} = \frac{\mathbf{rr}}{r^5} - \frac{\mathbf{I}}{3r^3}, \quad \mathbf{q} = \frac{3}{2}\int \mathbf{r}'\mathbf{r}'\rho(\mathbf{r}')\,d\mathbf{r}'. \tag{B.15}$$

This is not the standard form for the quadrapole, which is usually written as a tensor with zero trace, i.e., $\Sigma_i q_{ii} = 0$. We have the freedom to alter $\mathbf{q}$, as long as the changes yield the same value for ϕ as Eq. (B.14). First, we note that $\mathbf{T}$ does have a zero trace. The quadrapole tensor is written as

$$\mathbf{q}' = \mathbf{q} + C\,\mathbf{I} \tag{B.16}$$

where C is a constant. The "new" potential is

$$\phi(\mathbf{r}') = \frac{1}{4\pi\varepsilon_0}\,\mathbf{T}:\mathbf{q}' = \frac{1}{4\pi\varepsilon_0}\,[\mathbf{T}:\mathbf{q} + C\mathbf{T}:\mathbf{I}]. \tag{B.17}$$

Since

$$\mathbf{T}:\mathbf{I} = \sum_{ij} T_{ij}\,\delta_{ji} = T_{ii} = 0, \tag{B.18}$$

the potential is unaffected by this transformation. Defining

$$C = -\int \frac{1}{3}\, r'^2 \rho(\mathbf{r}')\,d\mathbf{r}', \tag{B.19}$$

then

$$\mathbf{Q} = \frac{3}{2}\int (\mathbf{r}'\mathbf{r}' - \frac{1}{3}\, r'^2\,\mathbf{I})\,\rho(\mathbf{r}')\,d\mathbf{r}' \tag{B.20}$$

is the quadrapole moment tensor, which has a zero trace. Furthermore, since $\mathbf{Q}$ has zero trace, then similarly $\mathbf{T}$ can be altered by adding or subtracting a multiple of $\mathbf{I}$. Inspection of Eq. (B.15) shows that $\mathbf{T}$ can be simplified by subtracting away the $-\mathbf{I}/3$ term. Hence the quadrapole contribution is given by

$$\phi(\mathbf{r}) = \frac{\mathbf{rr}}{4\pi\varepsilon_0 r^5} : \mathbf{Q}. \tag{B.21}$$

One can again ask under what conditions one can refer to "the quadrapole moment." It can be shown that $Q = 0$ and $\mathbf{p} = 0$ are sufficient conditions, i.e., the quadrapole term is the first nonvanishing term in the multipole expansion. However, the necissary conditions are more subtle and are discussed in Problem B.5b. Now consider the case where the charge distribution is of the form $\rho(\mathbf{r}') = q\delta(\mathbf{r}'-\mathbf{a})$ and calculate the moments up to the quadrapole term. First of all, note that the form of $\rho(\mathbf{r}')$ given does correspond to a charge density in the sense that

$$\int \rho(\mathbf{r}')\,d\mathbf{r}' = q$$

as required. Evaluating Eqs. (B.12) and (B.20) yields

$$\mathbf{p} = \mathbf{a}Q; \qquad \mathbf{Q} = \frac{3}{2}\,[\mathbf{aa} - \frac{1}{3}\,a^2\,\mathbf{I}]Q.$$

Since $Q \neq 0$, there is neither a well-defined dipole nor quadrapole moment. This is evident from the fact that both $\mathbf{p}$ and $\mathbf{Q}$ depend on the choice of the origin for the coordinates. Consider next the case of a vanishing total charge for $\rho(\mathbf{r}') = q\big(\delta(\mathbf{r}'-\mathbf{a}_+) - \delta(\mathbf{r}'+\mathbf{a}_-)\big)$ as shown in Figure B.2. In this case, it is easy to show that $Q=0$, $\mathbf{p} = q\big(\mathbf{a}_+ - \mathbf{a}_-\big)$ and

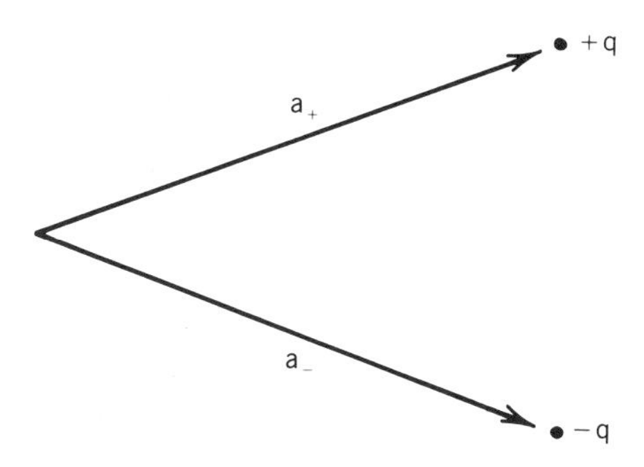

Figure B.2. Computation of dipole moment in case of charge neutrality.

$$\mathbf{Q} = \frac{3}{2}\,[\mathbf{a}_+\mathbf{a}_+ - \mathbf{a}_-\mathbf{a}_- - \frac{1}{3}\,\mathbf{I}\,(|\mathbf{a}_+|^2 + |\mathbf{a}_-|^2)]q.$$

Since the dipole moment is independent of the origin and depends only on the relative position of the charges, there is "a dipole moment." Note that $\mathbf{Q}$ depends on the choice of the origin and there is no "quadrapole moment." We finally consider the charge distribution shown in Figure B.3.

+q (-1,1)	-q (1,1)
-q (-1,-1)	+q (1,-1)

Figure B.3. Configuration for computing quadrapole moment. Charges are located at positions (x,y).

Simple application of the formulae leads to $Q=0$, $\mathbf{p}=(0,0,0)$ and $Q_{12}=6q$, $Q_{21}=6q$ and all other $Q_{ij}=0$.

We close this section by calculating the electric fields created by these multipole moments. Evaluating $\mathbf{E} = -\nabla\phi$ yields

monopole
$$\mathbf{E} = \frac{Q\mathbf{r}}{4\pi\varepsilon_0 r^3} \tag{B.22}$$

dipole
$$\mathbf{E} = \frac{3\mathbf{r}}{4\pi\varepsilon_0 r^5}(\mathbf{p}\cdot\mathbf{r}) - \frac{\mathbf{p}}{4\pi\varepsilon_0 r^3} \tag{B.23}$$

quadrapole
$$\mathbf{E} = \frac{5\mathbf{r}}{4\pi\varepsilon_0 r^7}\mathbf{rr} : \mathbf{Q} - \frac{2\mathbf{r}}{4\pi\varepsilon_0 r^5}\cdot\mathbf{Q} \tag{B.24}$$

B.3 MAGNETIC DIPOLE

The leading terms in the expression for the vector potential (Eq. (1.40)) are now expanded, i.e.,

$$\mathbf{A}(\mathbf{r}) = \frac{\mu_0}{4\pi}\int\frac{\mathbf{J}(\mathbf{r}')}{R}\,d\mathbf{r}'. \tag{B.25}$$

Using Eq. (B.9),

$$\mathbf{A}(\mathbf{r}) = \frac{\mu_0}{4\pi}\frac{1}{r}\int\mathbf{J}(\mathbf{r}')\,d\mathbf{r}' + \frac{\mu_0}{4\pi}\frac{\mathbf{r}}{r^3}\cdot\int\mathbf{r}'\mathbf{J}(\mathbf{r}')\,d\mathbf{r}' + \cdots. \tag{B.26}$$

Since current loops must close upon themselves, the first term is zero on physical grounds. This can also be verified mathematically.

We now search for a more convenient form for the second term. Consider

$$\mathbf{r}\times(\mathbf{r}'\times\mathbf{J}(\mathbf{r}')) = [\mathbf{r}\cdot\mathbf{J}(\mathbf{r}')]\mathbf{r}' - [\mathbf{r}\cdot\mathbf{r}']\mathbf{J}(\mathbf{r}'). \tag{B.27}$$

After some vector algebra and manipulation, this equation can be written as

$$\mathbf{r}\times\int[\mathbf{r}'\times\mathbf{J}(\mathbf{r}')]\,d\mathbf{r}' = -2\mathbf{r}\cdot\int\mathbf{r}'\mathbf{J}(\mathbf{r}')\,d\mathbf{r}'. \tag{B.28}$$

As a result,

$$\mathbf{A}(\mathbf{r}) = -\frac{\mu_0}{4\pi}\frac{\mathbf{r}}{r^3}\times\mathbf{m} \tag{B.29}$$

where

$$\mathbf{m} = \frac{1}{2}\int\mathbf{r}'\times\mathbf{J}(\mathbf{r}')\,d\mathbf{r}' \tag{B.30}$$

is the magnetic dipole moment.

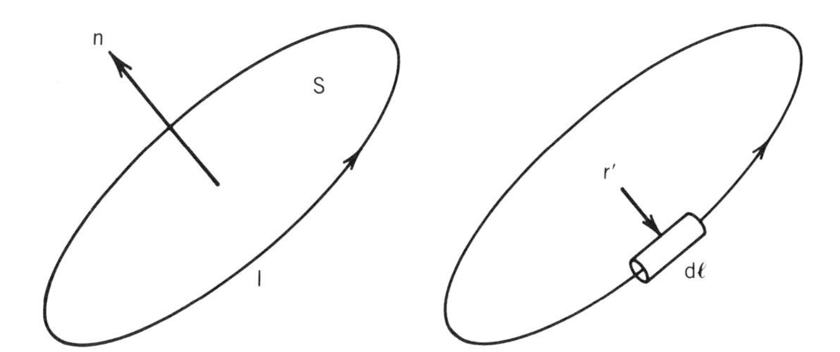

Figure B.4. Configuration and notation for computing a magnetic moment.

To see that Eq. (B.30) really describes a magnetic dipole moment, consider the simple current loop shown in Figure B.4. The usual definition of the magnetic moment found in elementary texts is

$$\mathbf{m} = I\mathcal{S}\mathbf{n}$$

where I is the current in the loop, **n** is the normal to the current loop and $\mathcal{S}$ is the area of the loop. We now attempt to rearrange Eq. (B.30) into this simple form. First we note that

$$\frac{1}{2}\int \mathbf{r}' \times \mathbf{J}(\mathbf{r}')dAd\ell = \frac{1}{2}\int \mathbf{r}' \times \boldsymbol{\ell}\ [J(\mathbf{r}')dA]$$

$$= \frac{1}{2}\, I \int \mathbf{r}' \times d\boldsymbol{\ell}.$$

The term $\mathbf{r}' \times d\boldsymbol{\ell}$ can be expressed in terms of a vector area element as

$$|\mathbf{r}' \times d\boldsymbol{\ell}| = |\mathbf{r}'|\ |d\boldsymbol{\ell}| = |dS|$$

so that

$$dS = \frac{1}{2}\int (\mathbf{r}' \times d\boldsymbol{\ell})$$

and therefore

$$\frac{1}{2}\int \mathbf{r}' \times \mathbf{J}(\mathbf{r}')\, dr' = I|\mathcal{S}|n = \mathbf{m}.$$

We close this section by calculating the magnetic field associated with a magnetic dipole. Evaluating Eq. (B.1) results in

$$B(\mathbf{r}) = \frac{\mu_0}{4\pi}\left(3(\mathbf{m}\cdot\mathbf{r})\,\frac{\mathbf{r}}{r^5} - \frac{\mathbf{m}}{r^3}\right). \tag{B.31}$$

Note the close analogy to the electric field of an electric dipole which has the same form.

B.4 HARMONIC DECOMPOSITION OF SOURCES AND FIELDS

In this book we are interested in fields that vary sinusoidally in time. Here we deal with only one harmonic component (characterized by the frequency ω) at a time. Hence the source fields are written in the form

$$\rho(\mathbf{r},t') = \frac{1}{2}\,\rho(\mathbf{r}',\omega)\,e^{-i\omega t'} + \text{cc} \tag{B.32}$$

$$J(\mathbf{r}',t') = \frac{1}{2}\,\mathcal{J}(\mathbf{r}',\omega)e^{-i\omega t'} + \text{cc}. \tag{B.33}$$

Substituting Eq. (B.32) in (B.6),

$$\phi(\mathbf{r},t) = \frac{1}{4\pi\varepsilon_0}\int d\mathbf{r}' \int dt'\,\frac{1}{2}\,[\rho(\mathbf{r}',\omega)e^{-i\omega t'} + \text{cc}]\,\frac{\delta(t-t'-R/c)}{R} \tag{B.34}$$

$$\phi(\mathbf{r},t) = \frac{1}{2}\,\phi(\mathbf{r},\omega)e^{-i\omega t} + \text{cc} \tag{B.35}$$

where integration over t' yields

$$\phi(\mathbf{r},\omega) = \frac{1}{4\pi\varepsilon_0}\int d\mathbf{r}'\,\rho(\mathbf{r}',\omega)\,\frac{e^{ikR}}{R}. \tag{B.36}$$

Similarly for

$$A(\mathbf{r},t) = \frac{1}{2}\,\mathcal{A}(\mathbf{r},\omega)e^{-i\omega t} + \text{cc}, \tag{B.37}$$

$$\mathcal{A}(\mathbf{r},\omega) = \frac{\mu_0}{4\pi}\int \mathcal{J}(\mathbf{r}',\omega)\,\frac{e^{+ikR}}{R}\,d\mathbf{r}'. \tag{B.38}$$

The electric and magnetic fields are also written in the form of Eq. (B.37) and the harmonic components of these fields are given by

$$\mathcal{B}(\mathbf{r},\omega) = \nabla \times \mathcal{A}(\mathbf{r},\omega) \tag{B.39}$$

and

$$\mathcal{E}(\mathbf{r},\omega) = -\nabla\phi(\mathbf{r},\omega) - i\omega\mathcal{A}(\mathbf{r},\omega). \tag{B.40}$$

Evaluating the electric field from Eq. (B.40) can be tedious because it requires that $\phi(\mathbf{r},\omega)$ be evaluated first. Instead it turns out to be easier to use Eq. (A.59) or Eq. (A.60), which gives the electric field directly in a gauge-independent fashion (see Problem B.9 for proof). Alternatively one can calculate

$\boldsymbol{E}(\mathbf{r},\omega)$ directly from $\boldsymbol{B}(\mathbf{r},\omega)$ by using Eq. (1.2). This procedure is also independent of the actual gauge used to compute the potentials. Assuming that the field point is outside the source region, Maxwell's equations give

$$\boldsymbol{E}(\mathbf{r},\omega) = \frac{ic}{k}\, \nabla \times \boldsymbol{B}(\mathbf{r},\omega). \tag{B.41}$$

B.5 MULTIPOLAR EXPANSION

One of the fundamental problems in electromagnetism is to calculate the fields generated by an oscillating charge distribution and the power radiated. By expanding the expression for the vector potential $\boldsymbol{A}(\mathbf{r},\omega)$, one can identify the various sources as electric dipole, magnetic dipole, electric quadrapole, etc. The price for this elegance is a fair amount of mathematical complexity if the complete fields are calculated.

The geometry for this problem is indicated in Figure 1.2, except that we denote the dimension of the medium by d. There are three characteristic dimensions in this problem, r the distance to the field point, d the extent of the source region and λ the wavelength of the electromagnetic field of frequency ω. As a result there are three "zones" in which the electromagnetic fields display different dependence on r:

1. Near (static zone) $d \ll r \ll \lambda$,
2. Intermediate (induction) zone $d \ll r \simeq \lambda$,
3. Far (radiation) zone $d \ll \lambda \ll r$.

We now proceed to expand the expression for the vector potential $\mathbf{A}(\mathbf{r})$. The problem is to expand $\exp(ikR)/R$ in powers of r'/r. Utilizing Eq. (B.9) gives

$$|\mathbf{r}-\mathbf{r}'| = r - \mathbf{r}'\cdot \nabla r + \frac{1}{2}\, \mathbf{r}'\mathbf{r}' : \nabla\nabla r + \cdots. \tag{B.42}$$

Hence

$$\frac{e^{ikR}}{R} = \frac{e^{ikr}}{r}\; \frac{e^{ik\{\mathbf{r}'\cdot\hat{\mathbf{r}} - \frac{1}{2} \mathbf{r}'\mathbf{r}' : (\frac{\mathbf{O}(\hat{\mathbf{r}})}{r}) + \cdots\}}}{\{1 - \frac{\mathbf{r}'\cdot\hat{\mathbf{r}}}{r} + \frac{1}{2}\, \mathbf{r}'\mathbf{r}' : (\frac{\mathbf{O}(\hat{\mathbf{r}})}{r^2})\}} \tag{B.43}$$

$$\frac{e^{ikR}}{R} \simeq \frac{e^{ikr}}{r}\, \{ 1 + (\frac{1}{r} - ik)\, \hat{\mathbf{r}}\cdot\mathbf{r}' + \cdots\}, \tag{B.44}$$

where $\mathbf{O}$ is defined in Eq. (1.48). Substituting into Eq. (B.38) gives

$$\boldsymbol{A}(\mathbf{r},\omega) = \frac{\mu_0}{4\pi}\, \frac{e^{ikr}}{r} \Big(\int \boldsymbol{J}(\mathbf{r}',\omega)\, d\mathbf{r}' + (\frac{1}{r} - ik) \int \hat{\mathbf{r}}\cdot\mathbf{r}' \boldsymbol{J}(\mathbf{r}',\omega)\, d\mathbf{r}' \Big). \tag{B.45}$$

Each term can be easily interpreted. The spherical wave nature of the fields emanating from the source is described essentially by the $\exp(ikr)/r$ term. The first term in the brackets gives rise to electric dipole fields and the second term is a combination of magnetic dipole and electric quadrapole fields. Successive terms which were neglected contain the magnetic quadrapole and electric octapole, etc. It is shown subsequently that it is the first nonvanishing term in the expansion of Eq. (B.45) which determines the observed radiation pattern.

B.6 MAGNETIC DIPOLE RADIATION

In practice the most important term in the expansion is that due to an electric dipole. This computation of electric dipole is carried out in Chapter 1, and it is left as an exercise to repeat the calculation using the notation of Fourier amplitudes. We now evaluate the electromagnetic fields and the power radiated by a magnetic dipole. Here we concentrate on the radiation zone in the detailed discussion. Since $kr \gg 1$ in the radiation region, the exponential phase term is the only one which varies rapidly with r in this region. Hence the curl operators implied by Eqs. (B.39) and (B.41) act only on the phase term. (The complete fields valid in all regions of space require derivatives of all terms which vary with r, not just the phase term.) The field is obtained from the second term of Eq. (B.45):

$$A(\mathbf{r},\omega) = \frac{\mu_0}{4\pi}\frac{e^{ikr}}{r}\left(\frac{1}{r} - ik\right)\int \hat{\mathbf{r}}\cdot\mathbf{r}'\mathcal{J}(\mathbf{r}',\omega)\,d\mathbf{r}'. \tag{B.46}$$

Some rearranging is necessary in order to express Eq. (B.46) in the standard form for a magnetic dipole. (It turns out that the electric quadrapole is also contained in Eq. (B.46).) First we consider the vector expansion

$$\mathbf{r}\times[\mathbf{r}'\times\mathcal{J}(\mathbf{r}',\omega)] = [\mathbf{r}\cdot\mathcal{J}(\mathbf{r}',\omega)]\mathbf{r}' + [\mathbf{r}\cdot\mathbf{r}']\mathcal{J}(\mathbf{r}',\omega) - 2\mathbf{r}\cdot\mathbf{r}'\mathcal{J}(\mathbf{r}',\omega), \tag{B.47}$$

which leads to

$$A(\mathbf{r},\omega) = \frac{\mu_0}{4\pi}\frac{e^{ikr}}{r^2}\left(\frac{1}{r} - ik\right)\left\{\frac{1}{2}\mathbf{r}\cdot\int[\mathbf{r}'\mathcal{J}(\mathbf{r}',\omega) + \mathcal{J}(\mathbf{r}',\omega)\mathbf{r}']\,d\mathbf{r}'\right.$$

$$\left. - \frac{1}{2}\mathbf{r}\times\int\mathbf{r}'\times\mathcal{J}(\mathbf{r}',\omega)\,d\mathbf{r}'\right\}. \tag{B.48}$$

Associating the second term with the magnetic dipole moment (Eq. (B.30)), the vector potential is given by

$$A(\mathbf{r},\omega) = \frac{\mu_0}{4\pi}ik\left(1 + \frac{1}{ikr}\right)\frac{e^{ikr}}{r}\,\hat{\mathbf{r}}\times\mathbf{m}_0 \tag{B.49}$$

where

$$\mathbf{m}_0 = \frac{1}{2}\int\mathbf{r}'\times\mathcal{J}(\mathbf{r}',\omega)\,d\mathbf{r}'. \tag{B.50}$$

The magnetic and electric fields are now calculated in the radiation zone ($kr \gg 1$). Using $B(\mathbf{r},\omega) = \nabla\times A(\mathbf{r},\omega)$,

$$B(\mathbf{r},\omega) = i\mathbf{k}\times A(\mathbf{r},\omega)$$

$$= -\frac{\mu_0}{4\pi}\frac{e^{ikr}}{r}k^2\,\hat{\mathbf{r}} \times (\mathbf{r} \times \mathbf{m}_0). \tag{B.51}$$

Note the similarity between the magnetic field of the magnetic dipole and electric field for the electric dipole. The corresponding electric field is given by

$$\boldsymbol{E}(\mathbf{r},\omega) = -\frac{k^2}{4\pi\varepsilon_0 c}\frac{e^{ikr}}{r}(\hat{\mathbf{r}} \times \mathbf{m}_0). \tag{B.52}$$

The power radiated is calculated in the same way as for the electric dipole case. For the Poynting vector,

$$\mathbf{S} = \frac{1}{8\pi^2 c\varepsilon_0}\frac{k^4}{r^2}|\mathbf{m}_0|^2\sin^2\theta\,\hat{\mathbf{r}}, \tag{B.53}$$

for the power radiated per unit solid angle,

$$P(\Omega) = \frac{1}{8\pi^2 c\varepsilon_0}k^4|\mathbf{m}_0|^2\sin^2\theta \tag{B.54}$$

and for the total power,

$$P = \frac{k^4}{3\pi\varepsilon_0 c}|\mathbf{m}_0|^2. \tag{B.55}$$

Comparison with the electric quadrapole and electric dipole is deferred until after the quadrapole discussion.

B.7 ELECTRIC QUADRAPOLE RADIATION

The vector potential due to a harmonic quadrapole is given by the first term in Eq. (B.48). (The second term is associated with the magnetic dipole.) Just as in the previous cases, some vector field manipulations are necessary in order to obtain the standard form for the quadrapole tensor. Again we restrict the discussion to the radiation zone for which

$$\boldsymbol{A}(\mathbf{r},\omega) = -\frac{\mu_0}{4\pi}ik\frac{e^{ikr}}{r}\frac{1}{2}\hat{\mathbf{r}} \cdot \int [\mathbf{r}'\boldsymbol{J}(\mathbf{r}',\omega) + \boldsymbol{J}(\mathbf{r}',\omega)\mathbf{r}']d\mathbf{r}'. \tag{B.56}$$

For this case, we expand the vector expression

$$\nabla \cdot [\boldsymbol{J}(\mathbf{r}',\omega)\mathbf{r}'\mathbf{r}'] = [\nabla \cdot \boldsymbol{J}(\mathbf{r}',\omega)]\mathbf{r}'\mathbf{r}' + \boldsymbol{J}(\mathbf{r}',\omega)\mathbf{r}' + \mathbf{r}'\boldsymbol{J}(\mathbf{r}',\omega), \tag{B.57}$$

eliminate the left-hand side by integrating over $d\mathbf{r}'$ and convert to a surface integral over a surface outside the source distribution. Applying the equation of continuity and identifying the result with Eq. (B.15),

$$A(\mathbf{r},\omega) = -\frac{\mu_0}{4\pi}\frac{k^2c}{2}\frac{e^{ikr}}{r}\hat{\mathbf{r}}\cdot\mathbf{q}_0. \tag{B.58}$$

It now remains to express Eq. (B.58) in the standard form for the quadrapole moment. Consider the transformation

$$\mathbf{q}'_0 = \mathbf{q}_0 + C\mathbf{I} \tag{B.59}$$

and the resulting vector potential

$$A'(\mathbf{r},\omega) = A(\mathbf{r},\omega) + f(r)\hat{\mathbf{r}}. \tag{B.60}$$

Since

$$B'(\mathbf{r},\omega) = \nabla\times A(\mathbf{r},\omega) + \nabla\times[f(r)\hat{\mathbf{r}}] = B(\mathbf{r},\omega), \tag{B.61}$$

the fields are unaffected by the transformation given by Eq. (B.59) and we are free to make $\mathbf{q}_0$ traceless. Defining

$$\mathbf{Q}_0 = \frac{3}{2}\int[\mathbf{r}'\mathbf{r}' - \frac{1}{3}r'^2\,\mathbf{I}]\,\rho(\mathbf{r}',\omega)\,d\mathbf{r}', \tag{B.62}$$

then

$$A(\mathbf{r},\omega) = -\frac{\mu_0 k^2 c}{12\pi}\frac{e^{ikr}}{r}\hat{\mathbf{r}}\cdot\mathbf{Q}_0. \tag{B.63}$$

The magnetic and electric fields are now calculated in the usual way. The following results are obtained in the radiation zone:

$$B(\mathbf{r},\omega) = -\frac{\mu_0}{12\pi}ik^3c\frac{e^{ikr}}{r}\hat{\mathbf{r}}\times(\hat{\mathbf{r}}\cdot\mathbf{Q}_0) \tag{B.64}$$

and

$$E(\mathbf{r},\omega) = \frac{1}{12\pi\varepsilon_0}ik^3\frac{e^{ikr}}{r}\{(\hat{\mathbf{r}}\hat{\mathbf{r}}:\mathbf{Q}_0)\hat{\mathbf{r}} - \hat{\mathbf{r}}\cdot\mathbf{Q}_0\}. \tag{B.65}$$

Again following standard procedures

$$\mathbf{S} = \frac{1}{72\pi^2\varepsilon_0}\frac{ck^6}{r^2}|\hat{\mathbf{r}}\times(\hat{\mathbf{r}}\cdot\mathbf{Q}_0)|^2\hat{\mathbf{r}}, \tag{B.66}$$

$$P(\Omega) = \frac{1}{72\pi^2\varepsilon_0}ck^6|\hat{\mathbf{r}}\times(\hat{\mathbf{r}}\cdot\mathbf{Q}_0)|^2 \tag{B.67}$$

and

$$P = \frac{ck^6}{90\pi\varepsilon_0}|\mathbf{Q}_0:\mathbf{Q}_0^*|. \tag{B.68}$$

B.8 COMPARISON OF MULTIPOLAR RADIATION

In the previous sections, the fields and power radiated by electric dipole, magnetic dipole and electric quadrapole sources were calculated. To summarize briefly for the total radiated power:

electric dipole $$P = \frac{1}{3\pi\varepsilon_0} ck^4 |\mathbf{p}_0|^2$$

magnetic dipole $$P = \frac{c^2}{3\pi\varepsilon_0} ck^4 |\mathbf{m}_0|^2$$

electric quadrapole $$P = \frac{1}{90\pi\varepsilon_0} ck^6 |\mathbf{Q}_0|^2$$

The total power radiated by both dipole distributions is proportional to k^4, whereas the quadrapole term varied as k^6. This is a general characteristic of multipoles in the sense that the dependence on k becomes stronger the higher the order of the pole.

It is instructive to compare the power radiated by various multipoles with the radiation in the visible region of the spectrum. For the magnetic dipole

$$|\mathbf{m}| \simeq |\mathbf{r}' \times \mathbf{J}| \simeq |\mathbf{r}' \times e\mathbf{v}| \simeq \frac{e}{m}(\mathbf{r}' \times m\mathbf{v}) \simeq \frac{e}{m}\mathbf{L}$$

where $\mathbf{L}$ is the angular momentum and $|\mathbf{L}| \simeq \hbar$ Hence $|\mathbf{m}_0| \simeq e\hbar/m$. The relative power radiated is

$$\frac{P(m_0)}{P(p_0)} \simeq \frac{1}{c^2}\frac{|m_0|^2}{|p_0|^2} \simeq \frac{e^2\hbar^2}{m^2c^2}\frac{1}{e^2a_0^2} \simeq \frac{\hbar^2}{m^2c^2}\frac{1}{a_0^2} \simeq \left(\frac{\lambda_c}{a_0}\right)^2$$

where λ_c is the Compton wavelength. Similarly

$$\frac{P(Q_0)}{P(p_0)} \simeq \frac{k^6c}{90}\frac{3}{ck^4}\frac{|\mathbf{Q}_0|^2}{|\mathbf{p}_0|^2} \simeq \frac{k^2a_0^2}{30} \simeq \left(\frac{a_0}{\lambda}\right)^2$$

where λ is the wavelength of the radiation. For a molecule or atom radiating in the visible ($\lambda \simeq 10^{-4}$cm, $a_0 \simeq 10^{-8}$ cm and $\lambda_c \simeq 10^{-11}$ cm)

$$\frac{P(m_0)}{P(p_0)} \simeq 10^{-6} \qquad \frac{P(Q_0)}{P(p_0)} \simeq 10^{-8}.$$

Hence if the radiation involves an electric dipole, its radiation dominates. If there is no electric dipole involved, magnetic dipole radiation is observed, with electric quadrapole only a few orders of magnitude further down.

ADDITIONAL READING

Multipole Expansion

Jackson, J.D. *Classical Electrodynamics*, 2nd Edition, Wiley, New York, 1975. (or 1st Ed. 1962). Moments, Chapter 4, Multipole Fields, Chapter 16.

PROBLEMS

B.1. Compute the electric and magnetic far field radiation pattern and the radiated power of an oscillating charge distribution of the form

$$\rho(x,t) = e[\delta(x-\zeta\cos\omega_a t)\ \delta(y-\zeta\sin\omega_a t)\ \delta(z) - \delta(x)\delta(y)\delta(z)].$$

B.2. Compute the far field radiation pattern of a charge distribution that oscillates as $\cos\omega_a t$ and has the properties that (1) it is symmetrical under all rotations about the z axis (cylindrical symmetry) and maintains its symmetry at all times; (2) it is invariant under 180° rotations about the x axis.

B.3. (a) Compute the vector potential of two pairs of charges placed as shown in Figure B.5. The pattern maintains its symmetry at all times and oscillates in the x dimension as $\cos\omega_a t$.

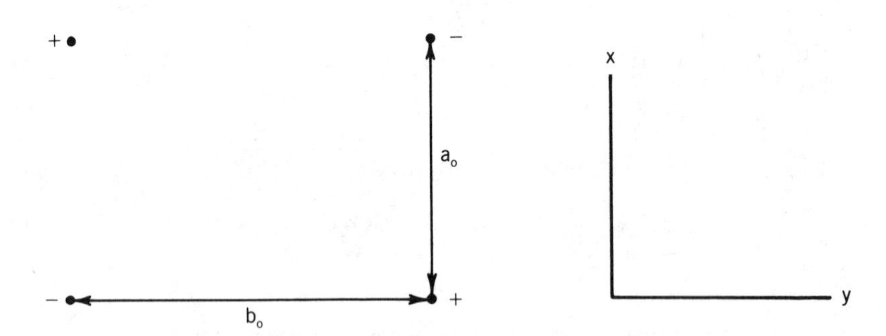

Figure B.5. Diagram for Problem B.3.

(b) Compute the vector potential of (a) by considering it as the interference pattern of the two dipole radiators located at $\mp b_0$. Assume $b_0 \ll \lambda$.

B.4. (a) Compute the dipole moment of the charge distribution in Eqs. (1.61) and (B.12). Show that the dipole moment is independent of the coordinate system.

(b) Compute the "magnetic moment" of the current distribution in Eqs. (1.62) and (B.30). Show that this depends on the choice of coordinate system.

(c) Consider a collection of charges of the form of Eq. (1.61) located at different points (x_0,y_0,z_0) with different displacements (ζ,η,ξ) (invent a notation). Show that the dipole moment of the collection of atoms is the sum of the individual dipoles.

(d) Show that the dot product of the magnetic and electric dipole moments of the set of charges in (c) is independent of the coordinate system used to compute it. This means that we can describe the system in terms of a magnetic moment that is independent of the coordinate system provided it is parallel to the dipole. This is important for optical activity, since it is frequently described as a magnetic effect.

(e) Show that the average dot product in (d) vanishes if the individual dipoles are located randomly with respect to each other (but not oriented randomly). In this case we cannot speak of a magnetic moment.

(f) Show that if the dipole computed in c is zero, then the magnetic dipole is independent of the coordinate system.

B.5. (a) Consider the set of dipoles in Problem B.4. Assume that the displacements (ζ,η,ξ) are much smaller than the distances between particles. Compute the quadrapole moment and show that it is a function of the coordinate system.

(b) Show that the 2x2 quadrapole tensor in the plane perpendicular to the total dipole of the set of charges is independent of the coordinate system. This result parallels the one in Problem B.4d since it shows that one can equally well discuss optical activity in terms of quadrapole moments.

(c) Show that the quadrapole tensor is independent of the coordinate system if the net dipole is zero.

B.6. (a) Show that in the far field, the time-averaged flux is

$$\mathbf{S} = \frac{1}{2}\left(\frac{\varepsilon_0}{\mu_0}\right)^{1/2} k^2\, \mathbf{O}(\hat{\mathbf{r}}):[\boldsymbol{A}(\mathbf{r})\boldsymbol{A}^*(\mathbf{r})] \tag{B.69}$$

(b) Sketch the far field radiation pattern for a quadrapole whose amplitude is proportional to the tensor $\hat{\mathbf{x}}\hat{\mathbf{y}}$.

B.7. Show that a negatively charged electron cloud that is spherically symmetrical and oscillates radially about a neutralizing positive charge at the center has no electric or magnetic moments at any order in the multipole expansion. This Raman active charge distribution is often misidentified as a quadrapole in the literature.

B.8. Consider a set of four charges that lie along a straight line. Let them be symmetrical with respect to a center of symmetry, and consider the situation in which the outermost pairs of charges are bound together with springs such that when one has a dipole $q_\alpha \boldsymbol{\ell}_\alpha$ the other has a dipole $-q_\alpha \boldsymbol{\ell}_\alpha$ (i.e., the charges oscillate from +--+ to -++-). Let the dipole pairs be separated by a distance $\mathbf{s}_\alpha$. Note that $\hat{\mathbf{s}}_\alpha$ is parallel to $\boldsymbol{\ell}_\alpha$ and defines the molecular axis. The molecules with this structure are assumed to lie with random orientation so that the medium is isotropic.

(a) Show that this set of charges has no magnetic dipole, i.e., it is a pure quadrapole.

(b) Show that the electric quadrapole reads

$$\mathbf{Q} = q_\alpha(3\boldsymbol{\ell}_\alpha \mathbf{s}_\alpha - s_\alpha \ell_\alpha\, \mathbf{I}).$$

(c) A field acts on this structure to break, slightly, the symmetry of oscillation. Treat the two dipoles as independent of each other and use the methods of Chapter 3 to show that the force on the structure reads

$$\mathbf{F}_\alpha = \ell_\alpha(\mathbf{E}(\mathbf{r},t)-\mathbf{E}(\mathbf{r}+\mathbf{s}_\alpha,t)) \simeq - \ell_\alpha \mathbf{s}_\alpha : \nabla\mathbf{E}.$$

(d) Show that one can derive this force from a potential

$$V_{int} = -\frac{1}{3}\,\mathbf{Q} : \nabla\mathbf{E}.$$

Try to justify this choice directly from the interaction of the structure with the field. Note; The second part of the exercise is nontrivial.

(e) Follow the methods of Section 3.2 to show that

$$Q_\alpha = -\frac{i\ell_\alpha \mathbf{s}_\alpha : \mathbf{k}\mathbf{E}}{M_\alpha D_\alpha(\omega)}.$$

Use Eq. (3.20) to show that the power absorbed by the ensemble of quadrapoles reads

$$P_{abs} = \frac{1}{2}\,\omega^2 \mathcal{N}\Gamma_\alpha \left\langle \frac{|\ell_\alpha \mathbf{s}_\alpha : \mathbf{k}\mathbf{E}|^2}{M_\alpha |D_\alpha(\omega)|^2} \right\rangle,$$

that under isotropic conditions the average absorption cross section reads

$$\sigma_{abs} = \frac{2\omega^4\Gamma_\alpha(\ell_\alpha s_\alpha)^2}{30 M_\alpha \varepsilon_0 c^3 |D_\alpha(\omega)|^2},$$

and that this absorption coefficient is $4(ks_\alpha)^2/45$ times smaller than the cross section that would be obtained from one dipole alone.

(f) Show that the macroscopic polarization resulting from the ensemble of quadrapoles can be written as

$$\boldsymbol{P} = -\,i\,\mathcal{N}\,\langle \ell_\alpha \mathbf{s}_\alpha \cdot \mathbf{k} Q_\alpha \rangle.$$

Use Eq. (4.55) to show that energy is conserved.

B.9. Let $\mathbf{A}$ and ϕ be explicit forms for the vector and scalar potentials that are solutions to Maxwell's equations.

(a) Show that if Φ is an arbitrary function and

$$\mathbf{A}' = \mathbf{A} + \nabla\Phi, \tag{B.70a}$$

$$\phi' = \phi - \frac{\partial\Phi}{\partial t}, \tag{B.70b}$$

then $\mathbf{A}'$ and ϕ' satisfy Eqs. (B.5), (1.18) and (B.3). Equations (B.70a) and B.70b) are a gauge transformation.

(b) Show that if $\mathbf{A}'$ and ϕ' describe plane-wave fields, then Eq. (A.59) gives a gauge independent answer for $\mathbf{E}$. Show the same for spherical waves in the far field using Eq. (A.60).

B.10. General interaction potentials are written

$$V_{int} = \sum_a \ell_a \left(-\phi(\mathbf{r}_a) + \mathbf{A}(\mathbf{r}_a)\cdot\mathbf{v}_a\right). \tag{B.71}$$

Consider a set of charges such that

$$\ell_\alpha = -\ell_{a+1} = \ell_a,$$

$$\mathbf{r}_{a+1} = \mathbf{r}_a - \hat{\boldsymbol{\ell}}_\alpha q_\alpha.$$

(a) Show that these charges have a polarization given by Eq. (5.1), an interaction potential given by Eq. (3.6), and a force given by Eq. (3.7).

(b) Use Eq. (B.71), (B.2) and the following generalization of Eq. (3.7) for velocity-dependent potentials which reads

$$F_\alpha = -\frac{\partial V_{int}}{\partial q_\alpha} + \frac{d}{dt}\frac{\partial V_{int}}{\partial \dot{q}_\alpha} \tag{B.72}$$

to obtain the same result.

B.11. (a) Use Eq. (B.71) with charges having a configuration

$$\ell_\alpha = \ell_{a+3} = -\ell_{a+2} = -\ell_{a+1} = \ell_a,$$

$$\mathbf{r}_{a+3} = \mathbf{r}_a + \mathbf{s}_\alpha - q_\alpha \boldsymbol{\ell}_\alpha,$$

$$\mathbf{r}_{a+2} = \mathbf{r}_a + \mathbf{s}_\alpha,$$

$$\mathbf{r}_{a+1} = \mathbf{r}_a - q_\alpha \boldsymbol{\ell}_\alpha.$$

to develop an interaction potential. Use Eq. (B.72) to show that

$$F_\alpha = -\frac{1}{2}(\mathbf{s}_\alpha \boldsymbol{\ell}_\alpha + \boldsymbol{\ell}_\alpha \mathbf{s}_\alpha) : \nabla\mathbf{E} - \frac{1}{2}(\mathbf{s}_\alpha \times \boldsymbol{\ell}_\alpha)\cdot\mathbf{B}$$

(b) Assume an isotropic medium. Develop this force from an interaction potential

$$V_{int} = -\frac{1}{3}\mathbf{Q} : \nabla\mathbf{E} + \mathbf{m}\cdot\mathbf{B}.$$

Where (show these)

$$\mathbf{Q} = q_\alpha \left(\frac{3}{2}(\boldsymbol{\ell}_\alpha \mathbf{s}_\alpha + \mathbf{s}_\alpha \boldsymbol{\ell}_\alpha) - \boldsymbol{\ell}_\alpha\cdot\mathbf{s}_\alpha \mathbf{I}\right),$$

$\mathbf{m} = -\frac{1}{2} q_\alpha (\mathbf{s}_\alpha \times \boldsymbol{\ell}_\alpha)$.

(c) Assume translational invariance and drop the subscript α for this part of the exercise. Use Eq. (1.4) to show that the force in part (a) reduces to $F = -\boldsymbol{\ell}_\alpha \mathbf{s}_\alpha \colon \nabla \mathbf{E}$. Write the charges a and a+1 in part (a) as one dipole sheet interacting with a field $\mathbf{E}(\mathbf{r}_a)$ and the charges a+2, a+3 as another sheet interacting with $\mathbf{E}(\mathbf{r}_a+\mathbf{s})$. Use $V_{int} = -\mathbf{P}\cdot\mathbf{E}$ to show $V_{int} = -q\boldsymbol{\ell}\mathbf{s}\colon\nabla\mathbf{E}$.

Hint: Part (a) and some of part (c) of this exercise are for the adventuresome. We have found Eqs. (A.16) (A.25) and (A.26) to be of value.

B.12. Show that the set of charges in Problem B.11 give a polarizability $\boldsymbol{\alpha} = \alpha\,\mathbf{I}$ such that

$$\alpha = -\left\langle\frac{(\boldsymbol{\ell}_\alpha\cdot\hat{\mathbf{e}})^2\,(\mathbf{s}_\alpha\cdot\hat{\mathbf{k}})^2 k E}{(\gamma/2+i(\omega_\alpha-\omega))mc}\right\rangle$$

Hence these charges are not optically active; rather they behave like a weak dipole.

B.13. The macroscopic polarization in Problem B.8 has a structure

$$P_w = \frac{1}{2}\,(\hat{\mathbf{k}}_p P_w e^{i(\mathbf{k}_p\cdot\mathbf{r}-wt)} + cc).$$

Let this polarization be a phased array in the sense of Section 10.4 (i.e., P_w is constant). Equations (10.7), (10.15), (10.16) and $\mathbf{O}(\hat{\mathbf{k}}_p)\cdot\hat{\mathbf{k}}_p = 0$ imply that the field radiated by this array is zero in all cases. Show that this conclusion is false for anisotropic crystals.

Hint: This exercise is logical rather than computational. The false conclusion does not involve a misuse of the SVEA; rather it results from a misconstruction of the weak polarization due to the neglect of nonradiative electromagnetic fields (longitudinal photons) which, under most circumstances, play no role in optics. This problem is discussed in more detail in Volume II.

Index